T0207403

Lecture Notes in Computer Science 14418

The series Lecture Notes in Computer Science (LNCS), including its subseries Lecture Notes in Artificial Intelligence (LNAI) and Lecture Notes in Bioinformatics (LNBI), has established itself as a medium for the publication of new developments in computer science and information technology research, teaching, and education.

LNCS enjoys close cooperation with the computer science R & D community, the series counts many renowned academics among its volume editors and paper authors, and collaborates with prestigious societies. Its mission is to serve this international community by providing an invaluable service, mainly focused on the publication of conference and workshop proceedings and postproceedings. LNCS commenced publication in 1973.

Vikram Goyal · Naveen Kumar ·
Sourav S. Bhowmick · Pawan Goyal ·
Navneet Goyal · Dhruv Kumar
Editors

Big Data and Artificial Intelligence

11th International Conference, BDA 2023
Delhi, India, December 7–9, 2023
Proceedings

Editors
Vikram Goyal
Indraprastha Institute of Information Technology
Delhi, India

Sourav S. Bhowmick
Nanyang Technological University
Singapore, Singapore

Navneet Goyal
Birla Institute of Technology and Science
Pilani, India

Naveen Kumar
University of Delhi
Delhi, India

Pawan Goyal
Indian Institute of Technology
Kharagpur, India

Dhruv Kumar
Indraprastha Institute of Information Technology
Delhi, India

ISSN 0302-9743 ISSN 1611-3349 (electronic)
Lecture Notes in Computer Science
ISBN 978-3-031-49600-4 ISBN 978-3-031-49601-1 (eBook)
https://doi.org/10.1007/978-3-031-49601-1

This Springer imprint is published by the registered company Springer Nature Switzerland AG
The registered company address is: Gewerbestrasse 11, 6330 Cham, Switzerland

Paper in this product is recyclable.

Preface

Data generation and capturing capabilities of humans is increasing rapidly with each passing year. As a consequence, the scope of Big Data Analytics is also expanding to cover all domains including healthcare, agriculture, social development, climate change, manufacturing, etc. Data feeds multiple downstream tasks which attempt to uncover the hidden information in data characterised by the 6Vs of Big Data. Big Data and AI are complementary technologies and their synergy is helping researchers solve many longstanding complex socio-economic and scientific problems.

The 11th International Conference on Big Data and Artificial Intelligence (BDA 2023) was held in physical mode during December 7–9, 2023 at the Indraprastha Institute of Information Technology Delhi (IIITD). The proceedings include 17 peer-reviewed research papers and 2 talks by distinguished keynote speakers. The program also included 2 workshops and 2 tutorial sessions related to Big Data and Artificial Intelligence. This edition of BDA covered a wide range of topics related to big data analytics, such as data science, deep learning, knowledge graphs, graph neural networks, healthcare analytics, information retrieval, LLMs, speech analytics, etc.

The conference received 67 submissions. The review process of research papers followed a single-blind two-tiered review system following the tradition of BDA. We employed Microsoft's Conference Management Toolkit (CMT) for the paper submission and review processes. The Program Committee (PC) consisted of competent data analytical researchers from academia and industry coming from several countries. Each submission was reviewed by at least three reviewers from the Program Committee (PC) and discussed by the PC chairs before a decision was made. Based on the above review process, the Program Committee accepted seventeen papers (acceptance ratio of 25%).

The two keynote talks were delivered by Constantine Dovrolis (The Cyprus Institute, Cyprus and Georgia Tech, USA) and Manish Gupta (Microsoft Bing, India). The two invited talks were delivered by Chetan Arora (IIT Delhi, India) and Sameep Mehta (IBM Research, India).

Two workshops were selected by the workshop co-chairs to be held in conjunction with BDA 2023: one workshop conducted by iHub Anubhuti-IIITD Foundation (TiH IIITD) and another conducted by the Ministry of Statistics and Programme Implementation (MosPI).

We want to thank all PC members for their hard work in providing us with thoughtful and comprehensive reviews and recommendations. Many thanks to the authors who submitted their papers to the conference. The Steering Committee and the Organizing Committee deserve praise for their support. Several volunteers contributed to the success of the conference. We thank all keynote speakers, invited speakers, workshop speakers, and panelists for their insights on the various themes of the conference. Lastly, we also acknowledge the Indraprastha Institute of Information Technology - Delhi for hosting the conference. We thank the Infosys Center for Artificial Intelligence (IIIT Delhi), Center for Design & New Media (IIIT Delhi), iHub Anubhuti-IIITD Foundation (TiH IIIT

Delhi) and Department of Computer Science & Engineering (IIIT Delhi) for sponsoring the conference. We also thank all partners for supporting the conference. We hope that the readers of the proceedings find the content interesting, contemporary, rewarding, and beneficial to their research.

December 2023

Vikram Goyal
Naveen Kumar
Sourav S. Bhowmick
Pawan Goyal
Navneet Goyal
Dhruv Kumar

Organization

Honorary Chairs

Ranjan Bose	IIIT Delhi, India
G. P. Samanta	Ministry of Statistics and Program Implementation, India

General Chairs

Vikram Goyal	IIIT Delhi, India
Naveen Kumar	University of Delhi, India

Steering Committee Chair

Sanjay Madria	Missouri S&T, USA

Steering Committee Members

Srinath Srinivasa	IIIT Bangalore, India
Mukesh Mohania	IIIT Delhi, India
Sharma Chakravarthy	University of Texas at Arlington, USA
Philippe Fournier-Viger	Shenzhen University, China
Masaru Kitsuregawa	University of Tokyo, Japan
Raj K. Bhatnagar	University of Cincinnati, USA
Vasudha Bhatnagar	University of Delhi, India
Ladjel Bellatreche	ISAE-ENSMA, France
H. V. Jagadish	University of Michigan, USA
Ramesh Kumar Agrawal	Jawaharlal Nehru University, India
Divyakant Agrawal	University of California at Santa Barbara, USA
Arun Agarwal	University of Hyderabad, India
Jaideep Srivastava	University of Minnesota, USA
Sanjay Chaudhary	Ahmedabad University, India
Jerry Chun-Wei Lin	Western Norway University of Applied Sciences, Norway
S. K. Gupta	IIIT Delhi, India
Krithi Ramamritham	IIT Bombay, India

Organizing Committee Chairs

R. K. Aggarwal	JNU Delhi, India
Md. Shad Akhtar	IIIT Delhi, India
Raghava Mutharaju	IIIT Delhi, India
Rajiv Ratn Shah	IIIT Delhi, India

Program Committee Chairs

Sourav S. Bhowmick	NTU, Singapore
Pawan Goyal	IIT Kharagpur, India

Program Committee Members

Abhilash Nandy	IIT Kharagpur, India
Anand Singh Jalal	GLA University, India
Anuj S. Saxena	National Defence Academy, India
Arfat Ahmad Khan	Khon Kaen University, Thailand
Arti Kashyap	IIT Mandi, India
Carlos Ordonez	University of Houston, USA
Rinkle Rani	Thapar University, India
Himanshu Gupta	IBM Research, India
Jainendra Shukla	IIIT Delhi, India
Karan Goyal	IIIT Delhi, India
Koustav Rudra	IIT Kharagpur, India
Krishna Reddy	IIIT Hyderabad, India
Ladjel Bellatreche	LIAS/ISAE-ENSMA, France
Mayank Singh	IIT Gandhinagar, India
Md. Shad Akhtar	IIIT Delhi, India
Naveen Kumar	University of Delhi, India
Navneet Goyal	BITS Pilani, India
Nazha Selmaoui-Folcher	ISEA/UNC, France
Neeraj Goel	IIT ROPAR, India
Poonam Goyal	BITS Pilani, India
Praveen Rao	University of Missouri-Columbia, USA
Pravin Nagar	IIIT Delhi, India
R. K. Agarwal	JNU Delhi, India
Raghava Mutharaju	IIIT Delhi, India
Rajeev Gupta	Microsoft, India
Ravindranath C. Jampani	Oracle Labs, India

Sadok Ben Yahia	Southern Denmark University, Denmark
Sahely Bhadra	IIT Palakkad, India
Sanjay Kumar Madria	Missouri S&T, USA
Shantanu Sharma	New Jersey Institute of Technology, USA
Sourav Kumar Dandapat	IIT Patna, India
Suresh Kumar	MRIIRS, India
Tzung-Pei Hong	National University of Kaohsiung, Taiwan
Uday Kiran Rage	University of Aizu, Japan
Vasudha Bhatnagar	University of Delhi, India
Venktesh Viswanathan	TU Delft, The Netherlands
Vikram Goyal	IIIT Delhi, India
Yash Kumar Atri	IIIT Delhi, India

Publication Chairs

Navneet Goyal	BITS Pilani, India
Dhruv Kumar	IIIT Delhi, India

Industry Track Chairs

Rajeev Gupta	Microsoft, India
Ashish Kundu	Cisco Research, India

Sponsorship Chairs

Sumit Bhatia	Adobe, India
Ankur Narang	Sigmoid Star, India

Workshop Chairs

Viswanath Gunturi	IIT Ropar, India
Arti Kashyap	IIT Mandi, India

Tutorial Chairs

Srikanta Bedathur	IIT Delhi, India
Jixue Liu	University of South Australia, Australia

Publicity Chairs

Neeraj Garg	Maharaja Agrasen Institute of Technology, India
Raj Sharma	Walmart, India

Diversity and Inclusion Chairs

Jyoti D. Pawar	Goa University, India
Om Pal	University of Delhi, India
Vinay Thakur	Ministry of Electronics and Information Technology, India

Student Chairs

Yash Kumar Atri	IIIT Delhi, India
Karan Goyal	IIIT Delhi, India
Akshit Jindal	IIIT Delhi, India

Web Chair

Yash Kumar Atri	IIIT Delhi, India

Panel Chairs

Vasudha Bhatnagar	Delhi University, India
Sameep Mehta	IBM Research, India

Keynote Lectures

Representation Learning for Dialog Models

Manish Gupta

Microsoft, Hyderabad, India
gmanish@microsoft.com

Abstract. Dialog models are useful in building chatbot systems where a bot generates a response given the conversation history, persona information, information about the topic of the conversation or any other context. The right representation of the input is crucial for effective and efficient response generation. In this talk, I will discuss ways to arrive at such input representations for dialog models in three settings. In the first setting, we only have text conversation history as input. Here we propose a novel pretraining objective called "Discourse Mutual Information" [3] that leverages the discourse-level organizational structure of dialog texts. In the second "comics" setting [1], we have a multimodal conversation history with persona information of characters as input. Here, we propose a multimodal embedding module that encodes both text and images using PersonaGPT and CLIP-ViT respectively. In the third setting, we have a rich input comprising text conversation history, persona information or information about the topic of the conversation. Here, we explore an efficient input representation to design prompts for in-context learning models and explore the tradeoff between model performance and cost [2]. Through various empirical results, I will show the effectiveness of the proposed dialog representation models in all the three settings.

Keywords: NLP · Dialog modeling · Large language models · Deep learning for NLP · Multimodal dialog models

Biography

Manish Gupta is a Principal Applied Researcher at Microsoft India R&D Private Limited at Hyderabad, India. He is also an Adjunct Faculty at International Institute of Information Technology, Hyderabad and a visiting faculty at Indian School of Business, Hyderabad. He received his Masters in Computer Science from IIT Bombay in 2007 and his Ph.D. from the University of Illinois at Urbana-Champaign in 2013. Before this, he worked for Yahoo! Bangalore for two years. His research interests are in the areas of deep learning, natural language processing, web mining and data mining. He has published more than 150 research papers in reputed refereed journals and conferences. He has also co-authored two books: one on Outlier Detection for Temporal Data and another one on Information Retrieval with Verbose Queries. He also runs a Youtube Channel called DataScienceGems on recent advancements in deep learning for NLP and vision.

References

1. Agrawal, H., Mishra, A., Gupta, M., Mausam: Multimodal persona based generation of comic dialogs. In: ACL 2023 (July 2023)
2. Santra, B., Basak, S., De, A., Gupta, M., Goyal, P.: Frugal prompting for dialog models. In: EMNLP (Findings) (December 2023)
3. Santra, B., et al.: Representation learning for conversational data using discourse mutual information maximization. In: NAACL (Jul 2022)

Sparsity, Modularity, and Structural Plasticity in Deep Neural Networks

Constantine Dovrolis[1,2]

[1] The Cyprus Institute, Nicosia, Cyprus
c.dovrolis@cyi.ac.cy

[2] School of CS, Georgia Institute of Technology, Atlanta, GA, USA
constantine@gatech.edu

Extended Abstract

There is a growing overlap between Machine Learning, Neuroscience, and Network Theory. These three disciplines create a fertile inter-disciplinary cycle: a) inspiration from neuroscience leads to novel machine learning models and deep neural networks in particular, b) these networks can be better understood and designed using network theory, and c) machine learning and network theory provide new modeling tools to understand the brain's structure and function, closing the cycle.

In this talk, we will "tour" this cross disciplinary research agenda by focusing on three recent works: a) the design of sparse neural networks that can learn fast and generalize well (PHEW), b) the use of structural adaptation for continual learning (NISPA), and c) the emergence of hierarchically modularity in neural networks (Neural Sculpting).

PHEW: Methods that sparsify neural networks at initialization enhance learning and inference efficiency. Our work leverages a decomposition of the Neural Tangent Kernel (NTK): the architecture-dependent Path Kernel. The Synflow-L2 algorithm, though optimal for convergence, leads to "bottleneck" layers in sub-networks. To address this, we introduce Paths with Higher-Edge Weights (PHEW), a data-agnostic method for constructing sparse networks based on biased random walks dependent on initial weights. PHEW retains path kernel properties akin to Synflow-L2 but produces wider layers, resulting in improved generalization and performance, outperforming data-independent SynFlow and SynFlow-L2 across various network densities.

NISPA: Continual learning (CL) aims to sequentially learn tasks, preserving performance on old tasks and efficiently adapting to new ones without requiring an expanding model or repeated training. Addressing this, we introduce the Neuro-Inspired Stability-Plasticity Adaptation (NISPA) architecture—a sparse neural network with constant density. NISPA uses stable paths to retain older task knowledge and connection rewiring to form plastic paths for new tasks, reusing established knowledge. Evaluations on various image datasets confirm NISPA's superior performance over state-of-the-art baselines, with up to tenfold fewer learnable parameters. We highlight sparsity as pivotal for effective continual learning.

Neural Sculpting: Natural tasks often display hierarchical modularity, allowing decomposition into simpler sub-functions with distinct inputs (inputseparability) that are reused higher in the hierarchy (reusability). Hierarchically modular neural networks, inherently sparse, promote learning efficiency, generalization, and transferability. This work seeks to identify such sub-functions and their structure within tasks. Specifically, we investigate Boolean functions to discern task modularity. Our approach employs iterative unit and edge pruning during training, paired with network analysis for module detection and hierarchy inference. We illustrate the effectiveness of this method on a variety of Boolean functions and vision tasks from the MNIST dataset.

Short Bio

Dr. Constantine Dovrolis is the Director of the center for Computational Science and Technology (CaSToRC) at The Cyprus Institute (CyI) as of 1/1/2023. He is also a Professor at the School of Computer Science at the Georgia Institute of Technology. He is a graduate of the Technical University of Crete (Engr.Dipl. 1995), University of Rochester (M.S. 1996), and University of Wisconsin-Madison (Ph.D. 2000). His research is highly inter-disciplinary, combining Network Theory, Data Mining and Machine Learning. Together with his collaborators and students, they have published in a wide range of scientific disciplines, including climate science, biology, and neuroscience. More recently, his group has been focusing on neuro-inspired architectures for machine learning based on what is currently known about the structure and function of brain networks. According to Google Scholar, his publications have received more than 15,000 citations with an h-index of 56. His research has been sponsored by US agencies such as NSF, NIH, DOE, DARPA, and by companies such as Google, Microsoft and Cisco. He has published at diverse peer-reviewed conference and journals such as the International Conference on Machine Learning (ICML), the ACM SIGKDD conference, PLOS Computational Biology, Network Neuroscience, Climate Dynamics, the Journal of Computational Social Networks, and others.

Contents

Artificial Intelligence in Healthcare

Tuberculosis Disease Diagnosis Using Controlled Super Resolution

P. V. Yeswanth(✉), Kunal Vijay Thool, and S. Deivalakshmi

Department of Electronics and Communication Engineering, National Institute of Technology, Tiruchirappalli, Tiruchirappalli, India
yeswanth2core@gmail.com, deiva@nitt.edu

Abstract. Tuberculosis (TB) is a chronic respiratory disease caused by a bacterial infection and has one of the highest mortality worldwide. Timely and precise TB detection is crucial as it can be dangerous if left untreated. To achieve accurate results, it is essential to have a high-resolution input. This paper introduces a Low and high level feature steering (LHFS) module, which reconstructs high-resolution images by a reference image that contains same information to the low-resolution input. Additionally, the Selective feature integration (SFI) module seamlessly integrates Ref image features into extracted features of LR image. The proposed model for factors 2, 4, and 6, attains super resolution metrics such as PSNR values of 30.225, 31.176, 33.836, and SSIM values of 0.8642, 0.8801, 0.9052 with classification metric accuracy values of 99.66, 98.96 and 98.32 respectively.

Keywords: Super Resolution · Selective feature integration · Low and high level feature steering · Tuberculosis detection · High resolution · Low resolution

1 Introduction

Tuberculosis (TB) is a contagious illness caused by the bacterium known as Mycobacterium tuberculosis. TB can impact not only the lungs but also various other body parts including the brain, spine and kidneys. The primary mode of TB transmission is through the air when an infected person releases bacteria-containing droplets into the environment through coughing, sneezing, or speaking. When these infected droplets are inhaled by another person, the bacteria can enter their respiratory system and lead to an active TB infection. Not all individuals infected with M. tuberculosis progress to develop active TB disease. In many cases, the immune system can contain bacteria, resulting in a latent TB infection. It also does not have any symptoms. It is not contagious, but the bacteria can remain in the body and become active in the future, leading to TB disease if the immune system weakens. TB presents several symptoms, including a persistent cough lasting over three weeks, loss of appetite and weight, chest pain, night sweats, high body temperature, fatigue, and general feelings of tiredness. Millions of people die every year due to TB. Detecting the disease early on can help with timely treatment and lower the chances of death caused by the illness.

V. Goyal et al. (Eds.): BDA 2023, LNCS 14418, pp. 3–15, 2023.
https://doi.org/10.1007/978-3-031-49601-1_1

Deep learning has gained prominence due to its ability to efficiently process and analyze massive volumes of data. Convolutional Neural Networks (CNNs) are among the most widely adopted deep neural networks. Deep learning has witnessed significant advancements and has become the preferred approach in the Radiology field. CNN has been instrumental in extracting features for detecting TB disease and classifying images. Researchers in medical field have integrated deep learning techniques into their work to automate disease detection.

Image super-resolution involves enhancing the resolution and quality of a low resolution image to produce a higher resolution version with improved details and sharpness. Image super-resolution methods are assessed by metrics like PSNR, SSIM, and perceptual quality assessments. These metrics are used to evaluate the functioning of methods objectively and measure factors such as image fidelity, similarity, and perceived quality. It is widely used for digital photography, medical imaging, surveillance systems, and video processing. The goal is to recover the detailed high frequency information that is lost during the process of downscaling an image. This process entails estimating the high-resolution image based on provided low-resolution input.

The paper presents the fundamental concept to utilize a Ref image that has the same information as LR input image to control the super-resolution process. In Sect. 2, the related work is discussed. In Sects. 3 and 4, the methodology and results are discussed. The conclusion is drawn out in Sect. 5.

2 Related Work

The increasing availability of computational power and advancements in computer hardware have led to the widespread adoption of deep learning techniques. Deep learning-based image reconstruction methods utilize nonlinear representation modules to learn intricate mapping relationships between images, resulting in higher and more abstract data representations.

Dong et al. introduced SRCNN [1] consisting of three layers, which were used to determine the nonlinear mapping between HR and LR images. Chen et al. [2] proposed attention in attention networks (A^2N). In this model, efficiency was quantified of different attention layers in a neural network to dynamically generate its weights. This allowed it to improve its capacity. Hu et al. [3] proposed a model that combined spatial and channel-wise attention to effectively modulate features for capturing more information for a longer period. Zhang et al. [4] incorporated an attention mechanism in residual blocks, introducing a highly deep network known as RCAN. This innovative approach significantly pushed forward the performance boundaries in SISR.

In contrast to SISR techniques, RefSR methods have the advantage of extracting more precise details from the reference image by various approaches such as patch matching or image alignment. In the CrossNet approach [5], optical flow was utilized to synchronize Ref and LR at various scales. The aligned images were then concatenated into their respective decoder layers. The effectiveness relies on the quality of alignment of Ref and LR. This alignment step significantly influences the overall effectiveness of approach. SRNTT (Super-Resolution using Non-Local Texture Transform) [6] applied to match on features patches of LR and Ref to swap texture features which are same, to enriched HR details. MASA [7] introduced a matching scheme that progresses from coarse to fine, along with a spatial adaptation module. This module is designed to dynamically remap distribution of features of Ref to align with the distribution of features of LR in a spatially adaptive manner. Yeswanth et. al proposed super resolution-based classification models to improve the classification accuracy [8–10, 13, 19, 21, 22].

Detecting TB poses significant challenges due to the diverse array of manifestations it can exhibit, including aggregation, cavities, focal lesions, and nodules in chest X-ray images. Chauhan et al. [11] developed an integrated system for tuberculosis (TB) detection using chest X-ray images. Their system is comprised of several modules that follow a stepwise process to classify the input image. The dataset is initially subjected to a pre-processing module based on denoising, followed by an extraction function. Subsequently, a model was constructed utilizing an SVM classifier. Liu et al. [12] presented a CNN-based approach to address the challenge of unbalanced and limited-category X-ray images in TB classification. They investigated efficiency of shuffle sampling in conjunction with validation for neural network training. They achieved a classification accuracy of 85.68% on a large dataset.

Ahsan et al. [20] accomplished accuracy of 80% and 81.25% without and with utilization of image augmentation respectively using pre-trained model. Main benefit of using a pre-trained CNN model was that it eliminated the need for developing complex segmentation algorithms, which can be tedious task. Norval et al. [14] detected pulmonary TB using CNNs with chest X-ray images. Their approach combined neural networks with conventional CAD method. The study involved simulations using both normal and abnormal images. The results of the simulations indicated that excellent outcomes could be achieved by utilizing a cropped interested region along with techniques to enhance contrast.

T. Rahman et al. [15] achieved reliable detection of TB by employing a combination of techniques, including various preprocessing methods, augmentation of data, as well as utilizing image segmentation and deep learning-based classification approaches. To achieve this, they employed different CNNs and applied transfer learning by utilizing pre-trained weights.

3 Methodology

Our Model consists of Low and high level feature steering (LHFS) module and Selective feature integration (SFI) module. LHFS module extracts and steers features of both low level and high level. SFI module assigns weights to features of Ref (Ground Truth) and LR image. SFI connects features of LR and Ref to fill in missing information in the LR.

First, to steer Ref and LR we down sample Ref. Both Ref and LR are then passed through convolutional layers, which extract feature maps at various levels for steering purposes. HR refers to high resolution image, LR refers to low resolution image, and Ref signifies the down-sampled reference image. Then, for finding similarity maps LHFS steers feature from low and high level. Later, Ref, LR and similarity map are given to swapping operator and replace patches of LR with corresponding Ref. This new image is feed to CIF module. To prevent transfer of unnecessary content to LR and maximize utilization of available contents in Ref, spatial and channel attention modules was devised. By applying these modules, features of LR and Ref are combined, enabling a selective connection of features of Ref to features of LR. Additionally, a residual block (RB) is employed to enhance high-frequency details in the final output.

3.1 Low and High Level Feature Steering (LHFS)

This sub-section primarily discusses Low and High level Feature Steering (LHFS). LHFS is employed to ensure precise steering between low level and high level features, mitigating any potential confusion caused by similar objects. The layers of high level contribute meaningful information, while layers of low level focus on capturing local features. To address the challenge of local region confusion in steering similar objects, we conduct steering of features of high level before proceeding to steering of features of low level (Fig. 1).

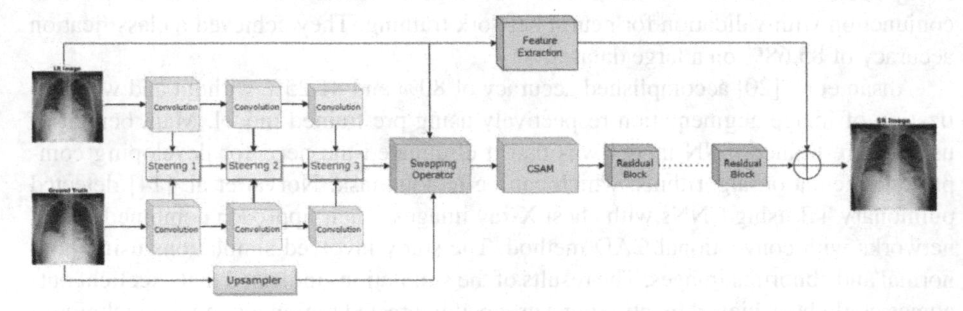

Fig. 1. Architecture of our proposed model

To facilitate feature steering of low and high levels, our approach involves using a down-sampled Ref and LR as inputs to convolutional layers. Our model aims to identify relevant textures in Ref by noting their positions at high level of network. We then query corresponding fine-grained patches in low level of network. Only patches of features of Ref that exhibit similarity to features of LR preserve corresponding positions in Ref. Irrelevant feature patches from Ref that are unrelated to the content of LR image are discarded. Next, a comparison is made between the corresponding features of LR and Ref, identifying the most similar patches based on similarity scores. For every patch, we identify the most relevant patch from Ref and obtain the corresponding position of the steered patch.

Once the similarity score and position of equivalent patches to LR are obtained, a swap operation is performed. In this operation, equivalent patches to position in Ref are substituted into LR, resulting in the generation of a HR image. Subsequently, based on the similarity score, the patch of generated HR is assigned a weight. This process ultimately yields a weighted HR image, which serves as Ref for the subsequent steps.

3.2 Selective Feature Integration

The primary objective of the Selective feature Integration (SFI) module is to integrate Ref features into LR features selectively. This is achieved through a combination of spatial and channel attention modules and use of a residual block (RB). By leveraging the attention mechanism, which effectively utilizes spatial and channel-wise features present in networks, we have designed CSAM. This module allows for the selective utilization of informative Ref features to enhance LR image (Fig. 2).

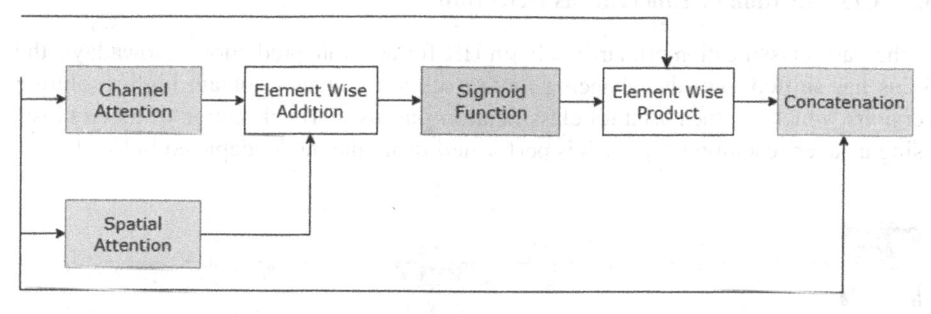

Fig. 2. CSAM Block

During LR reconstruction, we utilize EDSR [16] network structure to extract feature of HR from the image given as input. Next, the HR features are combined with the features that are swapped of Ref and fed into CSAM module. Within CSAM module, Ref features are selected using channel and spatial attention mechanisms. These selected Ref features are then connected with features of HR to enhance image given as input. Lastly, the feature transformation is accomplished by applying an RB (Residual Block) module, leading to the generation of HR. Our approach combines spatial and channel attention using extracted features from input. By connecting the Ref features, an attention map is obtained through feature fusion.

Channel Attention (CA) aids in capturing relevant information within different channels and spatial regions, which contributes in different manners to recovering of high-frequency details. Instead of using the average pooling method, we utilize the variance as pooling method for a more suitable representation. Since primary objective of SR is to restore high-frequency components, it is logical to determine attention maps based on high-frequency details within channels. The scaling and excitation processes in CA follow same approach of RCAN model [18].

Spatial Attention (SA) is employed to account for the diverse information present in input and feature maps across different spatial positions. Notably, regions with edges or textures typically contain higher-frequency information, which is crucial for restoring the details of the input image. By leveraging SA mechanism, network gains distinguish capabilities of various places greater emphasis on regions that are more significant and challenging to reconstruct and local regions. Moreover, the RB ensures that information is not compressed across channels, thereby preserving channel-specific characteristics.

The process of feature fusion entails combining attention mechanisms related to spatial and channel aspects through an addition-based operation. The resulting output is then passed through a sigmoid function for further processing. Afterward, the replaced features of Ref are multiplied by it. Next, features are linked to initial features preceding starting step, enabling integration of guidance information from Ref to populate corresponding positions determined by mechanism of attention. This fusion incorporates the necessary information from Ref to complement details of LR while avoiding errors due to same objects. By employing an attention mechanism, we achieve adaptive feature fusion, ensuring that only relevant and useful information is incorporated.

3.3 Classification of Tuberculosis Detection

In the past, classification primarily rely on HR for accurate predictions. Nowadays, the focus has shifted towards enhancing low-resolution images to obtain high-resolution versions, which are then used for classification purposes. The detection of tuberculosis using a super-resolution approach is performed using the model depicted in Fig 3.

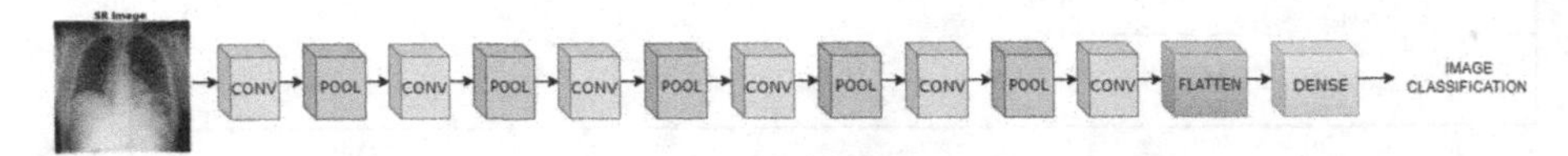

Fig. 3. Classification model

3.4 Dataset

On the X ray image dataset, which is open to the public, our high to low level feature steering model is assessed. Two classes make up this dataset. In these images, the backgrounds are all the same. There is also not much diversity in intensity (Table 1 and Fig. 4).

Table 1. Dataset information

Type of X ray	No of images
Normal	3500
Tuberculosis	700
Total images	4200

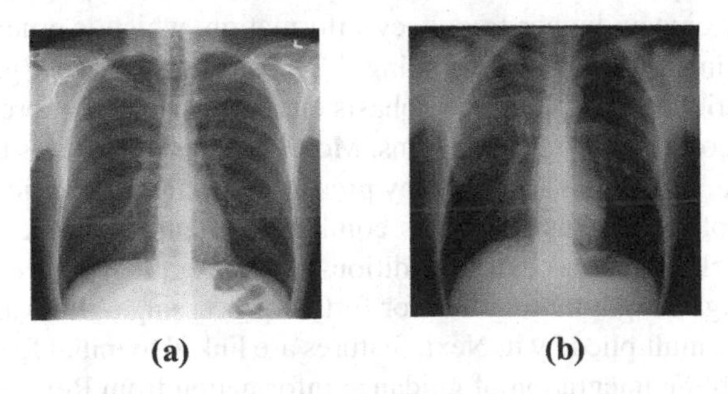

(a) **(b)**

Fig. 4. (a). Normal X ray (b). Tuberculosis X ray

The dataset is divided into three parts, with a split ratio of 80:10:10. Specifically, 80% of the dataset is allocated for training the model, while 10% is randomly selected for testing purposes. The remaining 10% is set aside for validation.

3.5 Experimentation

The experimentation involves a dataset of 4200 images, which are subjected to downsampling with factors 2, 4, and 6 to generate LR, as depicted in Fig. 5.

These Low-Resolution images will be used as input to our classification model. These are fed to the SR block to get HR. These images are then passed through the classification block in Fig. 3 to diagnose the disease.

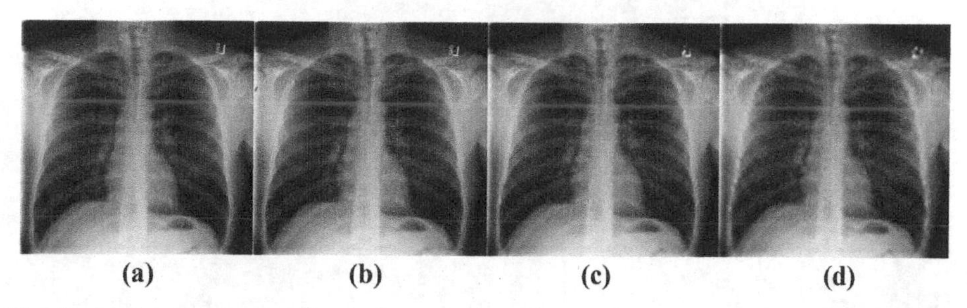

Fig. 5. (a) Original Image (b). LR image (Factor 2) (c). LR image (Factor 4) (d). LR image (Factor 6)

Glorot uniform for Initialization of weights, learning rate (α) value 0.001, batch size for training 32 are the hyper parameters are used for training SR, classification blocks.

Here, the measures PSNR and SSIM are used for training by the SR block. MSE (Mean Square Error) loss function was used for training.

Train cum validation losses versus the epochs are shown in Fig. 6 for various down sampling factors (Fig. 7).

4 Results

4.1 Qualitative Analysis

The qualitative analysis presented in Fig. 8 showcases SR images produced using different scaling factors from corresponding LR. Performance analysis of our model for these diverse scaling factors is provided in Tables 2 and 3. The results presented in Tables 2 and 3 demonstrate that proposed model achieves disease prediction accuracy exceeding 98.56%, with PSNR exceeding 30 and SSIM above 0.8.

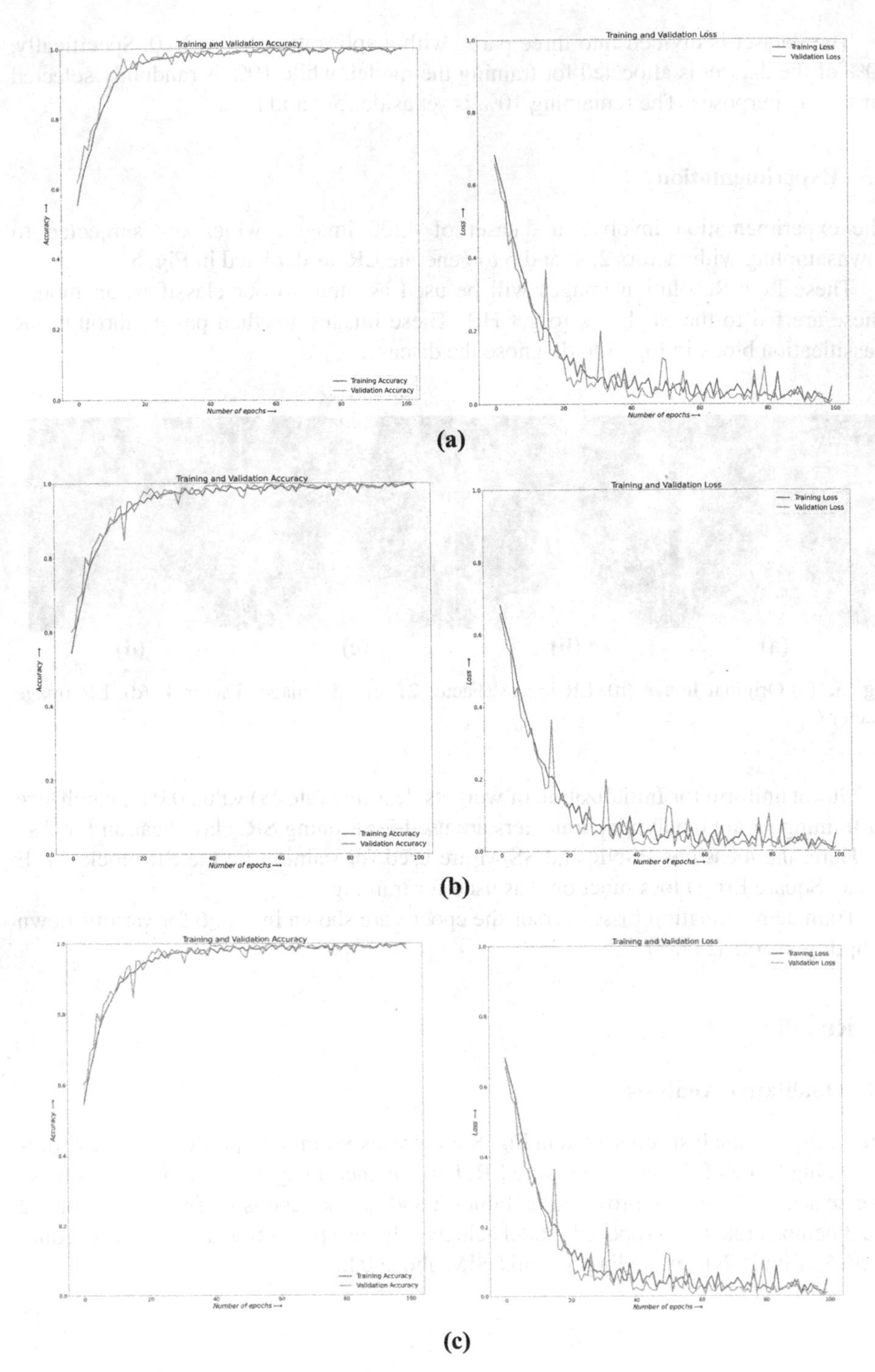

Fig. 6. (a). Accuracies and losses of training and validation vs. No. of epochs (factor 2) **(b).** Accuracies and losses of training and validation vs. No. of epochs (factor 4) **(c).** Accuracies and losses of training and validation vs No. of epochs (factor 6)

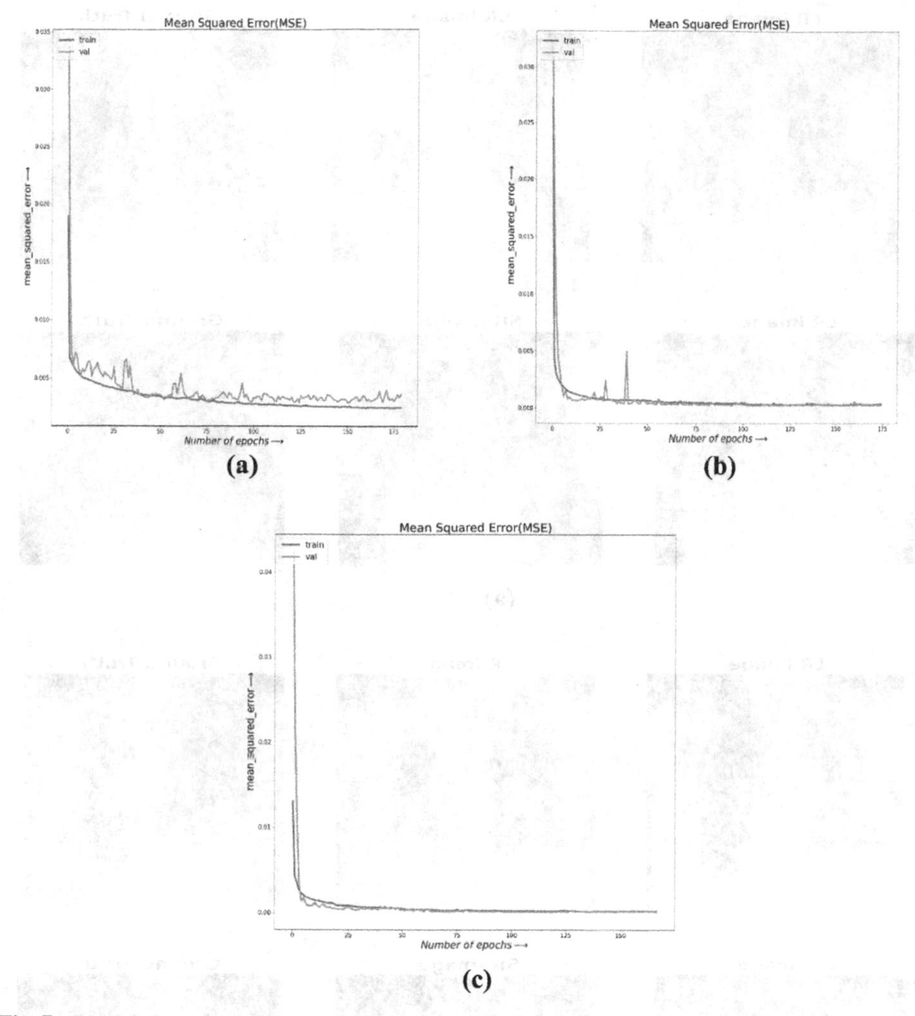

Fig. 7. Mean squared error versus the number of epochs of super resolution block **(a).** Super Resolution (factor 2) **(b).** Super Resolution (factor 4) **(c).** Super Resolution (factor 6)

4.2 Quantitative Analysis

See Table 4.

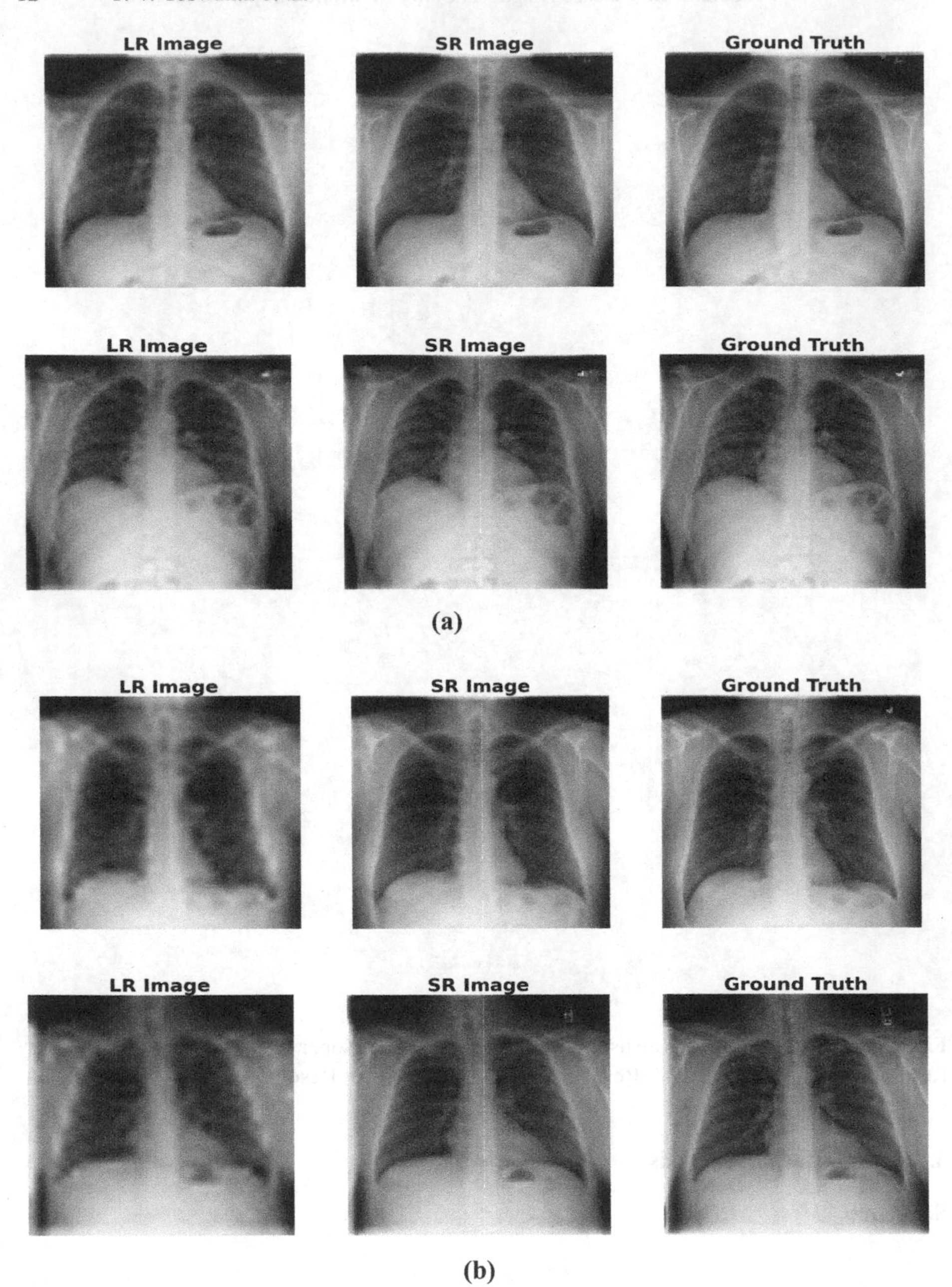

Fig. 8. **(a).** Super-resolution (factor 2) **(b).** Super-resolution (factor 4) **(c).** Super-resolution (factor 6)

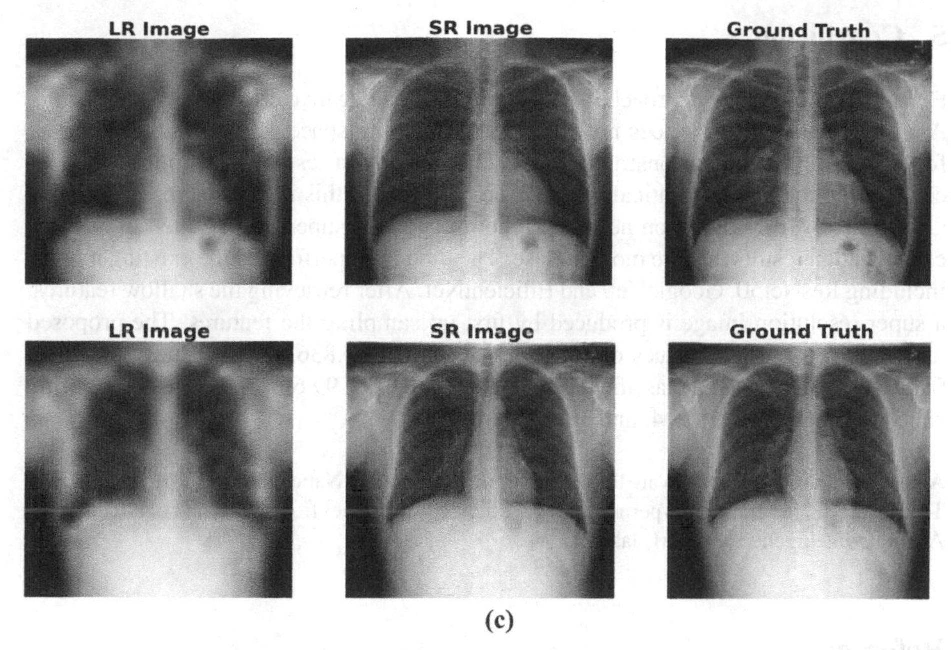

(c)

Fig. 8. (*continued*)

Table 2. Performance Evaluation of CSR-TDD

Metrics	SR image (factor 2)	SR image (factor 4)	SR image (factor 6)
PSNR	30.225	31.176	33.836
SSIM	0.8642	0.8801	0.9052

Table 3. Classification model Performance evaluation

Metrics	SR image (factor 2)	SR image (factor 4)	SR image (factor 6)
Accuracy	99.66	98.96	98.32

Table 4. Comparison with existing models

Metrics	ResNet50	GoogleNet	EfficientNet	CSR-TDD (Proposed)
Accuracy	94.31	96.15	97.51	99.66

5 Conclusion

Existing deep learning approaches fail to consider the negative effects of visual system. As a result, the network does not confine the mapping space. This is not appropriate for high-quality image reconstruction since it arbitrarily raises image resolution without considering the image's critical information. To address this issue, this research offers a controlled super-resolution network. Input images are super-resolutioned to produce classification results that are more precise. The model outperformed the existing models including ResNet50, GoogleNet, and EfficientNet. After retrieving the shallow features, a super-resolution image is produced by first up sampling the features. The proposed model achieves PSNR values of 30.225, 31.176, and 33.836, SSIM values of 0.8642, 0.8801, and 0.9052, and classification accuracy values of 99.66, 98.96, and 98.32 using super-resolution factors 2, 4, and 6.

Acknowledgement. Authors are thankful to the Director of the National Institute of Technology - Tiruchirappalli for granting us permission to use the GPU resources from the Center of Excellence – Artificial Intelligence (CoE-AI) lab.

References

1. Dong, C., Loy, C.C., He, K., Tang, X.: Image super-resolution using deep convolutional networks. IEEE Trans. Pattern Anal. Mach. Intell. **38**(2), 295–307 (2015)
2. Chen, H., Gu, J., Zhang, Z.: Attention in attention network for image super-resolution. arXiv preprint arXiv:2104.09497
3. Hu, Y., Li, J., Huang, Y., Gao, X.: Channel-wise and spatial feature modulation network for single image super-resolution. IEEE Trans. Circuits Syst. Video Technol. **30**(11), 3911–3927 (2019)
4. Dai, T., Zha, H., Jiang, Y., Xia, S.-T.: Image super-resolution via residual block attention networks, pp. 3879–3886 (2019). https://doi.org/10.1109/ICCVW.2019.00481
5. Zheng, H., Ji, M., Wang, H., Liu, Y., Fang, L.: CrossNet: an end-to-end reference-based super resolution network using cross-scale warping. In: Ferrari, V., Hebert, M., Sminchisescu, C., Weiss, Y. (eds.) ECCV 2018. LNCS, vol. 11210, pp. 87–104. Springer, Cham (2018). https://doi.org/10.1007/978-3-030-01231-1_6
6. Zhang, Z., Wang, Z., Lin, Z., Qi, H.: Image super-resolution by neural texture transfer. In: Proceedings of the IEEE/CVF Conference on Computer Vision and Pattern Recognition, pp. 7982–7991 (2019)
7. Lu, L., Li, W., Tao, X., Lu, J., Jia, J.: MASA-SR: matching acceleration and spatial adaptation for reference-based image super-resolution. In: IEEE/CVF Conference on Computer Vision and Pattern Recognition (CVPR), pp. 6368–6377 (2021)
8. Yeswanth, P.V., Deivalakshmi, S., George, S., Ko, S.-B.: Residual skip network-based super-resolution for leaf disease detection of grape plant. Circ. Syst. Signal Process. **42**(11), 6871–6899 (2023). https://doi.org/10.1007/s00034-023-02430-2
9. Yeswanth, P.V., Khandelwal, R., Deivalakshmi, S.: Super resolution-based leaf disease detection in potato plant using broad deep residual network (BDRN). SN Comput. Sci. **4**(2), 112 (2022)
10. Yeswanth, P.V., Deivalakshmi, S.: Extended wavelet sparse convolutional neural network (EWSCNN) for super resolution. Sādhanā **48**(2), 52 (2023)

11. Chauhan, A., Chauhan, D., Rout, C.: Role of gist and PHOG features in computer-aided diagnosis of tuberculosis without segmentation. PLoS ONE **9**, e112980 (2014). https://doi.org/10.1371/journal.pone.0112980
12. Liu, C., et al.: TX-CNN: detecting tuberculosis in chest X-ray images using convolutional neural network. In: 2017 IEEE International Conference on Image Processing (ICIP), Beijing, China, pp. 2314–2318 (2017). https://doi.org/10.1109/ICIP.2017.8296695
13. Yeswanth, P.V., Raviteja, R., Deivalakshmi, S.: Sovereign Critique Network (SCN) Based Super-Resolution for chest X-rays images. In: 2023 International Conference on Signal Processing, Computation, Electronics, Power and Telecommunication (IConSCEPT), Karaikal, India, pp. 1–5 (2023). https://doi.org/10.1109/IConSCEPT57958.2023.10170157
14. Norval, M., Wang, Z., Sun, Y.: Pulmonary Tuberculosis Detection Using Deep Learning Convolutional Neural Networks, pp. 47–51 (2019). https://doi.org/10.1145/3376067.3376068
15. Rahman, T., et al.: Reliable tuberculosis detection using chest X-ray with deep learning, segmentation and visualization. IEEE Access **8**, 191586–191601 (2020). https://doi.org/10.1109/ACCESS.2020.3031384
16. Lim, B., Son, S., Kim, H., Nah, S., Mu Lee, K.: Enhanced deep residual networks for single image super-resolution. In: Proceedings of the IEEE Conference on Computer Vision and Pattern Recognition Workshops, pp. 136–144 (2017)
17. Kim, J.-H., Choi, J.-H., Cheon, M., Lee, J.-S.: RAM: residual attention module for single image super-resolution. arXiv preprint arXiv:1811.12043 (2018)
18. Zhang, Y., Li, K., Li, K., Wang, L., Zhong, B., Fu, Y.: Image super-resolution using very deep residual channel attention networks. In: Ferrari, V., Hebert, M., Sminchisescu, C., Weiss, Y. (eds.) ECCV 2018. LNCS, vol. 11211, pp. 294–310. Springer, Cham (2018). https://doi.org/10.1007/978-3-030-01234-2_18
19. Yeswanth, P.V., Kushal, S., Tyagi, G., Kumar, M.T., Deivalakshmi, S., Ramasubramanian, S.P.: Iterative super resolution network (ISNR) for potato leaf disease detection. In: 2023 International Conference on Signal Processing, Computation, Electronics, Power and Telecommunication (IConSCEPT), Karaikal, India, pp. 1–6 (2023). https://doi.org/10.1109/IConSCEPT57958.2023.10170224
20. Ahsan, M., Gomes, R., Denton, A.: Application of a convolutional neural network using transfer learning for tuberculosis detection. In: Proceedings of the IEEE International Conference Electro Information Technology (EIT), pp. 427–433, May 2019
21. Yeswanth, P.V., Khandelwal, R., Deivalakshmi, S.: Two fold extended residual network based super resolution for potato plant leaf disease detection. In: Misra, R., Rajarajan, M., Veeravalli, B., Kesswani, N., Patel, A. (eds.) ICIoTCT 2022. LNCS, vol. 616, pp. 197–209. Springer, Singapore (2023). https://doi.org/10.1007/978-981-19-9719-8_16
22. Yeswanth, P.V., Rajan, J., Mantha, B.P., Siva, P., Deivalakshmi, S.: Self-Governing Assessment Network (SGAN) based super-resolution for CT chest images. In: 2023 International Conference on Computer, Electronics & Electrical Engineering & their Applications (IC2E3), Srinagar Garhwal, India, pp. 1–5 (2023). https://doi.org/10.1109/IC2E357697.2023.10262573

GREAT AI in Medical Appropriateness and Value-Based-Care

V Dinesh Datta[1], Sakthi Ganesh[2], Roland E. Haas[2,3], and Asoke K. Talukder[4,5,6,7](✉)

[1] Kakatiya Medical College, Rangampet Street, Nizampura, Warangal, Telangana 506007, India
[2] XAITeck GmbH, Heinrich-Otto-Straße 71, 73240 Wendlingen, Germany
sakthi.ganesh@xaiteck.ai, roland.haas@iiitb.ac.in
[3] International Institute of Information Technology, 26/C, Electronics City, Hosur Road, Bengaluru 560100, India
[4] BlueRose Technologies, 1-1, Langford Road, Shanti Nagar, Bengaluru 560027, India
asoke.talukder@bluerose-tech.com
[5] Computer Science & Engineering, National Institute of Technology, Surathkal, India
[6] XAITeck GmbH, Heinrich-Otto-Straße 71, 73240 Wendlingen, Germany
[7] SRIT, 113/1B ITPL Road, Brookfield, Bengaluru 560037, India

Abstract. Fee For Service, also known as Volume Based Care (VBC) model of healthcare encourages service volume – more service more reward. This model of care results in unnecessary, inappropriate, and wasted medical services. In the US, Fraud, Waste, and Abuse (FWA) ranges between $760 billion to $935 billion, accounting for approximately 25% of total healthcare spending. In India, the waste caused by FWA is estimated to be as high as 35%. This is due to a lack of smart digital health, absence of AI models, and lack of preventive vigilance against inappropriate medical interventions. Inappropriate medical intervention costs valuable resources and causes patient harm. This paper proposes GREAT AI (Generative, Responsible, Explainable, Adaptive, and Trustworthy Artificial Intelligence) in Medical Appropriateness. We show how GREAT AI is used to offer appropriate medical services. Moreover, we show how GREAT AI can function in vigilance role to curb FWA. We present two GREAT AI models namely MAKG (Medical Appropriateness Knowledge Graph) and RAG-GPT (Retrieval Augmented Generation – Generative Pretrained Transformer). MAKG is used as an autonomous coarse-grained medical-inappropriateness vigilance model for payers and regulators. Whereas RAG-GPT is used as a fine-grained LLM, with human-in-the-loop for medical appropriateness and medical inappropriateness model where the actor human-in-the loop can be anybody like providers, patients, payers, regulators, funders, or researchers.

Keywords: Medical Appropriateness · MAKG · RAG-GPT · GREAT AI · Generative AI · Responsible AI · Explainable AI · Adaptive AI · Trustworthy AI

1 Introduction

Fee For Service (FFS), also known as Volume Based Care (VBC) model of healthcare focuses on service volume – more service more reward. This model of care encourages defensive medicine [1]. Defensive medicine results in unnecessary interventions, waste,

V. Goyal et al. (Eds.): BDA 2023, LNCS 14418, pp. 16–33, 2023.
https://doi.org/10.1007/978-3-031-49601-1_2

and abuse, which is contrary to the outcome based or value-based care (VBC). Value in VBC, is defined as the ratio of quality of care to cost of care (Value = Quality/Cost). The underlying principle of "value-based care" is managing the cost of care for a patient population while aiming to improve result-oriented outcomes. The implications of wastage in healthcare resources are multifaceted, affecting not only financial aspects but also patient care. Inappropriate and unnecessary medical interventions, including overtreatment, overprescription, and medical misconduct pose risks to patient safety. Any medical service that is not value-based-care is considered as low-value-care. Addressing these issues is paramount to optimizing healthcare systems, enhancing patient care, and achieving efficient resource allocation.

About 20% to 40% of the total expenditure in healthcare is wasted due to fraud, waste, and abuse (FWA). In advanced economies where digital health and vigilance is in use, the estimated cost of waste is at the lower end of the spectrum. In the US healthcare system, this wastage ranges from $760 billion to $935 billion, accounting for approximately 25% of total healthcare spending [2]. In India, the waste due to FWA in Ayushman Bharat is estimated to be as high as 35% [3]. This is due to lack of smart digital health, no use of AI, and no appropriate vigilance against inappropriate medical interventions. In an extreme case of medical fraud, the FBI in the US caught a hematologist-oncologist for a crime that included administering medically unnecessary infusions to 553 individuals and submitting to Medicare and private insurance companies approximately $34 million in fraudulent claims [4].

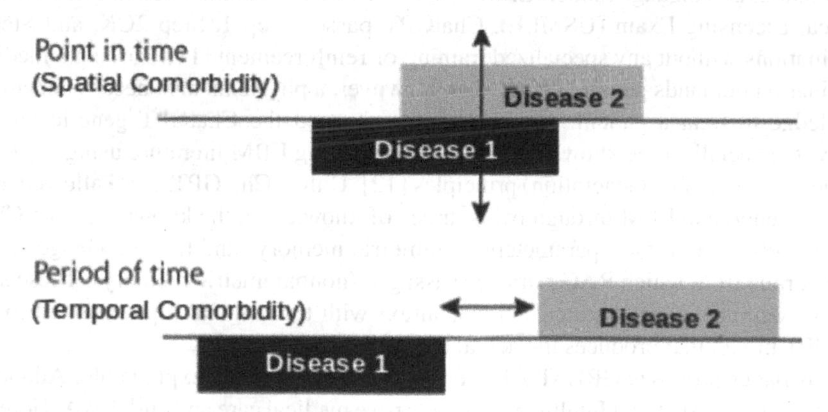

Fig. 1. Multimorbidity over space and time

Noncommunicable diseases (NCDs) cause 71% of all deaths globally [5]. NCDs are interrelated in space and time. This is known as comorbidity or multimorbidity at a point-in-time or over a period-of-time (see Fig. 1). Treating a symptom works well in infectious disease, however in case of NCD, where the symptoms are overlapping and interrelated, the patient needs to be treated as a whole. For the treatment of NCD, understanding the root cause of the symptom and the body system is critical. Also, in NCD, having a vigilance system for identifying appropriate or inappropriate intervention or a potential FWA case is a major challenge.

The concept of value-based care (VBC) was first attempted through the Inpatient Prospective Payment based on the DRG (Diagnosis Related Groups) system model [6]. The DRGs, group patients into classes that are clinically similar. In DRG, medical services offered by hospitals are considered as a "service product" like any other service industry that are governed by resource-based pricing and the principles of quality assurance. DRGs help determine how much the insurances company pays a hospital for each "product" – since patients within each DRG are clinically similar and are expected to consume the same level of hospital resources [7]. It is an attempt to move away from fee for service model to Merit-Based Incentive Payment System, to tie physician payments to quality and cost [8].

In this paper, we define the medical appropriateness in terms of evidence-based medicine (EBM). Evidence-based medicine (EBM) uses the scientific method to organize and apply current data (evidence) to improve healthcare decisions [9]. We used EBM in terms of DRG (Diagnosis Related Group) for patient classification, MDC (Major Diagnostic Categories) for body systems, ICD (International Classification of Disease) for disease categorization, HCPT (Healthcare Common Procedural Terminology) for surgery and procedures, and ATC (Anatomical Therapeutic Chemicals) for medicines and therapeutic drugs.

We constructed the Medical Appropriateness Knowledge Graph (MAKG) using Symbolic AI, Sub-symbolic AI, and Graphs database. We used Generative AI ChatGPT [10] for evidence based medical text understanding and appropriateness. A recent study found that large language model (LLM) ChatGPT passed examinations in United States Medical Licensing Exam (USMLE). ChatGPT passed Step 1, Step 2CK, and Step 3 examinations without any specialized training or reinforcement [11]. Passing a medical examination demands generic knowledge. However, a physician also needs specialized knowledge to treat a patient. Therefore, we enhanced the ChatGPT generic knowledge with specialist type knowledge through supporting EBM literature using the RAG (Retrieval Augmented Generation) principles [12]. Unlike ChatGPT, RAG allowed us to enhance generative LLM through two sources of knowledge: the knowledge that ChatGPT models store in their parameters (parametric memory) and the knowledge stored in the corpus from which RAG retrieves passages (nonparametric memory). These supporting documents are concatenated as context with the original input and fed to the ChatGPT model that produces the actual output with provenance.

This paper proposes GREAT AI (Generative, Responsible, Explainable, Adaptive, and Trustworthy Artificial Intelligence) to improve medical care and curb FWA. Here we show how RAG and medical appropriateness knowledge graph MAKG are constructed from medical ontologies and evidence-based medicine. To arrive at an accurate medical decision, we propose a human in the loop. From unlimited many-to-many possible relationships, RAG-GPT and MAKG helps identify the probable few options.

To the best of our knowledge, this is the first time a bouquet of AI technologies is being used to facilitate accurate medical decision making at the point-of-care. This bouquet of AI services also automates the adjudication process and vigilance of the medical service provided. This paper is divided into 6 sections. Following this Introduction section, we present the related previous works in Sect. 2. In Sect. 3 we describe the methods used for

GREAT AI. In Sect. 4 we present the Results including use cases. in Sect. 5 we discuss future work. We conclude the paper with concluding remarks in Sect. 6.

2 Related Prior Work

Medical appropriateness and patient safety have been discussed for a long time. 'Clinimetrics' term was introduced in the early 1980s by Alvan R. Feinstein for measuring the Medication Appropriateness Index (MAI) [13]. A study in Switzerland in 1990 and 1991 found that the level of inappropriate hospital utilization ranged between 8 and 15% in terms of length of stay (LOS) and was consistently higher in medical than in surgery [14]. Another study in 1996 in Canada found "that a substantial proportion of the healthcare delivered in Canada is inappropriate" [15]. Various adverse events or inappropriate medical interventions in healthcare puts a cost burden of $317.9 billion on hospitals across the US and Western Europe, adversely affecting 91.8 million patients and leading to 1.95 million deaths in 2016 [16].

To reduce medically inappropriate services and patient harm, digital health and EHR (Electronic Health Records) were introduced in the early 1990s. Technologies like CDS (Clinical Decision Support) were also introduced around the same time with the same objectives in mind. In recent times technologies like AI are also being used to address inappropriate medical interventions and FWA. Rasmy et al. used BERT (Bidirectional Encoder Representations from Transformers) for medical appropriateness [17]. Iqbal et al. in a review presented different AI methods that have been used to detect different types of fraud such as identify theft and kickbacks in healthcare [18].

Knowledge graphs were also used for medical appropriateness and FWA. Sun et al. constructed a medical knowledge graph from the China Food and Drug Administration data to identify inappropriate diagnosis/medications in the claims data [19]. Talukder et al. constructed an Drugomics knowledge graph that covers drug-drug interaction and drug-disease interaction for providers to prescribe the appropriate drug for polypharmacy and multimorbidity patients [20]. An AI-based comprehensive CDS was constructed by Talukder et al. for appropriate diagnosis [21]. Talukder et al. also created the Physicians' Brain Digital Twin that combined appropriate medical knowledge and stored in a knowledge graph [22].

Despite all these attempts, inappropriate medical interventions and patient harm is on the rise. The cost burden of patient harm is estimated to increase at a compound annual growth rate (CAGR) of 3.2%, which for 2022 was estimated to be $383.7 billion [16]. A recent review paper by Classen et al. investigated this subject under the notion of bending the patient safety curve [23].

3 Method

For medical appropriateness we need to validate that the Presented conditions (signs, symptoms, findings), Diagnosis, Procedures, Length of Stay (LOS) in the hospital, and Medication are all in-line with the evidence-based medicine (EBM) best practices. We used Generative AI in LLM (ChatGPT) and RAG (Retrieval Augmented Generation) for semantic EBM. We used Ontologies, Symbolic AI, and Knowledge Graphs for Responsible AI, Explainable AI, and Trustworthy AI. We used sub-symbolic Deep Neural Networks (DNN) and Transformers for Adaptive AI functions.

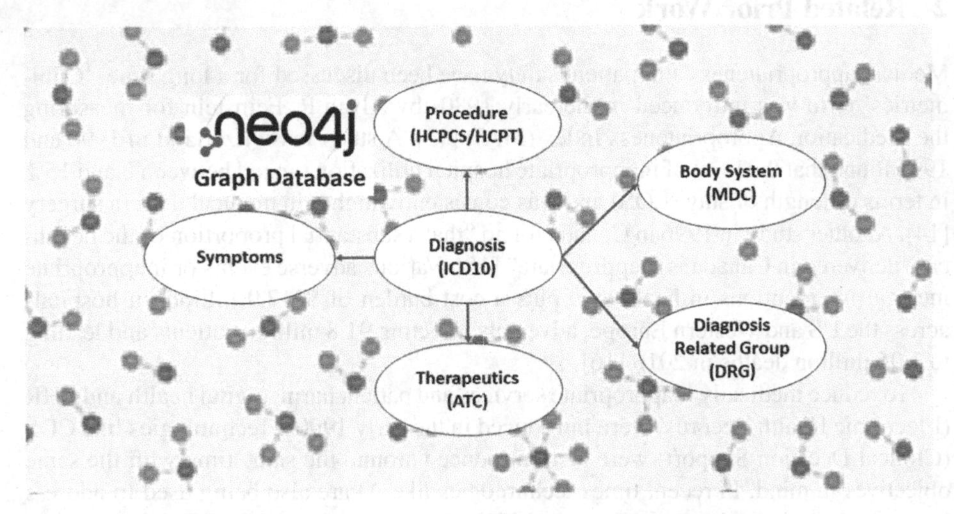

Fig. 2. Medical Appropriateness Knowledge Graph Architecture

We constructed four interconnected knowledge graphs. These knowledge graphs are constructed by means of controlled vocabularies and Symbolic AI. Symbolic AI is the collection of all methods in artificial intelligence that are based on high-level symbolic (human-understandable and human interpretable) representations of problems, logic, and search [24].

For these knowledge graphs we used following data (Fig. 2):

- The MDC codeset was downloaded from https://www-doh.state.nj.us/doh-shad/query/MDCtable.html in May 2023.
- The disease related ICD10 codeset was downloaded from https://www.cms.gov/medicare/coding-billing/icd-10-codes/2023-icd-10-cm in May 2023.
- The procedure, surgery, and hospitalization related HCPT (HCPCS II) codeset was downloaded from https://www.cms.gov/Medicare/Medicare-Fee-for-Service-Payment/HospitalOutpatientPPS/Addendum-A-and-Addendum-B-Updates in June 2023.
- The Drug or therapeutic related chemicals ATC ontology was downloaded from https://bioportal.bioontology.org/ontologies/ATC (July 2023).
- The Diagnosis Related Group (DRG) was taken from https://www.icd10data.com/ICD10CM/DRG (February 2023).

All these codes or ontologies are standalone codesets without any published relationship or crossmaps between any other codeset. The relationships were constructed in a semi-automated way. We mined PubMed, Wikipedia, and textbooks to build relationships between codesets. A team of qualified doctors went through literature and manually validated the semi-automated relationships between these ontologies under the supervision of author V Dinesh Datta, who is also a qualified doctor.

3.1 Major Diagnostic Categories

All possible principal diagnoses are grouped into 25 mutually exclusive diagnosis areas. These mutually exclusive diagnosis areas are called Major Diagnostic Categories (MDC). The diagnoses in each MDC correspond to a single organ system or etiology in the body and in general are associated with a particular medical speciality (Table 1).

Table 1. Major Diagnostic Category

0	Ungroupable
1	Nervous System
2	Eye
3	Ear, Nose, Mouth, And Throat
4	Respiratory System
5	Circulatory System
6	Digestive System
7	Hepatobiliary System and Pancreas
8	Musculoskeletal System and Connective Tissue
9	Skin, Subcutaneous Tissue, and Breast
10	Endocrine, Nutritional, and Metabolic System
11	Kidney and Urinary Tract
12	Male Reproductive System
13	Female Reproductive System
14	Pregnancy, Childbirth, and Puerperium
15	Newborn and Other Neonates (Perinatal Period)
16	Blood and Blood Forming Organs and Immunological Disorders
17	Myeloproliferative Diseases and Disorders (Poorly Differentiated Neoplasms)
18	Infectious and Parasitic Diseases and Disorders
19	Mental Diseases and Disorders
20	Alcohol/Drug Use or Induced Mental Disorders
21	Injuries, Poison, and Toxic Effect of Drugs
22	Burns
23	Factors Influencing Health Status
24	Multiple Significant Trauma
25	Human Immunodeficiency Virus (HIV) Infection

3.2 International Classification of Diseases, Tenth Revision (ICD10)

The diseases (or diagnoses) in the human body are represented in ICD10 codes. ICD10 is a globally used codification of diseases maintained by the World Health Organization (WHO). ICD10 disease codes are grouped into 22 Chapters as shown in Table 2.

Table 2. International Classification of Diseases

ICD10 Code	Disease Classification
A00-B99	Certain infectious and parasitic diseases
C00-D49	Neoplasms
D50-D89	Diseases of the blood and blood-forming organs and certain disorders involving the immune mechanism
E00-E89	Endocrine, nutritional and metabolic diseases
F01-F99	Mental and behavioral disorders
G00-G99	Diseases of the nervous system
H00-H59	Diseases of the eye and adnexa
H60-H95	Diseases of the ear and mastoid process
I00-I99	Diseases of the circulatory system
J00-J99	Diseases of the respiratory system
K00-K94	Diseases of the digestive system
L00-L99	Diseases of the skin and subcutaneous tissue
M00-M99	Diseases of the musculoskeletal system and connective tissue
N00-N99	Diseases of the genitourinary system
O00-O99	Pregnancy, childbirth and the puerperium
P00-P96	Certain conditions originating in the perinatal period
Q00-Q99	Congenital malformations, deformations and chromosomal abnormalities
R00-R99	Symptoms, signs and abnormal clinical and laboratory findings, not elsewhere classified
S00-T88	Injury, poisoning and certain other consequences of external causes
V00-Y99	External causes of morbidity
Z00-Z99	Factors influencing health status and contact with health services
U00-U99	Codes for special purposes

3.3 Anatomical Therapeutic Chemical (ATC)

The Anatomical Therapeutic Chemical (ATC) Classification System is a drug classification system that classifies the active ingredients of drugs according to the organ system on which they act and their therapeutic, pharmacological, and chemical properties. In ATC, drugs are classified into 14 groups as presented in Table 3.

Table 3. Anatomical Therapeutic Chemical Classification System

ATC	Code
A	Alimentary tract and metabolism
B	Blood and blood forming organs
C	Cardiovascular system
D	Dermatologicals
G	Genito urinary system and sex hormones
H	Systemic hormonal preparations, excl. Sex hormones and insulins
J	Anti-infectives for systemic use
L	Antineoplastic and immunomodulating agents
M	Musculo-skeletal system
N	Nervous system
P	Antiparasitic products, insecticides and repellents
R	Respiratory system
S	Sensory organs
V	Various

3.4 Healthcare Common Procedure Terminology (HCPT)

The Healthcare Common Procedure Terminology (HCPT) or Healthcare Common Procedure Coding System Level II (HCPCS II) is a controlled vocabulary for surgeries and procedures and resources used in a hospital. The Current Procedure Code (CPT) or HCPT or HCPCS II codes are divided into six main sections as listed in Table 4.

Table 4. Current Procedure Codes (CPT/HCPT/HCPCS II)

CPT Code	Section
99202–99499	Evaluation & Management
00100–01999	Anesthesia
10021–69990	Surgery—further broken into smaller groups by body area or system within this code range
70010–79999	Radiology Procedures
80047–89398	Pathology and Laboratory Procedures
90281–99607	Medicine Services and Procedures

3.5 DRG and Length of Stay (LOS)

Diagnosis-Related Group (DRG) is a patient classification or stratification scheme. The DRGs provide patient-disease classes that are clinically similar. One DRG comprises of multiple related ICD10 diagnoses. For a particular diagnosis or an illness, which is the input to the system, the output will be similar. The assumption is that for the same input and output, the service patterns and resources required will be consistent. One critical attribute of DRG is the length of stay (LOS) in a hospital. If two patients belong to the same DRG, the LOS will be statistically similar.

Let us look at Hospital Packages in India. Let us assume a male patient who presents with a case of gallbladder removal surgery. Many hospitals in India define these as package costs and publish the price list. In cases of hospital packages, hospitals already computed the prospective fee based on estimated LOS and resource utilization. However, such LOS are not available for all hospitalization cases.

Casemix, also known as patient mix, is a way to objectively quantify groups (cohorts) of statistically related patients and their average LOS. Casemix are generally computed by the public health departments of a country. All public health data in a country collected over many years are used for casemix analysis. We could not locate such data in the open domain. Therefore, we took some data from some payer and computed a limited average LOS. We used this data to construct a knowledge graph connecting ICD10 (diagnosis) to DRG to average LOS. Looking at the average LOS we can determine the outliers and potential fraudulent cases.

3.6 Medical Appropriateness Knowledge Graph (MAKG)

All ontologies and codesets presented in Sects. 3.1, 3.2, 3.3, 3.4, and 3.5 are standalone codesets and represent only a particular type of function. We considered all individual codes in these tables (Tables 1, 2, 3, and 4) as graph nodes or vertices. For example, there are 1361 diseases (codes) between A000 and B999 in Chapter 1 (Certain infectious and parasitic diseases) of ICD10 codeset. All these codes are represented as individual nodes (vertices) in the ICD10 graph. To construct a knowledge graph, the relationships amongst these vertices need to be established. However, only the ontology provides the parent child relationship within the objects in the ontology.

The core contribution of this paper is to construct a comprehensive network of MDC, ICD10, DRG, HCPT, and ATC with their interrelationships. This is done in a semi-automated fashion. Some interrelationships like DRG-ICD10-MDC are constructed algorithmically and validated by medical experts. In the case of ICD10, CPT, and ATC, the network was curated manually by domain experts. All nodes and relationships (crossmaps) related to MDC, ICD10, CPT, ATC, DRG, and LOS were considered. Cases where no substantial medical evidence of relationships was found were removed. Doubtful and conflicting cases (edges) were eliminated as well.

These nodes (vertices) and their relationships (edges) were converted into graph theoretic adjacency lists. In our network, there are multiple adjacency lists. Vertices generally contain the NodeID, NodeName, NodeDescription, and other attributes. Edges contain the NodeID and weights. The medical appropriateness knowledge graph (MAKG) is a multipartite, directed, weighted, probabilistic graph. For such graphs we needed a property Graph database.

The adjacency lists were loaded into a Neo4j Property Graph database to construct the medical appropriateness knowledge graphs (MAKG). We constructed four interconnected knowledge graphs as defined below:

1. Major Diagnostic Category (MDC) — International Classification of Diseases Revision Ten (ICD10)
2. International Classification of Diseases Revision Ten (ICD10) – Healthcare Common Procedure Terminology System (HCPT)
3. International Classification of Diseases Revision Ten (ICD10) – Anatomical Therapeutic Code (ATC)
4. International Classification of Diseases Revision Ten (ICD10) – Diagnostic Related Groups (DRG) — Length of Stay (LOS)

It may be noticed that the MDC, the ICD, and the ATC all are organized based on organ systems in the body. For symptom to disease, we used the open domain diseasomics knowledge graph [21]. If there is a surgery or procedure performed by the provider, we check the ICD10-HCPT knowledge graph for appropriateness – whether the procedure code (HCPT code) is associated with the diagnosis. Then we check the ICD10-ATC knowledge graph to check whether the medication or drug is the right drug for this diagnosis.

In the following (Results) sections we show how MAKG is used by a provider at the point-of-care for the medical appropriateness use case. We also show how the MAKG is used by a payer or vigilance team during claim adjudication to identify medical waste and medical abuse.

The graph we constructed is a directed graph. However, the direction of the edge in the MAKG does not imply causal relationship. It shows only associations between different standalone terminologies and helps reduce one-to-many relationships to one-to-few.

3.7 Retrieval Augmented Generation – Generative Pretrained Transformer (RAG-GPT)

Large language models (LLMs) can achieve impressive results of human like language understanding and language generation. This is possible through pretrained transformers. GPT-3 showed that without large-scale task-specific training (data collection) or model parameter updating it can perform in an impressive manner. ChatGPT can answer USMLE examination questions [11]. This implies that ChatGPT already has the basic medical knowledge built-in within its parameters to pass a medical examination. However, it may not have knowledge of a medical specialist. ChatGPT also does not have knowledge specific to a country or population. It also lacks the knowledge of specific medical pathways or protocols. For this purpose, we constructed RAG-GPT. This allowed us to enhance the ChatGPT knowledge with specialist knowledge.

We used the OpenAI "GPT-3.5-Turbo" APIs in the Azure Cloud for RAG-GPT development. We collected medical textbooks and specialist knowledge in electronic forms. These were in PDF files, MS-Words and Text documents. For certain use cases we added table data in CSV format. Instead of fine-tuning or retraining ChatGPT with this additional knowledge for few-shot learning, we embedded all these specialist medical knowledge and stored in the FAISS (Facebook AI Similarity Search) Vector store database in the Azure cloud within the Azure Blob Storage. We also stored the metadata and the embedding of the summary of the medical evidence as provenance.

When a user, which may be a provider, a patient, a payer, or a regulator asks a medical question, we embed the question, look for the best matching section within the documents in the FAISS vector store. Based on the similarity score, we select one or more documents (or sections). Using LangChain library in Python we retrieve the necessary text from the medical text and respond to the user's question through ChatGPT generator. Results of RAG-GPT are presented in the Results section. In some special cases we used prompt engineering to extract the medically appropriate answer.

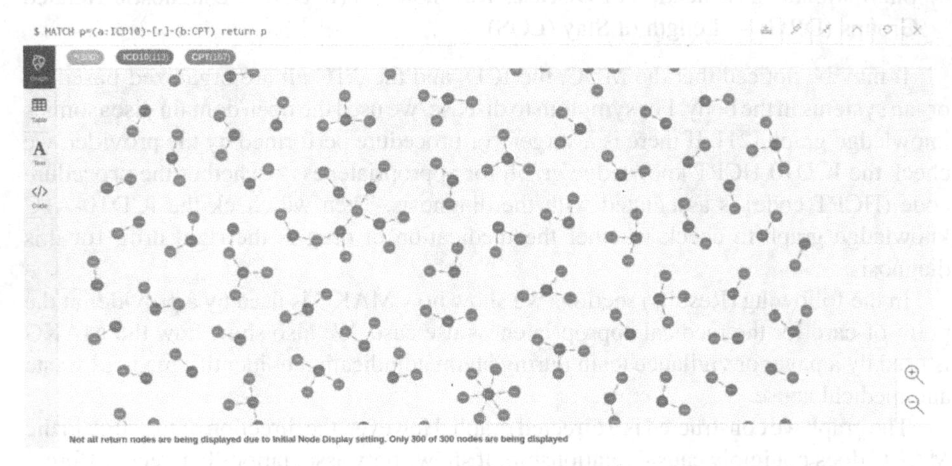

Fig. 3. ICD-CPT Knowledge graph on Neo4j browser for 300 nodes

4 Results

Here we present the results of our research. We show the results and use cases of both MAKG and RAG-GPT in this section. We presented the descriptive statistics of MAKG nodes and edges in Tables 5 and 6. In Fig. 3 we show the ICD-HCPT Knowledge graph visually on the Neo4j browser for 300 nodes. Figure 4 shows an API in Neo4j Cypher command to check the validity of a procedure.

4.1 MAKG Use Case

The MAKG knowledge base can be used either through the Neo4j browser in interactive mode, or through REST API. Vigilance teams or payers will access the MAKG programmatically through API. However, a researcher or a physician can use the MAKG through Neo4j graph browser.

Table 5. MAKG Nodes

Type of Node	Count
ATC	6692
CPT	9379
MDC	24
ICD10	62585
DRG	467

Table 6. MAKG Relationships (Edges)

Relationships	Count
MDC-ICD10 Relationships	62561
ICD10-CPT Relationships	9379
ICD10-ATC Relationships	5061

```
$ MATCH (a:CPT{CPT:'CPT:0001F'})--(b:ICD10) RETURN a.Name as CPT, a.Desc as Procedure,b.Name as ICD10,b.Desc as Disease
```

CPT	Procedure	ICD10	Disease
"0001F"	"Heart failure composite"	"I97131"	"Postprocedural heart failure following other surgery"
"0001F"	"Heart failure composite"	"I97130"	"Postprocedural heart failure following cardiac surgery"
"0001F"	"Heart failure composite"	"I509"	"Heart failure, unspecified"
"0001F"	"Heart failure composite"	"I5089"	"Other heart failure"
[illegible]	[illegible]	[illegible]	[illegible]
"0001F"	"Heart failure composite"	"I5083"	"High output heart failure"
"0001F"	"Heart failure composite"	"I5082"	"Biventricular heart failure"
"0001F"	"Heart failure composite"	"I50814"	"Right heart failure due to left heart failure"
"0001F"	"Heart failure composite"	"I50813"	"Acute on chronic right heart failure"
"0001F"	"Heart failure composite"	"I50812"	"Chronic right heart failure"
"0001F"	"Heart failure composite"	"I50811"	"Acute right heart failure"
"0001F"	"Heart failure composite"	"I50810"	"Right heart failure, unspecified"
"0001F"	"Heart failure composite"	"I5033"	"Acute on chronic diastolic (congestive) heart failure"
"0001F"	"Heart failure composite"	"I5032"	"Chronic diastolic (congestive) heart failure"
"0001F"	"Heart failure composite"	"I5031"	"Acute diastolic (congestive) heart failure"
"0001F"	"Heart failure composite"	"I5030"	"Unspecified diastolic (congestive) heart failure"

Started streaming 24 records after 39 ms and completed after 432 ms.

Fig. 4. The diseases associated with 'HCPT:0001' as an API

In the MAKG use case we take a case where a patient is presented with 'acute diarrhea' and 'dehydration'. The use case is demonstrated in Figs. 5, 6, and 7.

In Fig. 5, the physician entered the human understandable symptoms of 'acute diarrhea' and 'dehydration' to the Diseasomics [21] system. This system used diseasomics knowledge graph within the Physicians' Brain Digital Twin [22]. The human understandable n-gram symptoms are converted into machine interpretable UMLS (Unified Medical Language System) codes using NLP and Deep Neural Networks. The UMLS codes 'C0740441' and 'C0011175' can be seen on the upper right side of the Fig. 5. These codes are used to fetch the likely diagnoses as shown in the bottom part of Fig. 5. The likely diseases are 'cholera', 'acute diarrhea', and 'diarrhea'. The same response we

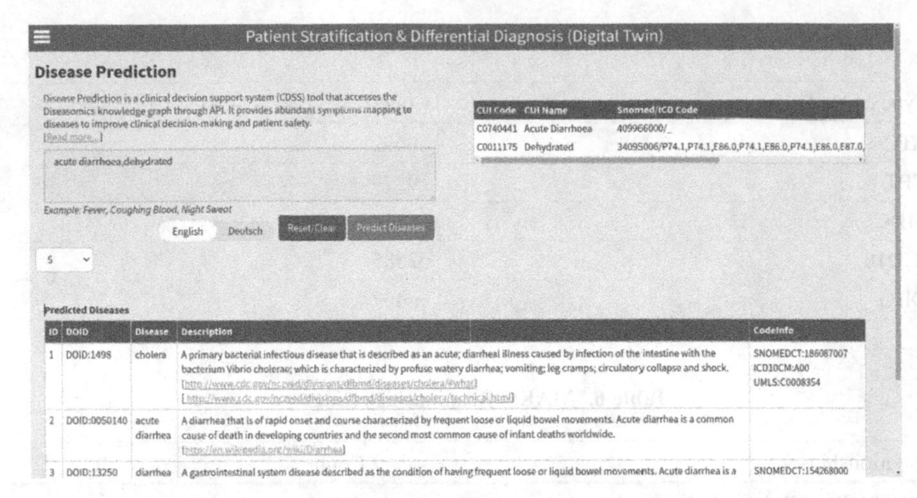

Fig. 5. The Knowledge Graph based Clinical Decision Support [21]

received when we interact with the Physicians' Brain Digital Twin through API. This is shown in Fig. 6.

```
$ WITH ["C0740441","C0011175"] AS SYMPID MATCH (s:SYMP)-[:SYMP]-(d:DO) WHERE s.Name IN SYMPID WITH d, size(SYMPID) as inputCnt, count(DISTINCT s) AS cnt W…
```

d.DOID	d.Name	d.Desc	d.xref
"DOID:13250"	"diarrhea"	"A gastrointestinal system disease described as the condition of having frequent loose or liquid bowel movements. Acute diarrhea is a common cause of death in developing countries and the second most common cause of infant deaths worldwide. The loss of fluids through diarrhea can cause severe dehydration which is one cause of death in diarrhea sufferers. Along with water; sufferers also lose dangerous amounts of important salts: electrolytes; and other nutrients. There are at least four types of diarrhea: secretory diarrhea; osmotic diarrhea; motility-related diarrhea; and inflammatory diarrhea. [http://en.wikipedia.org/wiki/Diarrhea#Types_of_diarrhea]"	"SNOMEDCT:154268000|UMLS:C0013369|"
"DOID:0050140"	"acute diarrhea"	"A diarrhea that is of rapid onset and course characterized by frequent loose or liquid bowel movements. Acute diarrhea is a common cause of death in developing countries and the second most common cause of infant deaths worldwide. [http://en.wikipedia.org/wiki/Diarrhea]"	"||"
"DOID:1498"	"cholera"	"A primary bacterial infectious disease that is described as an acute; diarrheal illness caused by infection of the intestine with the bacterium Vibrio cholerae; which is characterized by profuse watery diarrhea; vomiting; leg cramps; circulatory collapse and shock. [http://www.cdc.gov/nczved/divisions/dfbmd/diseases/cholera/#what; http://www.cdc.gov/nczved/divisions/dfbmd/diseases/cholera/technical.html]"	"SNOMEDCT:186087007|ICD10CM:A00|UMLS:C0008354|"

Fig. 6. API interface to Physicians Brain Digital Twin [22] with UMLS codes 'C0740441' and 'C0011175'

It may be noted that the API returned the same set of likely diagnoses in Fig. 6 as shown in Fig. 5. Once the diagnosis is complete, we can check for the appropriate or inappropriate medication. This is shown in Fig. 7.

In Fig. 7 we show two drug examples for the diagnosis of 'cholera'. In the upper panel, the Cypher command is unable to fetch any record connecting A000 (cholera) to L01CC (Colchicine). However, the Cypher command is able to fetch one record connecting A000 (cholera) to 'J01MA02' (Fluoroquinolones) as shown in the lower panel. In case of L01CC, the Cypher command failed to fetch any record, because there is no path between these two vertices. This implies that the drug 'L01CC' is inappropriate for cholera. Whereas, because there is a path between 'cholera' and J01MA02, this drug appropriate.

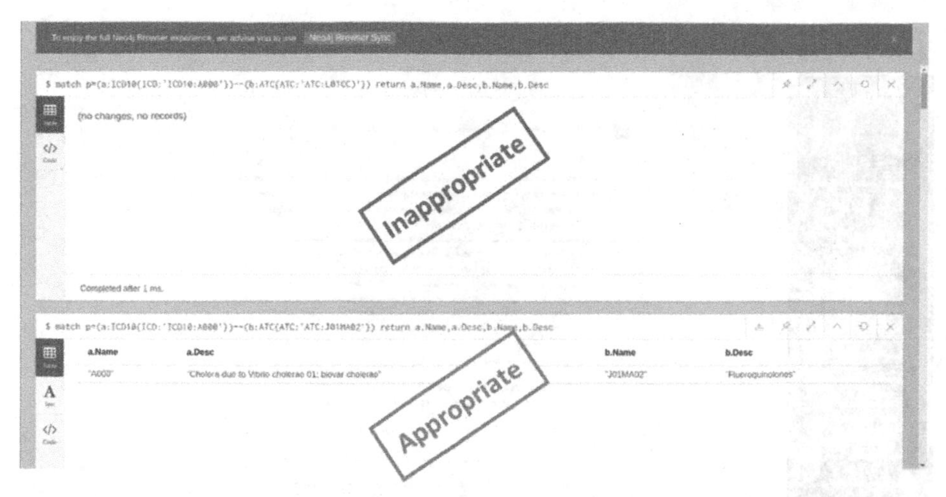

Fig. 7. The drug with ATC code 'L01CC' is inappropriate (upper panel) for Cholera. Whereas the drug 'J01MA02' (lower panel) is appropriate for Cholera.

4.2 RAG-GPT Use Case

Here we present a use case of RAG-GPT. A patient is presented with dengue fever somewhere in India. The physician wants to know the volume of fluid required for 24 h for a person weighing 30 kg. In Fig. 8 we show the native ChatGPT Web application where we asked a medical question "what is the volume of fluid required for 24 h for a 30 kg adult?". Though ChatGPT passed the medical exam in USA, its response is "For a 30 kg adult, 2 L might be a reasonable starting point, but individual needs can vary. Some experts suggest a daily fluid intake of 30–35 ml per kg of body weight, which for a 30 kg person would be around 900–1050 ml. However, this is just a rough estimate."

To answer the same question correctly, we enhanced the ChatGPT knowledge though RAG-GPT. For this exercise we used the "National Guidelines for Clinical Management of Dengue Fever" treatment protocol published in 2015 jointly by WHO and National Health Mission, Govt of India (downloaded from https://www.wbhealth.gov.in/uploaded_files/ticker/Dengue-National-Guidelines%20(Dec-2014).pdf).

The treatment protocol for dengue is a pdf document. This pdf file was embedded and given to ChatGPT as a RAG for enhanced country specific knowledge. We then asked the same question "what is the volume of fluid required for 24 h for a 30 kg adult?".

The answer given by ChatGPT with RAG is presented in Fig. 9. It responded with a detailed explanation "To calculate the volume of fluid required for a 30 kg adult, we need use the maintenance fluid formula. According to the given context, the maintenance fluid volume for 24 h for a person weighing more than 20 kg is 1500 + 20 ml/kg body weight exceeding 20 kg. Hence, for a 30 kg adult, the maintenance fluid volume for 24 h will be 1500 + 20 ml/kg x 10kg = 1700 ml. To maintain just adequate intravascular volume and circulation, we need to add 5% dehydration, which is equivalent 50 ml/kg. So, the total fluid required in 24 h for a 30 kg adult will be 1700 ml + (30 x 50) = 3200 ml. Therefore, a 30 kg adult requires 3200 ml of fluid in 24 h." The answer was not only accurate, but also without any hallucination.

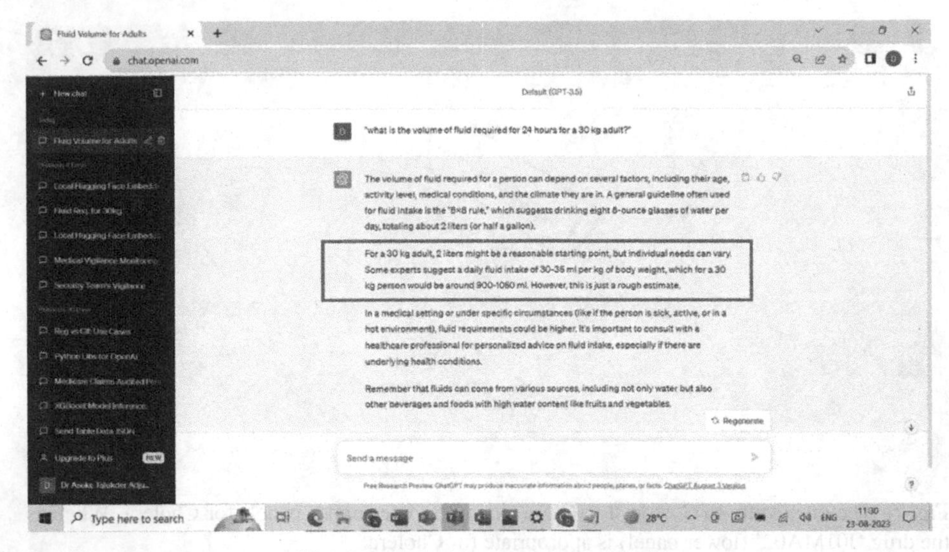

Fig. 8. Medical question to ChatGPT

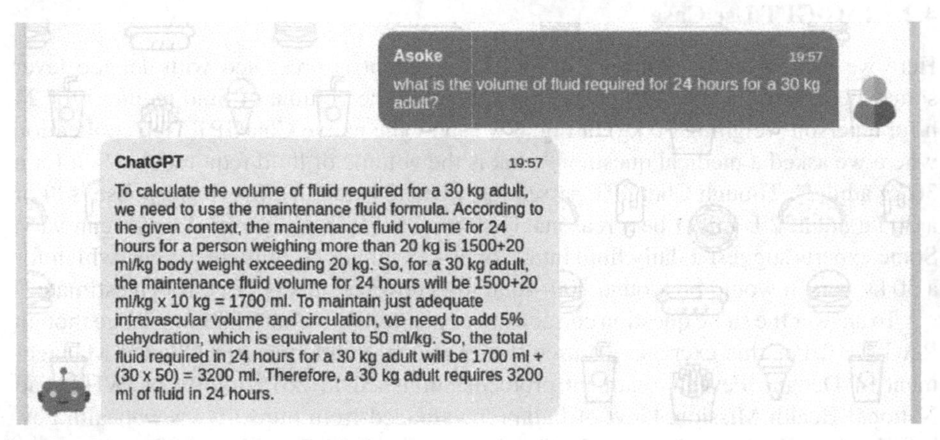

Fig. 9. RAG-GPT (ChatGPT with Retrieval Augmented Generation) from dengue treatment protocol

5 Future Work

The cliques between codesets are huge. In this paper we have limited out ICD-ATC to five leading characters only. We used the ATC ontology data from Bioportal. However we are in the process of getting a formal release from WHOCC [https://www.whocc.no/atc_ddd_index/]. This file has the DDD (Defined Daily Dose). This will be able to detect even the overprescription.

An intriguing aspect emerged from our research during the crossmapping process – the non-commutative nature of ICD to ATC and ATC to ICD mappings. For example, the case of typhoid fever (ICD10 ‘A010’), there are many ATCs which can be used depending upon the context of patient. What is noteworthy is that a single ATC, such as Paracetamol, can find utility across thousands of distinct conditions. So, the MAKGs mapped help reduce one-to-many relationships to one-to-few. These MAKGs will be constantly evolving using Adaptive AI and will be made more explainable and contextual in the future using better patient centric dynamic ontology tools.

6 Discussion

Any medical service that is inappropriate, is low-value-care. In this research paper we showed how GREAT AI can be used in medical appropriateness and value-based-care. This is achieved through both RAG-GPT and MAKG. RAG-GPT offers fine-grained generic and specialist medical knowledge. RAG-GPT is for use of providers, patients, payers, funders, and researchers alike. With additional data (knowledge) and prompt engineering, this medical knowledge can be refined further. RAG-GPT however, needs a human-in-the-loop to interpret and ingest the RAG-GPT output for medical appropriateness or inappropriateness.

MAKG in contrast, is coarse-grained medical knowledge. MAKG will be able to detect cases of medical inappropriateness in an autonomous fashion without any human-in-the-loop. Medically inappropriate cases are detected through graph traversal algorithm. If the path length between any two nodes (vertices) is more than 1 (see Fig. 1), they are medically inappropriate. In medically inappropriateness cases, there will not be any path between two vertices as we have seen in Fig. 7. For example, there is no path between ICD10:A000 (Cholera) and ATC:L01CC (Colchicine). This is because the diagnosis code ICD10:A000 is for Cholera whereas ATC:L01CC Colchicine is an anticancer drug and should not be used for cholera. Anything that does not have a path of traversal with pathlength not equal to 1, is medically inappropriate. This autonomous detection of inappropriate medical intervention features can be used for preventive FWA vigilance.

A notable observation that arises from this research is the fact that many diseases cataloged within the International Classification of Diseases (ICD) lack standardized treatment guidelines. In evidence-based medicine, the absence of established guidelines necessitates a multifaceted approach. The evaluation of relevant Randomized Controlled Trials (RCTs) in the PICO (Patient, Intervention, Comparison, Outcome) format, coupled with clinical judgment, plays a pivotal role in decision-making. Factors such as patient and payer costs, potential adverse effects, drug interactions, and more, are meticulously considered. In this context, AI becomes an indispensable tool, enabling the synthesis of complex medical data and expert insights.

The medical landscape is rapidly evolving, with new evidence and data emerging at an unprecedented pace. This exponential growth mandates the integration of AI to effectively manage and interpret dynamic patient-centric ontologies. AI serves as the bridge between vast amounts of data and actionable insights, thereby aiding medical professionals in rendering patient-centric care. Furthermore, the manually curated ICD to HCPT or ICD to ATC mapping holds potential beyond its initial scope of this research work. This crossmap or interrelationships could aid in the preventive identification of fraudulent, wasteful, and abusive (FWA) claims. Consider a scenario where a patient seeks insurance coverage for an anticancer drug, for a non-neoplastic condition. Such a claim raises suspicions of medical inappropriateness. The ICD to ATC crossmap will play a pivotal role in flagging such cases, thereby preventing FWA instances and optimizing the utilization of healthcare resources.

In essence, this research on GREAT AI underscores the symbiotic relationship between expert medical insight and AI-driven innovation. While AI accelerates data processing, pattern recognition, and evidence synthesis, the medical experts' judgment ensures the contextualization and appropriate application of these insights. Together, this partnership holds the promise of transforming healthcare practices, optimizing medical appropriateness, and ultimately improving patient outcomes.

References

1. Sekhar M.S., Vyas N.: Defensive medicine: a bane to healthcare. Ann Med Health Sci Res. 2013;3(2):295–6 (2013). https://doi.org/10.4103/2141-9248.113688
2. Shrank, W.H., Rogstad, T.L., Parekh, N.: Waste in the US Health Care System: Estimated Costs and Potential for Savings. JAMA **322**(15), 1501–1509 (2019). https://doi.org/10.1001/jama.2019.13978
3. Kamath, R., Brand, H.: A Critical Analysis of the World's Largest Publicly Funded Health Insurance Program: India's Ayushman Bharat. Int. J. Prev. Med. **14**, 20 (2023). https://doi.org/10.4103/ijpvm.ijpvm_39_22
4. Detroit Area Doctor Sentenced to 45 Years in Prison for Providing Medically Unnecessary Chemotherapy to Patients: https://www.justice.gov/opa/pr/detroit-area-doctor-sentenced-45-years-prison-providing-medically-unnecessary-chemotherapy
5. Noncommunicable Disease: https://www.who.int/news-room/fact-sheets/detail/noncommunicable-diseases
6. Inpatient Prospective Payment System Rule. https://www.facs.org/advocacy/regulatory-issues/payment-rules/inpatient-prospective-payment-system-rule/
7. Fetter, R.B., Freeman, J.L.: Diagnosis related groups: Product line management within hospitals. Acad. Manag. Rev. **11**(1), 41–54 (1986)
8. Ghosh, A.K., Ibrahim, S., Lee, J., Shapiro, M.F., Ancker, J.: Comparing Hospital Length of Stay Risk-Adjustment Models in US Value-Based Physician Payments. Qual. Manag. Health Care **32**(1), 22–29 (2023). https://doi.org/10.1097/QMH.0000000000000363
9. Sackett, D.L., Rosenberg, W.M., Gray, J.A., Haynes, R.B., Richardson, W.S.: Evidence based medicine: what it is and what it isn't. ClinOrthopRelat Res. **455**, 3–5 (2007)
10. Brown T.B., Mann B., Ryder N., Subbiah M., Kaplan J., Dhariwal P., Neelakantan A., Shyam P., Sastry G., Askell A., Agarwal S., Herbert-Voss A., Krueger G., Henighan T., Child R., Ramesh A., Ziegler D.M., Wu J., Winter C., Hesse C., Chen M., Sigler E., Litwin M, Gray S., Chess B., Clark J., Berner C., McCandlish S., Radford A., Sutskever I., Amodei D.: Language Models are Few-Shot Learners. (2020). https://arxiv.org/abs/2005.14165
11. Kung, T.H., Cheatham, M., Medenilla, A., Sillos, C., De Leon, L., Elepaño, C., et al.: Performance of ChatGPT on USMLE: Potential for AI-assisted medical education using large language models. PLOS Digit Health. **2**(2), e0000198 (2023). https://doi.org/10.1371/journal.pdig.0000198
12. Lewis P., Perez E., Piktus A., Petroni F., Karpukhin V., Goyal N., Küttler H., Lewis M., Yih W., Rocktäschel T., Riedel S., Kiela D.: Retrieval-Augmented Generation for Knowledge-Intensive NLP Tasks. https://arxiv.org/abs/2005.11401 (2020)
13. Hanlon, J.T., Schmader, K.E.: The Medication Appropriateness Index: A Clinimetric Measure. Psychother. Psychosom. **91**(2), 78–83 (2022)

14. Santos-Eggimann, B., Paccaud, F., Blanc, T.: Medical appropriateness of hospital utilization: an overview of the Swiss experience. Int. J. Qual. Health Care **7**(3), 227–232 (1995). https://doi.org/10.1093/intqhc/7.3.227
15. Lavis J.N., Anderson G.M.: Appropriateness in health care delivery: definitions, measurement, and policy implications. CMAJ. 154(3):321–8. PMID: 8564901 (1996)
16. Patient Safety in Healthcare, Forecast to 2022. https://store.frost.com/patient-safety-in-healthcare-forecast-to-2022.html
17. Rasmy, L., Xiang, Y., Xie, Z. et al.: Med-BERT: pretrained contextualized embeddings on large-scale structured electronic health records for disease prediction. npj Digit. Med. 4:86 (2021). https://doi.org/10.1038/s41746-021-00455-y
18. Iqbal, M.S., Abd-Alrazaq, A., Househ, M.: Artificial Intelligence Solutions to Detect Fraud in Healthcare Settings: A Scoping Review. Stud Health Technol Inform. **295**, 20–23 (2022). https://doi.org/10.3233/SHTI220649
19. Sun, H., et al.: Medical Knowledge Graph to Enhance Fraud, Waste, and Abuse Detection on Claim Data: Model Development and Performance Evaluation. JMIR Med. Inform. **8**(7), e17653 (2020). https://doi.org/10.2196/17653
20. Talukder, A.K., Selg, E., Fernandez, R., Raj, T.D.S., Waghmare, A.V., Haas, R.E.: Drugomics: Knowledge Graph & AI to Construct Physicians' Brain Digital Twin to Prevent Drug Side-Effects and Patient Harm. In: Roy, P.P., Agarwal, A., Li, T., Krishna Reddy, P., Uday Kiran, R. (eds) Big Data Analytics. BDA 2022. Lecture Notes in Computer Science, vol 13773. Springer, Cham. ((2022). https://doi.org/10.1007/978-3-031-24094-2_10
21. Talukder, A.K., Schriml, L., Ghosh, A., Biswas, R., Chakrabarti, P., Haas, R.E.: Diseasomics: Actionable machine interpretable disease knowledge at the point-of-care. PLOS Digit Health. **1**(10), e0000128 (2022). https://doi.org/10.1371/journal.pdig.0000128
22. Talukder, A.K., Selg, E., Haas, R.E.: Physicians' Brain Digital Twin: Holistic Clinical & Biomedical Knowledge Graphs for Patient Safety and Value-Based Care to Prevent the Post-pandemic Healthcare Ecosystem Crisis. In: Villazón-Terrazas, B., Ortiz-Rodriguez, F., Tiwari, S., Sicilia, MA., Martín-Moncunill, D. (eds) Knowledge Graphs and Semantic Web. KGSWC 2022. Communications in Computer and Information Science, vol 1686. Springer, Cham (2022). https://doi.org/10.1007/978-3-031-21422-6_3
23. Classen, D.C., Longhurst, C., Thomas, E.J.: Bending the patient safety curve: how much can AI help? npj Digit. Med. **6**, 2 (2023). https://doi.org/10.1038/s41746-022-00731-5
24. Garnelo, M., Shanahan, M.: Reconciling deep learning with symbolic artificial intelligence: representing objects and relations. Curr. Opin. Behav. Sci.Behav. Sci. **29**, 17–23 (2019). https://doi.org/10.1016/j.cobeha.2018.12.010

Large Language Models

KG-CTG: Citation Generation Through Knowledge Graph-Guided Large Language Models

Avinash Anand(✉), Mohit Gupta, Kritarth Prasad, Ujjwal Goel, Naman Lal, Astha Verma, and Rajiv Ratn Shah

Indraprastha Institute of Information Technology, Delhi, India
{avinasha,mohit22112,kritarth20384,ujjwal20545, asthav,rajivratn}@iiitd.ac.in

Abstract. Citation Text Generation (CTG) is a task in natural language processing (NLP) that aims to produce text that accurately cites or references a cited document within a source document. In CTG, the generated text draws upon contextual cues from both the source document and the cited paper, ensuring accurate and relevant citation information is provided. Previous work in the field of citation generation is mainly based on the text summarization of documents. Following this, this paper presents a framework, and a comparative study to demonstrate the use of Large Language Models (LLMs) for the task of citation generation. Also, we have shown the improvement in the results of citation generation by incorporating the knowledge graph relations of the papers in the prompt for the LLM to better learn the relationship between the papers. To assess how well our model is performing, we have used a subset of standard S2ORC dataset, which only consists of computer science academic research papers in the English Language. Vicuna performs best for this task with 14.15 Meteor, 12.88 Rouge-1, 1.52 Rouge-2, and 10.94 Rouge-L. Also, Alpaca performs best, and improves the performance by 36.98% in Rouge-1, and 33.14% in Meteor by including knowledge graphs.

Keywords: Citation Text Generation · Knowledge Graphs · Large Language Models · Natural Language Processing

1 Introduction

Generating text in the scientific domain is a complex task that demands a strong grasp of the input text and domain-specific knowledge. Citation Text Generation (CTG) is an NLP task that focuses on generating accurate citations or references to cited documents within a source document. To accomplish this, machine learning models must adeptly summarize the relationship between the original and the cited article in a given context. This involves analyzing the content of the documents, identifying their connections, and employing appropriate terminology and structure to convey this information with clarity and conciseness. The domain of Text Generation has gained significant attention in recent times,

V. Goyal et al. (Eds.): BDA 2023, LNCS 14418, pp. 37–49, 2023.
https://doi.org/10.1007/978-3-031-49601-1_3

largely attributed to the advancements in Transformer-based models [21]. These models have revolutionized the field, driving substantial progress in natural language processing and generation tasks.

CTG holds great potential for various applications, particularly in the field of scientific writing assistants. In the context of Education, CTG could be employed to teach students the correct approach to citing papers in academic writing. Another intriguing prospect is the ability to create summary sentences [1] from the referenced source papers, effectively summarizing the key ideas of the cited text. Next significant use is assisting researchers in paper writing by providing suggestions for suitable citations and generating the corresponding citation text. It can aid in plagiarism detection by comparing the generated citation text with the source material. CTG is a complementary task to citation recommendation and summarization, as it specifically focuses on explaining the relationships between documents rather than just summarizing their contents [8,23].

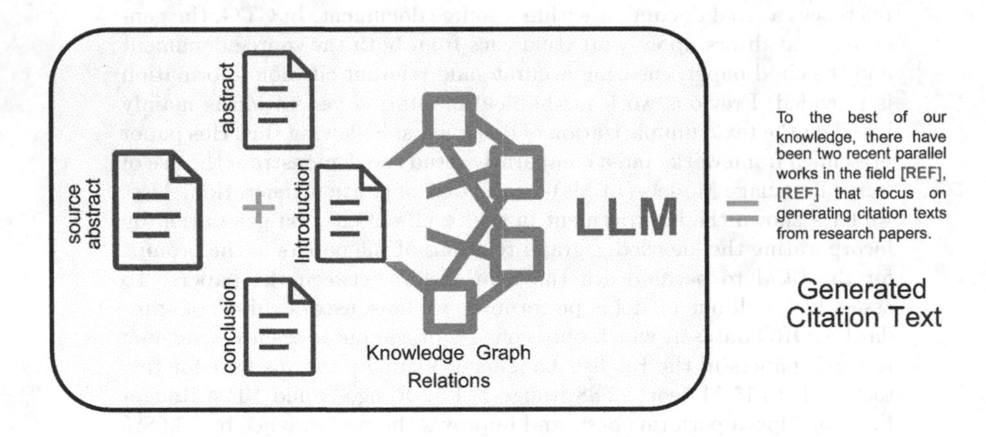

Fig. 1. Single-Sentence Citation Text Generation

Scientific texts are significantly longer than other domains typically studied in NLP. This is a challenging and unresolved problem for text generation models. This problem was solved by a few works like, Xing et al. [22] does automatic generation of citation text in scholarly articles. It makes use of a cross-attention multi-source pointer generating network. Luu et al. [16] introduced CTG using a pair of scientific documents as the source and cited papers. They have primarily used the text from the abstract of papers as input. It generated the citation text describing the relationship between the source and cited documents.

To our knowledge, researchers use subsets of the S2ORC[1] [15] dataset for the CTG task. We have only used the subset which contains research articles only in the domain of computer science in English Language. The original dataset contains *abstract* of paper, *body_text*, *paper_id*, etc. We have extracted the **introduction**, and **conclusion** from the *body_text*.

[1] https://github.com/allenai/s2orc.

In this paper, we proposed a methodology or technique to generate citation text using Large Language Models (LLMs). We have fine-tuned three LLMs for the task of generating citation text. The models are LLaMA [20], Alpaca [19], and Vicuna [7]. We have also performed the experiments of incorporating the Knowlegdge graph in the prompts for generating the citation, to provide better contextual understanding of the source and target paper to better learn the relationship between them. Incorporating the knowledge graphs shows the improvement in the performance and quality of text generation. To conclude, our main contributions in this paper is listed below:

- We propose the use of Large Language Model (LLMs) text generation capabilities in field of research writing. We fine-tuned three LLMs for generating the citation text given the content of citing and cited papers.
- We attempt to incorporate the knowledge graphs of the source and target papers in the prompts to make the model better understand the relationship between the papers. For creating the knowledge graphs, we have used **PL-Marker**[2] [24].
- We show the importance of incorporating the knowledge graphs to improve the performance. We achieve an increase of 33.14% in METEOR and 36.98% in Rouge-1 score for Alpaca on the S2ORC dataset.

The structure of the written work is as follows: Sect. 2 provides an overview of the related works on citation text generation. Section 3 describes the problem formulation, the utilized models, and their components. Section 4 outlines the processing and creation of the dataset from S2ORC. Section 5 presents the experimental setup, findings, and implementation details. Section 6 presents the evaluations conducted in the study. Section 7 summarizes the future aims of the paper. Lastly, Sect. 8 discusses the limitations of the proposed system.

2 Related Work

CTG is intricately connected to citation recommendation, scientific document understanding, and summarization. The task of citation recommendation complements CTG, as it provides references to pertinent publications for a particular document or text excerpt [3]. Additionally, citation recommendation systems [2] play a vital role in guiding researchers towards valuable sources of information. Summarization Systems [23] condense the information allowing scientists to understand the basic idea in a research section more quickly.

Citation information is also helpful for scientific paper summarization [18]. Previous work include the works of [11] in their literature review producing system for text summarization. The task of multi-document summarization in the scientific domain [5] and text generation for scientific documents is a particular case of multi-document scientific summarization [6].

[2] https://github.com/thunlp/PL-Marker.

Koncel-Kedziorski et al. [14] generated multi-sentence text from an information extraction system and improved performance using a knowledge graph. They did graph encoding using Graph Attention Network. Chen et al. [4] proposed a SciXGen dataset to solve the problem of context-aware text generation in a scientific domain. Zhu et al. [26] uses LLMs for constructing the knowledge graphs, and used them for reasoning. They proposed **AutoKG**, which uses LLMs for constructing and reasoning the knowledge graphs. Gosangi et al. [10] studied the significance of context in determining whether a sentence in an academic publication is worthy of reference. This paper can be considered complementary work with CTG tasks.

To the best of our knowledge, there have been two recent parallel works in the field that focus on generating citation texts from research papers. Luu et al. [16] were the first to introduce this task and successfully generated citation texts using the source and cited documents as input. On the other hand, Xing et al. [22] delved deeper into the relationship between scientific documents by leveraging a larger dataset. They employed an implicit citation extraction algorithm, utilizing GPT-2[3], which was trained on an annotated dataset to automatically enhance the training data.

3 Methodology

The goal of the task of citation generation, which tries to produce citation text in the context of both the source publication and the referenced paper, is outlined in this section. We leverage the advancements of Large Language Models (LLMs), which have demonstrated significant improvements in various text generation tasks [25].

- **Model Fine-Tuning:** We commence by fine-tuning three selected LLMs i.e. LLaMA [20], Alpaca [19], and Vicuna [7] specifically for the task of generating citation text. Through this process, we train the models on the subset of S2ORC dataset for citation generation and evaluate their performance and the quality of the generated text.
- **Incorporation of Knowledge Graphs:** To improve the model's comprehension of the context and relationship between the papers, we introduce a knowledge graph [14] derived from the papers as part of the input prompt. The knowledge graph provides structured information about the papers, including key concepts, entities, and their relationships. We explore the benefits of incorporating knowledge graphs with LLMs [17] in improving the performance of citation generation tasks.

By following this methodology, we show the effectiveness of fine-tuned LLMs for generating citation text. Additionally, we look into the effect of incorporating knowledge graphs on improving the model's performance in capturing the relationships and rich contextual information between the source paper and the cited paper.

[3] https://github.com/openai/gpt-2.

3.1 Large Language Models

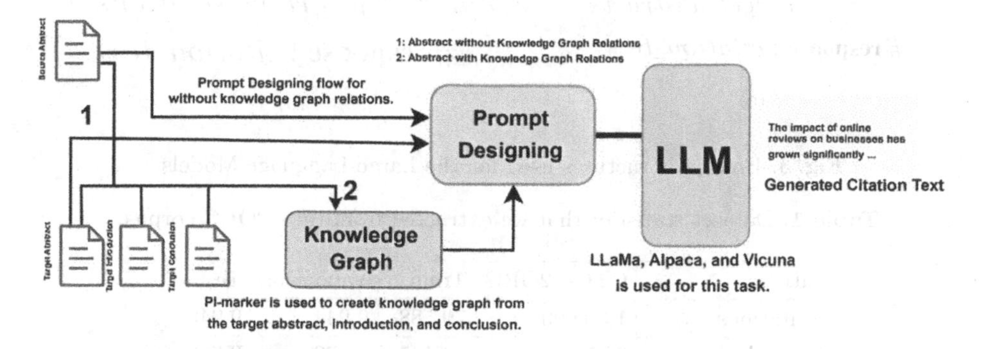

Fig. 2. Experimental setup workflow diagram depicting two workflows: **1**. Prompt creation without knowledge graph relations, and **2**. Relation extraction from abstracts followed by prompt creation.

In this subsection, we describe the Large Language Models (LLMs) used in our study for citation text generation. The LLMs investigated in our research include LLaMA [20], Alpaca [19], and Vicuna [7]. LLaMA is a transformer-based model available in multiple variations, such as 7B, 13B, 33B, and 65B parameters. For our study, we focused on LLaMA-7B. Alpaca, a variant of LLaMA, has been fine-tuned using 52k instructions from OpenAI's text-davinci-003 model. This targeted training allows Alpaca to specialize in generating instructional text. Vicuna is a supervised fine-tuned version of LLaMA, trained on 70K user-shared talks from ShareGPT.com. This variant of LLaMA captures the specific context and style of user-shared talks, enabling it to generate text aligned with conversational patterns.

To evaluate the performance of these LLMs on the task of citation text generation (CTG), we fine-tuned all models using our CTG dataset. This comparative analysis enables us to assess the strengths and weaknesses of each model in generating accurate and contextually appropriate citation text.

3.2 Knowledge Graphs and Prompting

The structured representation of knowledge is called knowledge graph that organizes information into entities and their relationships, enabling advanced data analysis and inference [12]. In our work, we constructed the knowledge graph of the source and target abstracts using a state-of-the-art tool called PL-Marker [24]. PL-Marker employs a novel packed levitated marker technique, combining a packing approach focused on the subject and the neighbourhood to obtain pair representations. The knowledge graph is constructed to capture the relationship and context between different entities within the abstracts of papers.

After generating the knowledge graph from the source abstract with the target abstract, introduction and conclusion, we then concatenated it with the

#input : *source_abstract*
target_abstract
#response : *citation_text*

(a)

#input : *source_abstract*
knowledge_graph_relations
#response : *citation_text*

(b)

Fig. 3. Prompt Structures used for the Large Language Models.

Table 1. Dataset statistics that we extracted from the S2ORC corpus.

Statistic	CTG-S2ORC	Train	Validation	Test
# citations	1,00,000	79,588	9,944	9,946
# unique papers	38,530	34,147	8,076	8,070
CITATIONS				
Avg # characters	171.3	171.34	171.47	170.87
Max # characters	3420	2986	1840	2170
SOURCE ABSTRACTS				
Avg # characters	1225.89	1185.74	1189.3	1190.75
Max # characters	56771	8945	7858	8024
TARGET ABSTRACTS				
Avg # characters	1065.64	1002.27	1001.54	993
Max # characters	93551	8059	7736	6647

source abstract. This final string is then passes in the *#input* section of the prompt. The structure of the prompt with knowledge graph relations is shown in Fig. 3b.

4 CTG Dataset

In our research, we utilized the S2ORC (Semantic Scholar Research Corpus) [15], which includes a large collection comprising of approximately 81.1 million English-language academic papers from various disciplines. This corpus encompasses diverse information such as abstracts, full paper texts, bibliographic references, and associated metadata for each paper.

To focus specifically on the field of Computer Science, we filtered the corpus by selecting papers with the "Field of Study" tag as Computer Science. Out of the total 81.1 million papers, we narrowed it down to approximately 6.0 million papers relevant to the computer science domain. However, not all papers in this subset had valid abstracts and body text. Some contained irrelevant or empty content, which we subsequently removed during the data cleaning process. As a result, our final dataset comprised approximately 1,00,000 samples.

The "body_text" of each paper consisted of paragraphs, including sections such as Introduction, Methodology, Conclusion, etc. Within these sections, we examined the presence of cite_spans, which are dictionaries containing citation information for referenced papers within each paragraph. Our analysis involved identifying these cite_spans within the body text and extracting the corresponding citation sentences. It is important to note that we excluded citations that referenced more than one paper within a single sentence. Additional statistical details about the dataset can be found in Table 1.

5 Experiments

We describe our experimental settings, evaluation metrics, and model comparisons in this section. We fine-tuned and evaluated three Large Language text generation models on our CTG dataset. Table 2 presents the results. Comparing LLaMA [20], Alpaca [19], and Vicuna [7], we observed Vicuna's superior performance.

Next, we integrate knowledge graphs constructed from the source and target papers using PL-Marker [24] and proceed to fine-tune the same set of models. The incorporation of knowledge graphs significantly enhances both the performance and the quality of the generated text. Notably, the Alpaca exhibits superior performance, as evidenced by notable increase in METEOR score by 33.14%, and 36.98% in Rouge-1. The Table 3 shows the outcomes for this configuration. These results affirm that the inclusion of knowledge graphs effectively guides the Large Language Models (LLMs) in text generation tasks.

5.1 Experimental Settings

In this research, we partitioned our CTG dataset into 79,588 training samples and 9,946 testing samples and 9,944 validation samples. For fine-tuning the Large Language Models (LLMs), we employed QLora [9] to minimize GPU usage. By back propagating gradients through a frozen, 4-bit quantized pretrained language model into Low Rank Adapters (LoRA), QLora is an effective method for maximizing memory utilization. We utilized the AdamW optimizer [13] with a Linear Scheduler. The learning rate was set to 3e-4, and we incorporated 100 warmup steps to gradually adjust the learning rate. By adopting this approach, we effectively trained the LLMs on the CTG dataset, allowing us to evaluate their performance on the respective testing samples.

$$k_i = \frac{1}{2}\left(Q_X\left(\frac{i}{2^n+1}\right) + Q_X\left(\frac{i+1}{2^n+1}\right)\right) \tag{1}$$

Where, $Q_x(.)$ is the quantile function of the standard normal distribution $N(0,1)$. For our experiments, we have used $n = 4$ as we are applying 4-bit quantization.

Table 2. Results of CTG Task using LLMs (Without Knowledge Graphs)

Model	METEOR	Rouge-1	Rouge-2	Rouge-L
LLaMA	12.83	11.26	1.36	9.59
Alpaca	10.53	9.22	1.21	7.81
Vicuna	**14.15**	**12.88**	**1.52**	**10.94**

Evaluation Metrics: For the text generation and summarization tasks, we employed common evaluation metrics as METEOR, ROUGE-N, and ROUGE-L. ROUGE-L evaluates the longest common subsequence between the text generated and the reference, while the overlap of n-grams between the two is measured using ROUGE-N. ROUGE-N receives additional information from METEOR, which takes word similarity into account during stemming.

Table 3. Results of CTG Task using LLMs (With Knowledge Graphs)

Model	METEOR	Rouge-1	Rouge-2	Rouge-L
LLaMA	11.61	10.61	0.99	9.01
Alpaca	**14.02**	12.63	**1.54**	10.71
Vicuna	13.80	**12.87**	1.48	**10.96**

6 Evaluations

Our proposed work highlights the use case of Large Language Models (LLMs) in the domain of generating citation text for scientific papers. Furthermore, our study emphasizes the significance of knowledge graphs generated from both the source and target papers, as they facilitate capturing deeper relationships and structured contextual data between these papers. Through our research, we have showcased the effectiveness of the Alpaca LLM in generating citation text, which outperforms LLaMA and Vicuna in terms of both the results obtained and the quality of the generated text. Figure 4 and Fig. 5 illustrate citation examples generated by the best model during the inference process, providing compelling evidence of the exceptional quality achieved. These findings underscore the value of leveraging LLMs and incorporating knowledge graphs to enhance the generation of accurate and contextually appropriate citation text for scientific papers.

7 Conclusion

This paper explores the task of generating citation texts in research papers. To accurately understand and capture relevant features from scientific papers, we leverage the synthesis of knowledge graphs. We present a compelling use case for employing Large Language Models (LLMs) in the domain of citation text generation, demonstrating their impressive performance when given the source abstract and target abstract, introduction, and conclusion. The efficiency of LLMs is substantiated through automatic evaluations employing various metrics. Our experiments also emphasize the significance of utilizing knowledge graphs as prompts to guide the model's generation process. Looking ahead, we plan to further enhance the capabilities of LLM models by incorporating Chain-of-Thoughts prompting, which will improve their reasoning abilities and enable the generation of more plausible and higher-quality citations.

8 Limitations

While our proposed solution excels in generating single-sentence citations, its effectiveness is primarily constrained in scenarios where authors frequently employ multiple citations within a single paragraph. To overcome this limitation, we can enhance our model by incorporating multi-citation examples into our dataset.

Another limitation of our proposed work is the presence of certain keywords in the target citation text that are not found within the section constrained by the token limit of the source and target papers. This discrepancy adversely affects the performance of the models, resulting in a decrease in overall effectiveness.

Acknowledgements. Rajiv Ratn Shah is partly supported by the Infosys Center for AI, the Center for Design and New Media, and the Center of Excellence in Healthcare at IIIT Delhi.

A Appendix

This section shows the inference examples used to test the fine-tuned model and checking the generated text quality, and context.

Figure 4 illustrates an example of the generated citation text obtained from the fine-tuned Vicuna model. The provided source_abstract and target_abstract have resulted in high-quality generated citation text that aligns well with the context of both the source and target papers.

In Fig. 5, the generated citation text demonstrates a higher level of context richness due to the incorporation of knowledge graph relations. These relations enable a better understanding of the connections between words in the source and target abstracts, resulting in more contextually relevant generated text.

source_abstract

Recovering missing values plays a significant role in time series tasks in practical applications. How to replace the missing data and build the dependency relations from the incomplete sample set is still a challenge. The previous research has found that residual network (ResNet) helps to form a deep network and cope with degradation problem by shortcut connection. Gated recurrent unit (GRU) can improve network model and reduce training parameters by update gate which takes the place of forgetting gate and output gate in long short-term memory (LSTM). Inspired by this finding, we observe that shortcut connection and mean of global revealed information can model the relationship among missing items, the previous and overall revealed information. Hence, we design an imputation network with decay factor for shortcut connection and mean of the global revealed information in GRU, called decay residual mean imputation GRU (DRMI-GRU). We introduce a decay residual mean unit (DRMU), which takes full advantage of the previous and global revealed information to model incomplete time series; and the decay factor is applied to balance the previous long-term dependencies and all non-missing values in the sample set. In addition, a mask unit is designed to check the missing data existing or not. An extensive body of empirical comparisons with other existing imputation algorithms over real-world data and public dataset with different ratio of missing data verifies the performance of our model.

target_abstract

In this work we propose a novel interpretation of residual networks showing that they can be seen as a collection of many paths of differing length. Moreover, residual networks seem to enable very deep networks by leveraging only the short paths during training. To support this observation, we rewrite residual networks as an explicit collection of paths. Unlike traditional models, paths through residual networks vary in length. Further, a lesion study reveals that these paths show ensemble-like behavior in the sense that they do not strongly depend on each other. Finally, and most surprising, most paths are shorter than one might expect, and only the short paths are needed during training, as longer paths do not contribute any gradient. For example, most of the gradient in a residual network with 110 layers comes from paths that are only 10-34 layers deep. Our results reveal one of the key characteristics that seem to enable the training of very deep networks: Residual networks avoid the vanishing gradient problem by introducing short paths which can carry gradient throughout the extent of very deep networks.

target_citation

REF propose an enlightening analysis of the Residual Networks (ResNets), and suppose that ResNet **REF** can be looked upon as a set of shallow networks and helps to form a deep network and copes with degradation problem by shortcut connection.

predicted

Recovering missing values plays a significant role in time series tasks in practical applications. How to replace the missing data and build the dependency relations from the incomplete sample set is still a challenge. The previous research has found that residual network (ResNet) helps to form a deep network and cope with degradation problem by shortcut connection. Gated recurrent unit (GRU) can improve network model and reduce training parameters by update gate.

Fig. 4. Example of Generated Citation text from **Vicuna** without knowledge graph relations

source_abstract
Recovering missing values plays a significant role in time series tasks in practical applications. How to replace the missing data and build the dependency relations from the incomplete sample set is still a challenge. The previous research has found that residual network (ResNet) helps to form a deep network and cope with degradation problem by shortcut connection. Gated recurrent unit (GRU) can improve network model and reduce training parameters by update gate which takes the place of forgetting gate and output gate in long short-term memory (LSTM). Inspired by this finding, we observe that shortcut connection and mean of global revealed information can model the relationship among missing items, the previous and overall revealed information. Hence, we design an imputation network with decay factor for shortcut connection and mean of the global revealed information in GRU, called decay residual mean imputation GRU (DRMI-GRU). We introduce a decay residual mean unit (DRMU), which takes full advantage of the previous and global revealed information to model incomplete time series; and the decay factor is applied to balance the previous long-term dependencies and all non-missing values in the sample set. In addition, a mask unit is designed to check the missing data existing or not. An extensive body of empirical comparisons with other existing imputation algorithms over real-world data and public dataset with different ratio of missing data verifies the performance of our model.

relations
decay factor **USED-FOR** shortcut connection . update gate **USED-FOR** Gated recurrent unit (GRU) . identity skip-connections **USED-FOR** residual layers . Gated recurrent unit (GRU) **USED-FOR** network model . public dataset **EVALUATE-FOR** imputation algorithms . unraveled view **USED-FOR** residual networks . Gated recurrent unit (GRU) **USED-FOR** training parameters . update gate **USED-FOR** training parameters . AlexNet **HYPONYM-OF** architectures . residual network (ResNet) **USED-FOR** deep network . ensemble-like behavior **FEATURE-OF** they . VGG **HYPONYM-OF** architecture . Residual networks **USED-FOR** vanishing gradient problem . forgetting gate **CONJUNCTION** output gate . missing data **USED-FOR** dependency relations . biological systems **HYPONYM-OF** architectures . imputation network **USED-FOR** shortcut connection . real-world data **USED-FOR** imputation algorithms . gradient **PART-OF** residual network . ensemble-like behavior **FEATURE-OF** paths . architecture **USED-FOR** computer vision systems . AlexNet **CONJUNCTION** biological systems . redundancy **FEATURE-OF** they . architectures **USED-FOR** complex concepts . mask unit **USED-FOR** missing data . decay factor **FEATURE-OF** imputation network . forgetting gate **USED-FOR** update gate . residual networks **USED-FOR** gradient flow . incomplete sample set **USED-FOR** dependency relations . residual networks **USED-FOR** vanishing gradient problem . decay residual mean unit (DRMU) **USED-FOR** incomplete time series . shortcut connection **USED-FOR** degradation problem . decay residual mean imputation GRU (DRMI-GRU) **HYPONYM-OF** imputation network . Recovering missing values **USED-FOR** time series tasks . residual network (ResNet) **USED-FOR** degradation problem . they **USED-FOR** deep networks . output gate **PART-OF** long short-term memory (LSTM) . decay factor **USED-FOR** long-term dependencies .

target_citation
REF propose an enlightening analysis of the Residual Networks (ResNets), and suppose that ResNet **REF** can be looked upon as a set of shallow networks and helps to form a deep network and copes with degradation problem by shortcut connection.

predicted
We propose a novel approach to address the issue of replacing missing data and building dependency relations from an incomplete sample set in time series tasks using residual networks (ResNets) and gated recurrent units (GRUs). Our proposed method, called Decay Residual Mean Imputation GRU (DRMI-GRU), incorporates a decay residual mean unit (DRMU) and a mask unit to handle missing data. The DRMU models the incomplete time series by.

Fig. 5. Example of Generated Citation text from **Alpaca** with knowledge graph relations

References

1. Agarwal, N., Reddy, R.S., Kiran, G., Rose, C.: Scisumm: a multi-document summarization system for scientific articles. In: Proceedings of the ACL-HLT 2011 System Demonstrations, pp. 115–120 (2011)
2. Bhagavatula, C., Feldman, S., Power, R., Ammar, W.: Content-based citation recommendation. arXiv preprint arXiv:1802.08301 (2018)
3. Bornmann, L., Mutz, R.: Growth rates of modern science: a bibliometric analysis based on the number of publications and cited references. J. Am. Soc. Inf. Sci. **66**(11), 2215–2222 (2015)
4. Chen, H., Takamura, H., Nakayama, H.: Scixgen: a scientific paper dataset for context-aware text generation. arXiv preprint arXiv:2110.10774 (2021)
5. Chen, J., Zhuge, H.: Summarization of scientific documents by detecting common facts in citations. Futur. Gener. Comput. Syst. **32**, 246–252 (2014)
6. Chen, J., Zhuge, H.: Automatic generation of related work through summarizing citations. Concurrency Comput. Pract. Experience **31**(3), e4261 (2019)
7. Chiang, W.L., et al.: Vicuna: an open-source chatbot impressing GPT-4 with 90%* chatgpt quality, March 2023. https://lmsys.org/blog/2023-03-30-vicuna/
8. Cohan, A., Goharian, N.: Scientific article summarization using citation-context and article's discourse structure. arXiv preprint arXiv:1704.06619 (2017)
9. Dettmers, T., Pagnoni, A., Holtzman, A., Zettlemoyer, L.: Qlora: efficient finetuning of quantized llms. arXiv preprint arXiv:2305.14314 (2023)
10. Gosangi, R., Arora, R., Gheisarieha, M., Mahata, D., Zhang, H.: On the use of context for predicting citation worthiness of sentences in scholarly articles. arXiv preprint arXiv:2104.08962 (2021)
11. Jaidka, K., Khoo, C., Na, J.-C.: Imitating human literature review writing: an approach to multi-document summarization. In: Chowdhury, G., Koo, C., Hunter, J. (eds.) ICADL 2010. LNCS, vol. 6102, pp. 116–119. Springer, Heidelberg (2010). https://doi.org/10.1007/978-3-642-13654-2_14
12. Ji, S., Pan, S., Cambria, E., Marttinen, P., Yu, P.S.: A survey on knowledge graphs: representation, acquisition, and applications. IEEE Trans. Neural Netw. Learn. Syst. **33**(2), 494–514 (2022). https://doi.org/10.1109/tnnls.2021.3070843
13. Kingma, D.P., Ba, J.: Adam: a method for stochastic optimization. arXiv preprint arXiv:1412.6980 (2014)
14. Koncel-Kedziorski, R., Bekal, D., Luan, Y., Lapata, M., Hajishirzi, H.: Text generation from knowledge graphs with graph transformers. arXiv preprint arXiv:1904.02342 (2019)
15. Lo, K., Wang, L.L., Neumann, M., Kinney, R., Weld, D.S.: S2orc: the semantic scholar open research corpus. arXiv preprint arXiv:1911.02782 (2019)
16. Luu, K., Koncel-Kedziorski, R., Lo, K., Cachola, I., Smith, N.A.: Citation text generation. ArXiv abs/2002.00317 (2020)
17. Pan, S., Luo, L., Wang, Y., Chen, C., Wang, J., Wu, X.: Unifying large language models and knowledge graphs: a roadmap (2023)
18. Qazvinian, V., Radev, D.R.: Scientific paper summarization using citation summary networks. arXiv preprint arXiv:0807.1560 (2008)
19. Taori, R., et al.: Stanford alpaca: an instruction-following llama model (2023)
20. Touvron, H., et al.: Llama: open and efficient foundation language models, Feb 2023. 10.48550/arXiv. 2302.13971
21. Vaswani, A., et al.: Attention is all you need. In: Advances in Neural Information Processing Systems, vol. 30 (2017)

22. Xing, X., Fan, X., Wan, X.: Automatic generation of citation texts in scholarly papers: a pilot study. In: Proceedings of the 58th Annual Meeting of the Association for Computational Linguistics, pp. 6181–6190 (2020)
23. Yasunaga, M., Kasai, J., Zhang, R., Fabbri, A.R., Li, I., Friedman, D., Radev, D.R.: ScisummNet: a large annotated corpus and content-impact models for scientific paper summarization with citation networks. In: Proceedings of the AAAI Conference on Artificial Intelligence, vol. 33, pp. 7386–7393 (2019)
24. Ye, D., Lin, Y., Li, P., Sun, M.: Packed levitated marker for entity and relation extraction. In: Muresan, S., Nakov, P., Villavicencio, A. (eds.) Proceedings of the 60th Annual Meeting of the Association for Computational Linguistics (Volume 1: Long Papers), ACL 2022, Dublin, Ireland, 22–27 May 2022, pp. 4904–4917. Association for Computational Linguistics (2022). https://aclanthology.org/2022.acl-long.337
25. Zhao, W.X., et al.: A survey of large language models (2023)
26. Zhu, Y., et al.: LLMs for knowledge graph construction and reasoning: recent capabilities and future opportunities (2023)

SciPhyRAG - Retrieval Augmentation to Improve LLMs on Physics Q&A

Avinash Anand(✉), Arnav Goel, Medha Hira, Snehal Buldeo, Jatin Kumar, Astha Verma, Rushali Gupta, and Rajiv Ratn Shah

Indraprastha Institute of Information Technology, Delhi, India
{avinasha,arnav21519,medha21265,snehal22074,jatin20206, asthav,rajivratn}@iiitd.ac.in

Abstract. Large Language Models (LLMs) have showcased their value across diverse domains, yet their efficacy in computationally intensive tasks remains limited in accuracy. This paper introduces a comprehensive methodology to construct a resilient dataset focused on High School Physics, leveraging retrieval augmentation. Subsequent finetuning of a Large Language Model through instructional calibration is proposed to elevate outcome precision and depth. The central aspiration is reinforcing LLM efficiency in educational contexts, facilitating more precise, well-contextualized, and informative results. By bridging the gap between LLM capabilities and the demands of complex educational tasks, this approach seeks to empower educators and students alike, offering enhanced support and enriched learning experiences. Compared to Vicuna-7b, the finetuned retrieval augmented model **SciPhy-RAG** exhibits a **16.67% increase** in BERTScore and **35.2%** increase on ROUGE-2 scores. This approach has the potential to be used to reshape Physics Q&A by LLMs and has a lasting impact on their use for Physics education. Furthermore, the data sets released can be a reference point for future research and educational domain tasks such as **Automatic Evaluation** and **Question Generation**.

Keywords: Document Retrieval · Neural Text Generation · Large Language Models · Natural Language Processing · Question-Answering

1 Introduction

The rising popularity of transformer-based [4] Large Language Models can be attributed to their exceptional performance in tasks such as text generation, question answering, and document summarization [1]. Recent advancements in language model architectures, such as the GPT-3.5 [2], PaLM [30], and LLAMA [8] have showcased their remarkable ability to comprehend and generate human-like text. However, when it comes to domain-specific computational tasks, such as solving physics problems, language models often struggle to achieve the desired level of accuracy.

V. Goyal et al. (Eds.): BDA 2023, LNCS 14418, pp. 50–63, 2023.
https://doi.org/10.1007/978-3-031-49601-1_4

Datasets have played an instrumental role in pushing forward performance on domain-specific tasks. In this research paper, we address the challenge of improving the accuracy of large language models (LLMs) for computational tasks by finetuning the model on a high school physics dataset we designed. We also introduce a high-quality corpus containing high school physics concepts from the NCERT textbook, considered a higher-education physics standard. The corpus is annotated manually to ensure its reliability and investigate the effectiveness of retrieval-augmentation techniques in further enhancing the model's performance.

Physics problem-solving requires an in-depth understanding of all underlying concepts and the ability to apply them sequentially and logically. Traditional approaches to computational tasks rely on rule-based algorithms or symbolic manipulation [32,33]. However, recent advancements in deep learning and language modelling have presented an opportunity to leverage the power of LLMs for these tasks. We believe that the efficacy of LLMs in domain-specific computational tasks would be a pivotal step in their usage for education and learning.

Language Models (LMs) rely on their inherent parameters to generate responses based on their knowledge accumulated via training of a huge corpus. With thousands of parameters holding the vast information, instances arise where domain-specific expertise, particularly prominent in complex fields such as physics question answering, demands enhanced support [15]. The notable complexities of physics queries suggest a potential advantage in incorporating relevant passages within the prompt. Given the complex and knowledge-intensive nature of physics problems, such an approach becomes imperative. In light of this, retrieval mechanisms offer a viable avenue to explore.

This paper explores retrieval augmentation techniques to enhance the finetuned LLM's performance using good-quality support passages. Retrieval enhancement involves incorporating a retrieval-based model to provide additional context and guidance to improve the model's reasoning capabilities.

The contributions of this research paper are twofold. First, we present a high school physics dataset corpus using NCERT textbook content, with good-quality annotations for mathematical equations specifically curated to enhance the precision of LLMs in computational tasks and serve as a benchmark for evaluating their performance. Second, we prepare a novel retrieval pipeline with a self-annotated document corpus to enhance model performance.

By addressing the limitations of LLMs in computational problem-solving and leveraging the potential of finetuning and retrieval augmentation, we aim to push the boundaries of language models' accuracy for high school physics tasks. The insights gained from this research have the potential to revolutionize educational technologies, enabling intelligent tutoring systems and personalized learning experiences that support students on their journey to master complex scientific concepts.

The written work is structured as follows: Sect. 2 addresses the related works on math and scientific problem solvers, and Sect. 3 explains the process behind data collection, annotation and augmentation. Section 4 explains how we performed our experiments, and Sect. 5 describes our evaluation metrics used and the results obtained. The paper's future scope and conclusion are summarised in Sect. 6 respectively.

2 Related Work

2.1 Mathematics and Science Solvers

We encountered several Mathematics and Science domain Q&A datasets in our exploration of datasets. However, we encountered a significant absence of high-quality and challenging physics question-answering datasets.

AQuA-RAT [11] is a dataset comprising more than 100K algebraic word problems with natural language rationales. Each data point in the dataset provides four values, i.e. the question, the options, the rationale behind the solution and the correct option.

The **MathQA** [12] was created using a novel representation language to annotate AQuA-RAT and correct the noisy-incorrect data. The limitation of this approach is its exclusive focus on multiple-choice questions. However, we found the rationale parameter intriguing as it empowers the finetuned model to offer logical explanations, aiding users in deeper comprehension of the problem and its solution.

GSM8K [6] claims to contain questions that can be solvable by a bright middle school student. Therefore, it cannot be used to finetune a model for complex reasoning and computational tasks. **MATH** [7] is a dataset of 12.5K challenging mathematics problems in competition. Recently, there has also been an emphasis on evaluating the capabilities of models in answering open-domain science questions.

ScienceQA [14], a dataset collected from elementary and high school science curricula, contains 21K multimodal multiple-choice science questions. **ASDiv (Academia Sinica Diverse MWP Dataset)** [18] is a notable dataset where each math word problem(MWP) is associated with a problem type and grade-level.

The **SCIMAT (Science and Mathematics Dataset)** [5] is a valuable resource as it provides chapter-wise questions for class 11 and 12 Mathematics and Science subjects, together with numerical answers. However, a notable limitation of the dataset is the absence of explicit explanations for the solutions or the underlying formulas related to the topics. This absence of contextual information hinders the dataset's effectiveness in fully supporting comprehensive learning and understanding.

2.2 Document Retrieval in Science Q&A

In Large Language Models, the learned knowledge about the outside world is implicitly stored in the parameters of the underlying neural network. Finding out what knowledge is stored in the network and where becomes challenging as a result [15]. The network's size also affects how much data can be stored; to capture more global information, one must train larger and larger networks, which might be prohibitively slow or expensive. To bridge this gap, adding context to a given query in the input of a Large Language Model is helpful.

[13] mentions how generative models for question answering can benefit from passage retrieval. It retrieves support text passages from an external source of knowledge, such as Wikipedia. Then, a generative encoder-decoder model produces the answer, conditioned on the question and the retrieved passages.

This approach has obtained state-of-the-art results on the **Natural Questions** [16] and **TriviaQA** [17] open benchmarks. This idea can be extended to domain-specific question answering. Given the complex nature of Physics and Math Questions, we assume that presenting the model with a set of relevant passages containing domain knowledge and required formulas can help the generative model answer the questions more accurately.

Most of the research in retrieval-augmented generation has been directed towards open-domain question answering. This involves extracting relevant information from external sources like Wikipedia to answer questions across various topics [15] [13]. The wide range of information in Wikipedia greatly aids open domain question answering.

Previously, domain-specific knowledge has been used to improve the performance of information retrieval systems. [34] created a data set by collecting question and answer pairs from the internet in the insurance domain. Our assumption in this work is that a corpus containing information in the physics domain would increase a model's performance on physics question-answering tasks.

3 Datasets

To further advance the application of language models in physics, we propose the creation of a comprehensive and challenging dataset. This study thus releases two open-source datasets :

1. **PhyQA** consists of 9.5K high school physics questions and answers with step-wise explanations. The dataset is diverse and consists of topics studied by high school physics students in the range of 15–19 years of age are included in the dataset. The list of topics includes: Alternating Current; Atoms and Nuclei; Communication Systems; Electric Charges, Fields and Current; Electromagnetic Induction; Electromagnetic Waves; Capacitors; Dynamics and Rotational Mechanics; Units, Dimensions and Kinematics; Ray and Wave Optics; Thermodynamics and Heat; Gaseous State; Waves, Sound and Oscillations. Each topic is divided into several subtopics. Both of these datasets are meticulously formatted to make it easy to train and test on open-source language models such as: **LLAMA** [8], **Alpaca** [20] and **Vicuna** [10].
2. The **RetriPhy Corpus** is a curated collection of Physics chapter content from NCERT books for 11th and 12th grades. It is annotated manually with LaTeX representations of equations and examples. The dataset includes 14 chapters from each grade, 11th and 12th, respectively. When prompted with a question, the combined corpus of all chapters creates context passages using the retrieval pipeline.

The proposed data sets serve multiple important purposes. Firstly, they enable the training of language models on complex reasoning and computational tasks specific to physics, which requires a deeper understanding of scientific principles and relationships between variables. Second, they could serve as a valuable benchmark to assess the performance of existing language models in physics problem-solving, allowing for a meaningful comparison of different models and driving further advancements in natural language understanding within the physics domain. Third, our retrieval corpus can be used to prepare and benchmark retrieval systems and to retrieve high-quality passages for physics Q&A (Fig. 1).

3.1 PhyQA

The dataset comprises 9.5K Physics questions, with each chapter having nearly equal representation. Each data point in the dataset is associated with two keys, i.e. "instruction" and "output", to organize the information. They are described as follows:

- **Instruction**: Key containing the question which needs to be answered by the model
- **Output**: Key containing the corresponding numerical answer and a detailed explanation of how that answer was obtained.

An example of a question from the chapter Newton's Laws of Motion is given below:

Instruction: An aircraft of mass 176 kg executes a horizontal loop at a speed of 249 m/s with its wings banked at 80 degrees. What is the radius of the loop?

Output: Answer: *35162.514 m* <sep> Explanation: *To calculate the radius, we use the formula*

$$\frac{(v^2)}{(g * \tan(\frac{param*\pi}{180}))}$$

where v is the speed (249*m*/*s*), *g is the acceleration due to gravity* (10*m*/*s*^{2}), *and param is the angle of banking* 80°. *Substituting the values,*

$$R = \frac{249 \cdot 249}{10 \cdot \tan(\frac{80*\pi}{180})} = 35162.514m$$

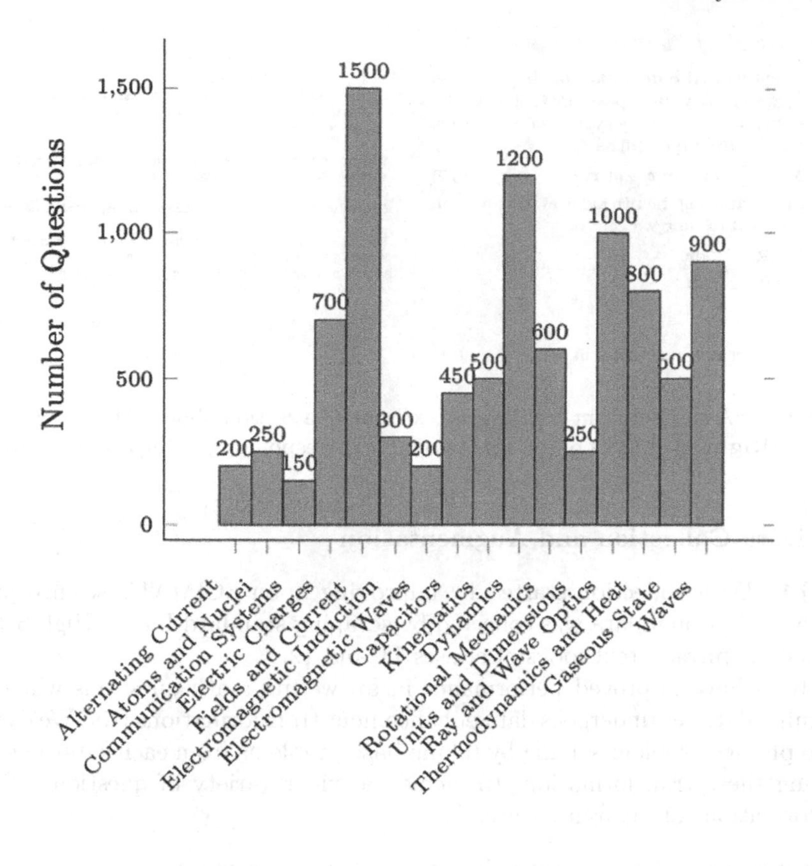

Fig. 1. Insight into the number of questions at chapter-wise granularity in PhyQA

3.2 RetriPhy Corpus

The RetriPhy Corpus comprises content extracted from NCERT books of Physics subjects for 11th and 12th grades. These NCERT (National Council of Educational Research and Training) books are known for their concise, accurate, and easily understandable presentation of concepts. The corpus contains theorems, equations, solving numerical problems, and explanations spanning various topics such as electric charges, gravitation, optics, atoms and nuclei, etc.

Containing material from all 14 chapters of the 11th and 12th grade NCERT Physics books, the corpus comprises **28 documents**, each corresponding to a chapter. Each chapter contains around **30–32 paragraphs**, with approximately 400 tokens per paragraph. The corpus extracts a total of **927 paragraphs**. We have also included an overlapping of 20 tokens in these paragraphs to ensure consistency (Fig. 2).

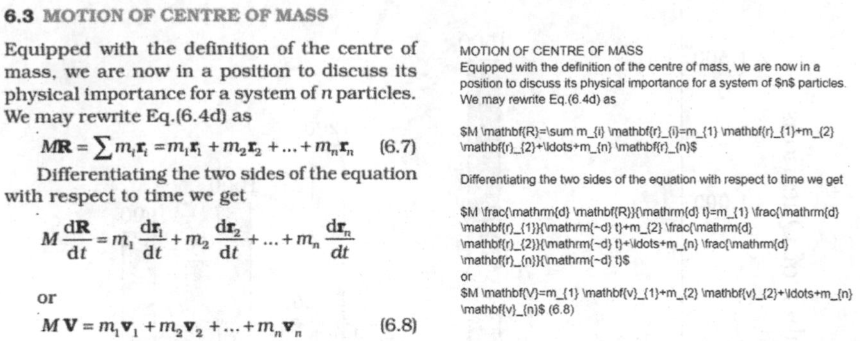

6.3 MOTION OF CENTRE OF MASS

Equipped with the definition of the centre of mass, we are now in a position to discuss its physical importance for a system of n particles. We may rewrite Eq.(6.4d) as

$$M\mathbf{R} = \sum m_i \mathbf{r}_i = m_1 \mathbf{r}_1 + m_2 \mathbf{r}_2 + \ldots + m_n \mathbf{r}_n \quad (6.7)$$

Differentiating the two sides of the equation with respect to time we get

$$M\frac{d\mathbf{R}}{dt} = m_1 \frac{d\mathbf{r}_1}{dt} + m_2 \frac{d\mathbf{r}_2}{dt} + \ldots + m_n \frac{d\mathbf{r}_n}{dt}$$

or

$$M\mathbf{V} = m_1 \mathbf{v}_1 + m_2 \mathbf{v}_2 + \ldots + m_n \mathbf{v}_n \quad (6.8)$$

```
MOTION OF CENTRE OF MASS
Equipped with the definition of the centre of mass, we are now in a
position to discuss its physical importance for a system of $n$ particles.
We may rewrite Eq.(6.4d) as

$M \mathbf{R}=\sum m_{i} \mathbf{r}_{i}=m_{1} \mathbf{r}_{1}+m_{2}
\mathbf{r}_{2}+\ldots+m_{n} \mathbf{r}_{n}$

Differentiating the two sides of the equation with respect to time we get

$M \frac{\mathrm{d} \mathbf{R}}{\mathrm{d} t}=m_{1} \frac{\mathrm{d}
\mathbf{r}_{1}}{\mathrm{~d} t}+m_{2} \frac{\mathrm{d}
\mathbf{r}_{2}}{\mathrm{~d} t}+\ldots+m_{n} \frac{\mathrm{d}
\mathbf{r}_{n}}{\mathrm{~d} t}$
or
$M \mathbf{V}=m_{1} \mathbf{v}_{1}+m_{2} \mathbf{v}_{2}+\ldots+m_{n}
\mathbf{v}_{n}$ (6.8)
```

Fig. 2. Left: A snippet from the Physics textbook of a section about "Motion of Centre of Mass"; **Right**: A snippet of the annotation file corresponding to the textbook snippet

3.3 Data Collection and Augmentation

PhyQA: Data collection started by improving upon SCIMAT's science problems [5]. Additional data was collected by scraping standard Indian High School open-source physics textbooks of classes 11 and 12.

Solvers have improved performance in answering math questions when the finetuning data set undergoes data enhancement transformations [19]. We extend this to physics problem-solving by taking base problems from each sub-topic and applying these transformations to include a wider variety of questions. These transformations are two in nature:

- Substitution: Changing the values of constants in a question.
- Paraphrasing: Paraphrasing the problem q using a model to generate N candidate questions that differ from the one in which they were written.

RetriPhy: We have used NCERT textbook content to create the RetriPhy (Retrieval-based Physics) corpus. Our motivation for using NCERT textbook content is the concise and easy-to-understand explanation of concepts in these books. Also, PhyQA consists of problems related to the topics of grade 11th and 12th physics; hence, the focus is on the content from the 11th and 12th grade NCERT Physics textbooks.

These chapter-wise documents on physics are accessible on the official NCERT website. Our methodology involved using these documents to retrieve textual content from the chapters. To ensure an accurate representation of mathematical symbols and equations, we employed LaTeX annotations, thereby eliminating the potential for ambiguity in text interpretation.

3.4 Data Annotation

PhyQA: Our team consisted of five dataset annotators, each of whom had graduated high school and studied physics until class 12. Upon self-evaluation on a scale of 5, the annotators rated themselves as 3, 3, 4, 4 and 5. We verified this self-evaluation by giving them a small test on basic questions to check fundamental understanding.

They used this in-depth understanding of physics concepts to annotate the solutions and provide relevant formulae and concepts. The annotated questions were then shuffled amongst the remaining annotators to evaluate the dataset's quality. Upon this, the annotators who rated themselves as 3 could solve only 55% problems given to them, while the one rated 5 could solve 80% of the problems. This shows that the dataset has high-quality questions and can be challenging for humans, too.

RetriPhy: The annotation process for the Retriphy corpus is aimed at the accurate representation of mathematical symbols and equations present in the content of NCERT textbooks. For every mathematical notation or equation in the text, we have used LaTeX for its annotation. To identify the start and end of the LaTeX content in the text, we have added a $ symbol as the start and end delimiters (Fig. 2). In the annotation process of RetriPhy, our team comprised three annotators, each contributing to approximately one-third of the annotations. A shared segment of annotations was distributed to all annotators to validate the annotations, allowing for cross-evaluation. This experiment revealed an impressive accuracy rate of 87% among the annotations.

Furthermore, the accuracy of annotations for all chapters is verified by both the annotators and an expert in the domain.

3.5 Inter-Annotator Agreement for Data Validation

In the data validation process, a team of five data annotators, all proficient college students in the domain of high school Physics, was employed. The dataset was equally divided among the annotators to ensure a balanced workload. Within this team, one annotator possessed expert-level knowledge, while the remaining four were classified as having intermediate expertise.

Rigorous attention to detail was exercised throughout the data annotation process to uphold the accuracy of the annotations. A meticulous approach was taken, whereby each segment annotated by one annotator underwent a verification stage involving assessment by two other annotators. This multiple-layer validation strategy was adopted to enhance the reliability of the annotated data.

The **Fleiss' Kappa** score of this annotation process was **0.65**. By combining the expertise of the annotators, the cross-validation process, and the Fleiss' Kappa coefficient application, a robust framework for data validation was established, ensuring the accuracy and integrity of the annotated high school Physics dataset.

4 Experiments

We describe the **SciPhy** retrieval experiment that uses both data sets we released. Language models' performance in question-answering tasks improves with finetuning on data sets with questions rephrased as instruction-following data points [3]. Following this, we finetune an open-source large language model with **PhyQA**.

We then incorporate retrieval by using our high-quality retrieval document dataset as a database for retrieving documents. We prepare a retrieval pipeline with our document vectorbase and use that to provide context on the test questions to evaluate the accuracy of the model answers on evaluation metrics. We elaborate on this approach in forthcoming subsections.

4.1 Fine-Tuning Using LoRA

Model and Hyperparameters: Vicuna [10] is a large language model prepared by finetuning a Llama base model on 70k user-shared conversations. We used a Vicuna model with 7 billion parameters as our baseline. Natural language processing consists of pre-training language models on general text and finetuning model parameters on domain-specific data. However, as the model size increases, it is computationally expensive to finetune models, which involves retraining all parameters fully. We thus adopt the **L**ow **R**ank **A**daptation (LoRA) [9], which proposes freezing model weights and injecting lower rank matrices into transformer layers that can be trained. This reduces training time and the hardware needed to keep model accuracy intact. We hypothesize that finetuning **Vicuna-LoRA** on our annotated **PhyQA** physics dataset will greatly improve the model's capabilities to answer physics questions.

Experiment: For this task, the PhyQA dataset is split into *8000 training samples* and *1500 test samples*. The training set is used to finetune the model. The Vicuna-LoRA model is run with different model weight representations, i.e. an **8-bit** representation and a **16-bit** representation. The LoRA rank r is set to **8** with a LoRA-dropout of **0.05** for preparing the 8-bit finetuned model and is set to **16** for preparing the 16-bit finetuned model. The finetuning is run for **3 epochs** with a batch size of **128** and the learning rate equal to **3e-4** i.e. the Karpathy constant on an NVIDIA RTX A6000 GPU.

4.2 Rationale Behind Retrieval:

Our second experiment hypothesizes that providing relevant context to our language model about the question as input will greatly improve the explanation and precision of the answers. This is based on physics being driven by concepts and their interpretation rather than simply applying formulae. Given a query q, the retrieval-based system is prompted to find N relevant passages. Each passage is retrieved and appended with a $< sep >$ token. Query q is prepended to the N retrieved passages to form the final user query q_f, which can be described as:

$$q_f = q + < sep > + \sum_{i=1}^{n} (N_i + < sep >)$$

q_f is then prompted to the finetuned Vicuna LoRA 7-billion model described in Sect. 4.1 to get the answer.

4.3 Retrieval System Design:

Combined with Vicuna-7b, LoRA finetuned on **PhyQA**, the retrieval system is called **SciPhy-RAG**. The RetriPhy Corpus creates passages of **400 tokens** each.

After creating the passages, the next step is to perform indexing, which addresses the challenge of memory storage. We retrieve relevant passages by using similarity matches between the indexed passages and the user query. However, as the corpus grows with more passages, this task becomes progressively more time-consuming. To efficiently store and index the passages, we adopt a method of representing them as dense vectors [22].

VectorStores, like Pinecone, are used for indexing and storing vector embeddings of text data for fast retrieval. It uses Approximate Nearest Neighbour (ANN) search in higher dimensions [27], which allows for handling large numbers of queries. ANN proposes, when given a set P of n points (in this case, a set of n queries), a metric ball $B_D(q|r)$ in metric space (X, D). It creates a data structure S such that for any query $q \in$ X, S returns a point p that satisfies:

$$D(p, q) \leq r$$

$$\forall p \in B_D(q, cr) \cap P$$

It minimises this for some $c \geq 1$ and returns the point p at the minima. In our approach, each passage is converted into a 384-dimensional dense vector embedding using the **all-MiniLM-L6-v2 model** [12]. These passages are then stored in the VectorStore described above.

Experimentation: When prompted with a user query q, the system applies ANN search to identify top K relevant passages where the user specifies K. The passages are returned by the system and appended to q. This final prompt with the passages q_f is prepared. We use the technique described in Sect. 4.1 to finetune Vicuna-7b using LoRA on PhyQA. The prompt q_f is inputted into the model specified above to obtain our final output.

5 Results

5.1 Evaluation Metrics

We perform a two-tier evaluation of the fine-tuned models. We first choose the evaluation metrics such as BERT-Score [21], METEOR [24] and ROUGE-L, ROUGE-1 and ROUGE-2 [25]. These help us assess the quality and correctness of the explanations generated by the models. We then sample 100 questions from each chapter and prompt the model to give a “One-Word Answer”. This gives the numeric answer, and we treat it as a classification task measuring the accuracies with the ground-truth answers of the test set. We call this metric as **Final Answer Accuracy (FAA)**. We repeat this with ten randomly chosen samples and report the lowest accuracies achieved out of the 10.

5.2 Experimental Results and Analysis

Upon finetuning the baseline Vicuna-7B model and attaching the retrieval system, we get two models from the experimental setups described above. Table 1 shows their results on the evaluation metrics described above. Our finetuned SciPhy-RAG 16-bit model (i.e. our retrieval pipeline + Vicuna-LoRA 16-bit) shows a **16.67%** increase on BERT-F1 scores over the base Vicuna-7b model. This shows that finetuning with **PhyQA** and using **RetriPhy** as our retrieval corpus increases the quality of explanations by a significant amount over the base model.

Table 1. Table showing the BERT, ROUGE and METEOR scores of the two finetuned Vicuna-LoRA models

Metric	Vicuna-7B	SciPhy-RAG (8-bit)	SciPhy-RAG (16-bit)
BERT (F1)	0.768	0.887	**0.899**
BERT (Precision)	0.744	**0.876**	0.865
BERT (Recall)	0.784	0.886	**0.895**
METEOR	0.285	0.347	**0.352**
ROUGE-L	0.321	0.371	**0.389**
ROUGE-1	0.315	0.358	**0.363**
ROUGE-2	0.147	0.181	**0.195**

On METEOR, our 8-bit SciPhy-RAG shows a **22.8%** increase. On the ROUGE evaluation metrics (ROUGE-L, ROUGE-1 and ROUGE-2), 16-bit SciPhy-RAG shows a much higher improvement (**19.4%**, **22.2%** and **35.3%** respectively) over the base Vicuna-7b model. We hypothesize that METEOR scores are calculated based on unigram matching between the reference and candidate sentences [23], and the retrieval models drive the model output generation slightly away from the ground truth explanation. The same hypothesis holds for lesser improvements in ROUGE scores. However, the increase in BERT scores validates that the explanations are semantically similar and high quality.

Table 2 shows the final answer accuracy (FAA) for the base-Vicuna-7b model (i.e. before) and the SciPhy-RAG (16-bit) model after finetuning and applying retrieval. Note that the lowest scores have been reported for both models, and this shows a massive increase across chapters, showing improvement despite skewness in the training data. Due to resource constraints, the lower accuracies can be attributed to running these experiments on smaller models. However, our hypothesis can be extended to larger models with sizes of parameters ¿ 50 B and newer architectures.

Table 2. Final Answer Accuracy **before and after Fine-Tuning with PhyQA**

Chapter	Before	After
Alternating Current	21.3	26.2
Atoms and Nuclei	20.8	27.1
Communication Systems	15.3	21.2
Electric Charges	22.9	26.5
Fields and Current	23.5	28.6
Electromagnetic Induction	22.2	29.1
Electromagnetic Waves	23.8	26.3
Capacitors	25.1	28.3
Dynamics & Rotational Mechanics	24.7	29.2
Units, Dimensions & Kinematics	22.4	27.3
Ray and Wave Optics	23.6	26.5
Thermodynamics and Heat	21.5	27.2
Gaseous State	19.6	24.3
Waves, Sound and Oscillations	20.6	28.4

6 Conclusion and Future Work

As we enter the future, we envision several promising avenues for further exploration. One such direction involves the incorporation of the **Chain of Thought** [28] and **Tree of Thought** [29] prompting techniques. Augmenting our datasets to incorporate these prompting techniques could lead to richer explanations and higher accuracies. A custom annotated retrieval dataset could also be prepared for preparing the vector database to improve the quality of retrieved texts.

Another extension area is curating benchmarks for other STEM fields, such as Chemistry and Biology, which will give rise to Multimodal models and further advancements in using LLMs and Artificial Intelligence to hasten up research and improve the quality of learning and education in these fields.

The symbiosis of **PhyQA** and **RetriPhy** with advanced LLM finetuning techniques and retrieval augmentation marks a significant step towards empowering AI-driven physics education. We envision a physics-based Q&A system catering to a student. As we embark on continuous improvement and exploration, we anticipate that our research will pave the way for more intelligent, interactive, and personalized learning experiences in physics and beyond. Additionally, these datasets will help drive research in AI-driven education tasks such as Automatic Evaluation and provide a foundation for making similar datasets.

Acknowledgments. We want to acknowledge the contribution of our data annotators, Aryan Goel, Ansh Varshney, Siddhartha Garg and Saurav Mehra. Rajiv Ratn Shah is partly supported by the Infosys Center for AI, the Center for Design and New Media, and the Center of Excellence in Healthcare at IIIT Delhi.

References

1. Goel, A., Hira, M., Anand, A., Bangar, S., Shah, D.R.R.: Advancements in scientific controllable text generation methods (2023)
2. Brown, T., et al.: Language models are few-shot learners (2020)
3. Chung, H., et al.: Scaling instruction-finetuned language models (2022)
4. Vaswani, A., et al.: Attention is all you need (2017)
5. Chatakonda, S.K., Kollepara, N., Kumar, P.: SCIMAT: dataset of problems in science and mathematics. In: Srirama, S.N., Lin, J.C.-W., Bhatnagar, R., Agarwal, S., Reddy, P.K. (eds.) BDA 2021. LNCS, vol. 13147, pp. 211–226. Springer, Cham (2021). https://doi.org/10.1007/978-3-030-93620-4_16
6. Cobbe, K., et al.: Training verifiers to solve math word problems. ArXiv Preprint ArXiv:2110.14168 (2021)
7. Hendrycks, D., et al.: Measuring mathematical problem solving with the MATH dataset (2021)
8. Touvron, H., et al.: Open and efficient foundation language models. In: LLaMA (2023)
9. Hu, E., et al.: Low-rank adaptation of large language models. In: LoRA (2021)
10. Chiang, W.L., et al.: Vicuna: an open-source chatbot Impressing GPT-4 with 90 (2023)
11. Ling, W., Yogatama, D., Dyer, C., Blunsom, P.: Learning to solve and explain algebraic word problems. In: Program induction by rationale generation (2017)
12. Miao, S., Liang, C., Su, K.: A diverse corpus for evaluating and developing English math word problem solvers (2021)
13. Izacard, G., Grave, E.: Leveraging passage retrieval with generative models for open domain question answering (2021)
14. Lu, P., et al.: Multimodal reasoning via thought chains for science question answering, learn to explain (2022)
15. Guu, K., Lee, K., Tung, Z., Pasupat, P., Chang, M.: Retrieval-augmented language model pre-training. In: REALM (2020)
16. Kwiatkowski, T., et al.: Natural questions: a benchmark for question answering research. In: Transactions Of The Association For Computational Linguistics, vol. 7, pp. 453–466 (2019)
17. Joshi, M., Choi, E., Weld, D., Zettlemoyer, L.: Triviaqa: a large scale distantly supervised challenge dataset for reading comprehension. ArXiv Preprint ArXiv:1705.03551. (2017)
18. Miao, S., Liang, C., Su, K.: A diverse corpus for evaluating and developing English math word problem solvers. ArXiv. abs/2106.15772 (2020)
19. Kumar, V., Maheshwary, R., Pudi, V.: Data augmentation for math word problem solvers. In: Practice Makes a Solver Perfect (2022)
20. Taori, R., et al.: Hashimoto stanford alpaca: an instruction-following LLaMA model (2023). https://github.com/tatsu-lab/stanford_alpaca
21. Zhang, T., Kishore, V., Wu, F., Weinberger, K., Artzi, Y. Evaluating text generation with BERT. In: BERTScore (2020)
22. Karpukhin, V., et al.: Dense passage retrieval for open-domain question answering (2020)
23. Saadany, H.: Constantin orăsan and BLEU, METEOR, BERTScore: evaluation of metrics performance in assessing critical translation errors in sentiment-oriented text. In: Proceedings Of The Translation And Interpreting Technology Online Conference TRITON 2021 (2021). https://doi.org/10.26615/978-954-452-071-7_006

24. Banerjee, S., Lavie, A.: METEOR: an automatic metric for MT evaluation with improved correlation with human judgments. In: Proceedings of The ACL Workshop On Intrinsic And Extrinsic Evaluation Measures For Machine Translation And/or Summarization, pp. 65–72 (2005)
25. Lin, C.: Rouge: a package for automatic evaluation of summaries. In: Text Summarization Branches Out, pp. 74–81 (2004)
26. Wang, W., et al.: Deep self-attention distillation for task-agnostic compression of pre-trained transformers. In: MiniLM (2020)
27. Andoni, A., Indyk, P., Razenshteyn, I.: Approximate nearest neighbor search in high dimensions (2018)
28. Wei, J., et al.: Chain-of-thought prompting elicits reasoning in large language models (2023)
29. Yao, S., et al.: Deliberate problem solving with large language models. In: Tree of Thoughts (2023)
30. Chowdhery, A., et al.: Scaling language modeling with pathways. In: PaLM (2022)
31. Deng, J., Dong, W., Socher, R., Li, L., Li, K., Fei-Fei, L.: ImageNet: a large-scale hierarchical image database. In: 2009 IEEE Conference On Computer Vision And Pattern Recognition, pp. 248–255 (2009)
32. Kojima, T., Gu, S., Reid, M., Matsuo, Y., Iwasawa, Y.: Large language models are zero-shot reasoners. Adv. Neural. Inf. Process. Syst. **35**, 22199–22213 (2022)
33. He-Yueya, J., Poesia, G., Wang, R., Goodman, N.: Solving math word problems by combining language models with symbolic solvers (2023)
34. @miscfeng2015applying, title=Applying Deep Learning to Answer Selection: A Study and An Open Task, author=Minwei Feng and Bing Xiang and Michael R. Glass and Lidan Wang and Bowen Zhou, year=2015, eprint=1508.01585, archivePrefix=arXiv, primaryClass=cs.CL

Revolutionizing High School Physics Education: A Novel Dataset

Avinash Anand[1(✉)], Krishnasai Addala[1], Kabir Baghel[1], Arnav Goel[1], Medha Hira[1], Rushali Gupta[2], and Rajiv Ratn Shah[1]

[1] Indraprastha Institute of Information Technology, Delhi, India
{avinasha,krishnasai20442,kabir20564,
arnav21519,medha21265,rajivratn}@iiitd.ac.in
[2] Banaras Hindu University, Varanasi, India

Abstract. Despite the growing capabilities of Large Language Models (LLMs) in various domains, their proficiency in addressing domain-specific high-school physics questions remains an unexplored area. In this study, we present a pioneering data set curated from NCERT exemplar solutions strategically designed to facilitate the use of LLMs to solve school physics questions. Originally comprising 766 questions accompanied by LaTeX representations, the dataset underwent a sophisticated augmentation process that expanded its scope to an impressive 7,983 questions. The augmentation employed innovative techniques which effectively broaden the dataset's coverage. The dataset, prioritizing text-based questions, is formatted as JSON objects detailing instructions, inputs, and outputs. Post evaluation, we noted significant scores: **METEOR at 0.282** and **BERTScore F1 at 0.833**, indicating a close alignment between generated and reference texts.

Keywords: Large Language Models · High School Education · Dataset · Chain of Thought · Machine Learning · Artificial Intelligent · Physics Questions

1 Introduction

The advent of transformer-based Large Language Models [28] has revolutionized various sectors, ranging from finance to healthcare. Recently, these technologies have begun to make significant inroads into the field of education. The capacity of large language models to personalize learning, adapt to individual student needs, and provide scalable solutions makes them particularly well-suited for educational applications. This paper explores the application of controllable text generation in high school education, with a focus on physics.

In the realm of education, there exists a growing demand for effective and innovative teaching methodologies. Traditional educational content often struggles to cater to the diverse needs and learning styles of students. Moreover, as the world becomes increasingly digital and connected, there is a pressing need to harness the power of technology to enhance the learning experience.

V. Goyal et al. (Eds.): BDA 2023, LNCS 14418, pp. 64–79, 2023.
https://doi.org/10.1007/978-3-031-49601-1_5

While this study makes significant strides in the creation and augmentation of a high school-level physics question dataset, it is important to acknowledge the challenges and limitations encountered. We grappled with issues such as achieving comprehensive topic coverage, balancing dataset size with quality, and addressing the structural nuances needed for AI model interpretability and educational platform integration. Additionally, the reliance on NCERT exemplar solutions as a primary source brought its own set of limitations, which are elaborated in Sect. 6.

In this study, we evaluated the efficacy of our dataset by fine-tuning the Llama2-7B model, a state-of-the-art machine learning algorithm designed for natural language processing tasks. The model was fine-tuned on our high school-level physics question dataset, allowing us to assess its suitability for solving physics problems. To mitigate the computational challenges associated with fine-tuning, the LoRA method was employed. Additionally, we conducted experiments using both the Chain-of-Thought (COT) and non-COT models to offer a more comprehensive assessment of their performance.

Physics, being a fundamental subject in high school education, presents its own set of challenges. It involves abstract concepts that can be difficult to grasp, and students often require personalized explanations to comprehend complex principles. This is where the potential of large language models comes into play. These models can assist students in learning concepts and answering questions, potentially providing a more dynamic education experience. This paper addresses the absence of a high-quality physics question dataset by contributing to the creation and augmentation of a dataset of high school-level physics questions. The original dataset consisted of 807 questions, This dataset was cleaned, by removing undesirable questions and questions which contained multimodal instructions. In this manner, we obtained a base dataset of 766 high school level physics questions. We elected to utilize NCERT questions in the construction of the dataset, as it is a well-respected institution of education throughout the whole of India. Leveraging controllable text generation, we augmented this base dataset to approximately 8000 questions. Our goal is to enrich the existing repository of educational content and facilitate the training of AI models that can potentially enhance the learning experience in high school physics.

The contribution of this research is the creation and augmentation of a dataset of high school-level physics questions, at a high level of quality, such that any model fine-tuned on the data is able to solve and work through physics problems with a high level of accuracy. We also introduce a dataset of COT prompts, to investigate the potential change in performance. Potential applications for this contribution include being used for the finetuning of QA models used in High-School level educational contexts.

The written work is structured as follows: Sect. 2 addresses the related works, Sect. 3 explains how we curated this dataset, data annotation, and data augmentation techniques used. In Sect. 4, we have explained how we performed our experiments, and then Sect. 4.1 shows the evaluation metrics and evaluations. Section 4.3 discusses the analysis and study about this research, and the paper's future scope, its limitations, and conclusion are summarized in Sect. 5, 6, and 7 respectively.

2 Related Work

The application of AI in education has been an area of active research in recent years. AI-driven systems have been developed for a variety of educational tasks, including intelligent tutoring, automated grading, and adaptive learning.

Specifically, in the realm of question generation and augmentation, previous work has demonstrated the potential of AI models. For instance, Kumar et al. [8] proposed a data augmentation technique that we have adopted in our research. Their methodology forms the backbone of our augmentation process.

There exist several comparable datasets in this domain, such as GSM8K [9], SCIMAT2 [8], and JEEBench [10]. However, there still exists a lack of high-quality, high-school-level physics questions. JEEBench consists of approximately 450 pre-engineering questions in the domain of mathematics, physics, and chemistry. These questions tend to be above the difficulty of standard Indian high school physics problems. GSM8K [9] contains questions that are purported to be of middle school difficulty which is less than optimal for finetuning a model with the intention of solving high school level physics problems. SciQ [17] contains over 13,000 questions focused on science, however, these questions are entirely multiple-choice questions (MCQ), the dataset we introduce includes both MCQ style questions and questions which require more intermediate steps to solve. Other similar datasets are mainly focused on mathematics, such as AquaRat [15] or GeoQA [16].

Bryant [4] shows the efficacy of using LLMs (in particular, GPT-4) to assist in academic pursuits related to physics. Although the study does point towards some potential issues in this domain, it is plausible that some of these problems, such as issues of replication, can be mitigated using prompt engineering.

Few-shot prompting, as proposed by Liu et al. [13], is a technique that has shown promise in enhancing the capabilities of language models. By providing a few examples of the desired output format or context, the model can be guided to generate text that adheres to the specified style or content. This approach is particularly relevant in our case, as it allows us to fine-tune the model to generate physics questions and explanations that align with the curriculum and difficulty level of high school physics.

Chain-of-thought prompting [3] presents a valuable technique for enhancing language models' utility in education, involving the generation process guidance through sequential prompts that break down the desired output into coherent intermediate steps. Applied to high school physics question augmentation, this method enables the model to produce well-structured content by following a logical progression. By providing step-by-step prompts, the model generates questions and explanations that encompass essential content while illustrating problem-solving methodologies, aligning effectively with educational goals of imparting both knowledge and problem-solving strategies crucial for high school physics comprehension.

The advent of transformer-based models, has further opened the avenue for controllable text generation. These models have shown remarkable success in generating human-like text, making them a promising tool for educational content generation. Our work builds upon these prior studies but with a focus on high school physics. To the best of our knowledge, this is the first study that applies controllable text generation for the augmentation of a physics question dataset.

3 Dataset Description

3.1 Dataset Description

Our original dataset consists of 766 questions generated from physics problems. After augmentation, our dataset consists of 7983 questions. Table 1 shows the distribution of questions across question types and topics. A majority are subjective questions, as augmenting subjective problems was simpler. The questions in the dataset cover a wide range of topics in physics, including mechanics, electromagnetism, thermodynamics, optics, and atomic physics. The questions are designed to test students' understanding and problem-solving skills in physics.

3.2 Sample Question in Physics Dataset

Instruction: A circular current loop of magnetic moment M is in an arbitrary orientation in an external magnetic field B. The work done to rotate the loop by 300° about an axis perpendicular to its plane is:

Input: (a) MB, (b) $\frac{\sqrt{\mathbf{MB^2}}}{2}$, (c) $\frac{\text{MB}}{2}$, (d) Zero
Output: The work done to rotate the loop by 300° about an axis perpendicular to its plane is zero, as the correct answer is, **(d) Zero**.

3.3 Data Collection

We adopted a semi-automated process to construct a dataset of over 8000 physics problems:

1. Indian high school textbook question banks (i.e. NCERT Exemplar) were scraped and formatted using various open-source OCR tools. The formatting was maintained as standard TeX formatting.
2. The formatted LateX text was formatted into an appropriate key-value pair format.
3. Data augmentation was performed using a combination of Self-Instruct [25] and a custom Python script to generate variations for select questions.

Our base dataset is derived from the NCERT exemplar solutions for high school physics. This choice is motivated by the comprehensiveness of the NCERT exemplars, which provide a wide range of questions covering all key areas of the

Table 1. Distribution of Questions by Chapter

Chapter	# Questions
Units and Measurements	95
Motion in a Straight Line	121
Motion	183
Laws of Motion	336
Work, Energy and Power	148
System of Particles and Rotational Motion	176
Gravitation	260
Mechanical Properties of Solids	176
Mechanical Properties of Fluids	169
Thermal Properties of Matter	527
Thermodynamics	624
Kinetic Theory	1227
Oscillations	739
Waves	933
Electric Charges and Fields	214
Electrostatic Potential and Capacitance	216
Current Electricity	427
Moving Charges and Magnetism	116
Magnetism and Matter	323
Electromagnetic Induction	129
Alternating Current	110
Electromagnetic Waves	78
Ray Optics and Optical Instruments	105
Wave Optics	90
Dual Nature of Radiation and Matter	96
Atoms	127
Nuclei	101
Semiconductor Electronics: Materials, Devices and Simple Circuits	137
Communication Systems	107
Total	7983

high school physics curriculum. Each question in the JSON file is structured as an object comprising an instruction (the question text), an input (options for multiple-choice questions or blank otherwise), and an output (the solution).

The augmentation of this dataset was achieved through the application of controllable text generation, leveraging the methodology provided by Kumar et al [8]. Through this process, we expanded the dataset from 766 questions to a total of 7983 questions.

The above graph (Fig. 1) illustrates the distribution of question lengths throughout the corpus. The average question prompt length of approximately

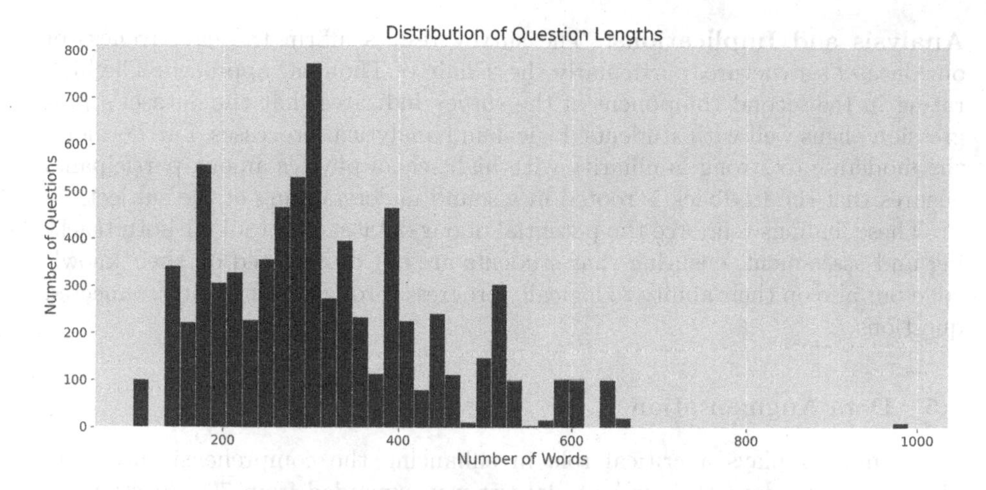

Fig. 1. Question length distribution

350 words allows for LLMs to hold the question prompts in their Context Window without much issue.

3.4 Student Survey

To gauge the effectiveness of our dataset and its alignment with the 'Chain of Thought' approach, we conducted a survey involving around 500 participants. The survey was designed to understand students' familiarity with high school physics and evaluate how well the first three questions in a set prepared them for the final question.

Survey Design. The survey contained the following components:

A question to rate the participants' familiarity with high school physics on a scale from 1 (Weak) to 5 (Strong). Five sets of questions where participants were shown an image (representing a series of physics questions) and asked to rate on a scale of 1 to 5 how well the first three questions in the set provided a basis for answering the final question.

Survey Findings. Familiarity with High School Physics: The majority of participants (approximately 65%) rated their familiarity with high school physics between 3 and 4, indicating a moderate to strong grasp of the subject.

Basis for Answering the Final Question: On average, participants rated the effectiveness of the first three questions in providing a basis for the final question at 3.8. This suggests that most students found the initial questions effective in preparing them for the concluding question.

Analysis and Implications. The survey results affirm the effectiveness of our dataset's structure, particularly the 'Chain of Thought' approach. The high rating in the second component of the survey indicates that the dataset's progression aligns well with students' logical and analytical processes. Furthermore, the moderate to strong familiarity with high school physics among participants ensures that the feedback is rooted in a sound understanding of the subject.

These findings reinforce the potential of our dataset as a tool for both teaching and assessment, ensuring that students are not only tested on their knowledge but also on their ability to logically progress through a set of interconnected questions.

3.5 Data Augmentation

Augmentation plays a critical role in enhancing the comprehensiveness and diversity of the dataset. The base dataset was expanded from 766 questions to 7983 questions through the augmentation process. The technique adopted from Kumar et al. [8] data augmentation method was used as a foundation, forming the backbone of the augmentation process.

Custom augmenter scripts were developed for each chapter in the physics curriculum. These scripts employed a structured approach to generate questions tailored to the specific topics and concepts within each chapter. The scripts followed a template-based approach, where predefined question and answer templates were populated with randomly selected elements from predefined lists of options. To ensure the augmented data remains contextually relevant without compromising its scientific accuracy, the scripts introduced variations in multiple aspects:

- **Values Variation:** Numerical values and specific data points within the questions were randomized, ensuring that the model doesn't get biased towards specific solutions.
- **Conceptual Variation:** Questions were generated to approach the same core concept from different angles or perspectives, offering a comprehensive understanding.
- **Structural Variation:** The linguistic structure of the questions, including phrasing and order, was diversified to prevent the model from memorizing specific patterns.

This approach not only expanded the dataset but also ensured a wide range of physics problems across different difficulty levels, promoting a model that can generalize well across diverse query types and scenarios. By generating such extensive variability in the questions, the chances of the model overfitting to the original dataset are minimized. The augmentation process was specifically designed to create a balance, ensuring the model is exposed to varied data while still maintaining the core principles of the physics curriculum.

The dataset currently focuses on the Indian context, but there is potential for expansion to other educational contexts. Future plans include benchmarking the dataset on various state-of-the-art models, such as Falcon, GPT-4, or Vicuna, to further evaluate the effectiveness of the augmentation techniques.

3.6 Construction of a COT Dataset

At the heart of our experimental approach lies the "Chain-of-Thought" (CoT) [3] methodology, which aims to guide the model's reasoning by providing a sequence of related questions leading up to the main query. The underlying philosophy is that by priming the model with questions and their answers that are semantically close to the main query, we guide its reasoning in a desired direction. This CoT is achieved using BERT [27] embeddings and cosine similarity. The methodological steps are detailed below:

Data Loading. The dataset, replete with questions and their associated details, is loaded into the Python environment. Each question in the dataset acts as a potential link in our chain of thought.

BERT Initialization. To effectively capture the semantic essence of each question, we utilize the 'bert-base-uncased' variant from the Hugging Face Transformers library. This choice ensures that the CoT is built on rich, contextual embeddings.

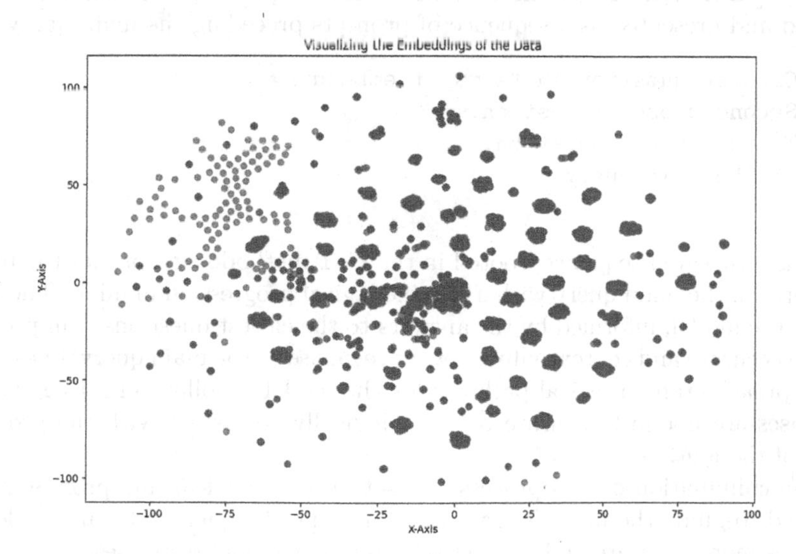

Fig. 2. t-SNE visualization of BERT embeddings for the Dataset (*t-SNE* [29] *was utilized for its proficiency in mapping high-dimensional data to a lower-dimensional space, particularly preserving local data point relationships, hence enabling an effective visualization of potential clusters.*)

Embedding Generation. Each question is converted into a BERT embedding, capturing its semantic essence. These embeddings serve as the foundation for determining the logical and contextual proximity between questions.

Figure 2 is a visualization of the BERT Embeddings of the dataset. Each point represents a question from the dataset. The visualization illustrates the diversity of the dataset, while also maintaining enough samples to form clusters, which represent similar questions in the corpus. Areas with many points represent questions with a similar semantic structure, which may also represent similar concepts (i.e., similar questions about Classical Mechanics are likely to be clustered together).

Similarity Calculation. For a given query question, its BERT embedding is generated. Cosine similarity then measures the degree of alignment between this query and all other questions in our dataset, determining which questions are most semantically proximate.

Retrieval of Similar Questions. Harnessing the power of cosine similarity, we identify the top three questions that are most closely aligned with the query in terms of their chain of thought. These questions act as the guiding prompts for the model.

Prompt Generation. With the three closest questions in hand, they are ordered and presented as a sequence of prompts preceding the main query:

```
q1: Closest question in terms of embedding.
q2: Second closest question.
q3: Third closest question.
query: The main query.
```

This structured sequence, rooted in the CoT methodology, primes the model to approach the main query with a specific logical progression in mind. The hope is that the model, informed by the answers to the closest questions, will produce a more accurate and contextually relevant response to the main query. In essence, our approach crafts a logical pathway for the model to follow, ensuring that its responses are not just accurate but also logically consistent with the provided chain of thought.

The culmination of this process is a set of few-shot learning prompts, each designed to guide the model's reasoning for a specific query. This methodology promises more accurate and contextually relevant model responses.

4 Experiments

The dataset was evaluated by finetuning the Llama2 [26] model with 7 billion parameters. Llama2-7B model is a state-of-the-art machine learning algorithm

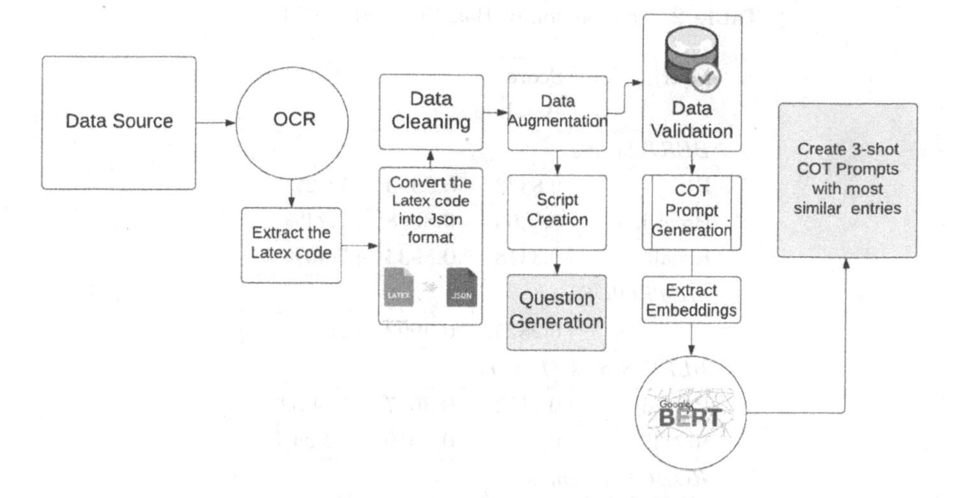

Fig. 3. Visualization of the Prompt Generation Pipeline

designed for natural language processing tasks. It is based on the transformer architecture and has been pre-trained on a large corpus of text data. By fine-tuning the model on our dataset, we were able to evaluate the efficacy of the dataset for solving physics problems at a high school level. Fine-tuning a model such as Llama2 is computationally taxing, so we employed the LoRA [11] method to make the task less demanding (Fig. 3).

4.1 Evaluation of COT

In addition to our primary evaluation, we also investigated the performance of the COT model, as discussed in a reference paper which highlighted its strong performance [3]. We conducted experiments using both the COT and non-COT models and compared their outputs with the ground truth to provide a more comprehensive assessment of their capabilities.

Results. After fine-tuning on the Llama2-7B model, we evaluated the model using various metrics including BERTScores [21], METEOR score [22], ROUGE scores [23], and BLEU scores [24]. The scores obtained are detailed in the following sections.

4.2 Evaluation Metrics

METEOR: The METEOR scores obtained are **0.282** and **0.221** which indicates that the generated content aligns to some extent with the reference data. However, certain domain-specific terms or phrasings common in physics might still be missing (Table 2).

Table 2. Comparison of Baseline and CoT Results

Metric Score	Score		% Change
	Baseline	CoT	
BERT Scores			
F1	0.8333	0.8943	+7.32%
Precision	0.8363	0.8968	+7.24%
Recall	0.8318	0.8934	+7.40%
METEOR Score			
	0.2825	0.4992	+76.68%
BLEU Scores (1-gram)			
Precision	0.1123	0.3647	+224.89%
Recall	0.1571	0.3019	+92.24%
ROUGE-1 Scores			
F1	0.2258	0.4845	+114.65%
Precision	0.2581	0.5528	+114.18%
Recall	0.3746	0.5372	+43.44%
ROUGE-L Scores			
F1	0.2732	0.5307	+94.23%
Precision	0.2980	0.5880	+97.32%
Recall	0.4118	0.5757	+39.79%

BERTScore: BERTScores compute BERT-based embeddings of the ground truth answer in the annotated dataset and the answer generated by the model. The scores obtained indicate that the model's generated text is relatively close to the reference text in terms of content and meaning.

BLEU: BLEU scores measure the overlap of n-grams between the generated content and reference content. The evaluation suggests that certain n-gram overlaps are missing, which means the generated content might lack some important multi-word physics terms or phrasings.

ROUGE: ROUGE scores assess the quality of the summaries generated by the model. The various ROUGE scores show the degree of overlap between the generated summaries and reference summaries in terms of n-grams and sequences. The F1 scores suggest that while there's a decent overlap of terms with the reference, there is still room for improvement.

4.3 Discussion

Taking into account the relatively modest scale of the evaluated model, it is reasonable to anticipate that employing a more substantial and advanced model could yield improved results. The consistency in strong BERT Score implies that our model's ability to convey semantics is notable, even with current limitations. The variability in BLEU and ROUGE scores highlights specific areas for growth, particularly in capturing longer n-grams and linguistic subtleties. A larger model could potentially address these challenges and lead to enhanced performance across a wider spectrum of linguistic contexts.

Evaluation metrics, in general, can be divided into those that focus on token-level matches and those that emphasize semantic-level understanding. While BLEU and METEOR combine both these perspectives, it's important to interpret their scores with this duality in mind.

BERT Score-: The BERT scores indicate a consistent and positive impact from utilizing Chain-of-Thought (CoT) Data compared to the baseline. Across all three metrics (F1, Precision, and Recall), there are improvements of +7.32%, +7.24%, and +7.40% respectively. This uniform increase suggests that the CoT Data enhances the model's ability to retain semantic meaning and precision in its responses, without sacrificing recall. It's noteworthy that the BERT Score, which is based on cosine similarity, can be influenced by good grammar. Therefore, a model that generates grammatically correct sentences can achieve a high BERT Score even if the semantic content isn't fully aligned with the reference.

METEOR Score-: A significant enhancement is observed in the METEOR score, which leaps from 0.2825 to 0.4992, constituting a +76.68% increase when employing CoT Data. METEOR, considering both precision and recall and being particularly sensitive to synonymy, indicates that the CoT-enhanced model might be effectively utilizing synonyms and aligning phrases in a semantically relevant manner, thus achieving a higher score.

BLEU Score-: For BLEU scores, particularly focusing on 1-gram, the Precision metric experiences an extraordinary increase of +224.89%, while Recall also experiences a significant boost of +92.24%. This suggests that CoT Prompting aids in producing responses that are considerably more aligned with reference sentences, especially in utilizing appropriate 1-grams. It's inherent in the nature of BLEU that unigram matches are the most frequent, while bigram and higher n-gram matches become progressively rarer. This points towards the need for further analysis into higher n-grams to provide insights into more intricate linguistic alignments.

ROUGE Scores-: In the context of ROUGE-1 scores, there is a remarkable increase across all metrics, with F1, Precision, and Recall improving by +114.65%, +114.18%, and +43.44% respectively. This indicates a substantial enhancement in producing unigram overlaps with reference sentences, which is crucial for maintaining semantic and informational fidelity.

Similarly, ROUGE-L (longest common subsequence) scores also witness substantial improvements, where F1, Precision, and Recall are boosted by +94.23%, +97.32%, and +39.79% respectively, when utilizing CoT Data. This underscores the model's enhanced capability to maintain coherent and informationally equivalent responses over longer subsequences of text, indicating a potential strength in preserving sentence-level informational and structural fidelity.

It's worth noting that LLMs typically struggle with some of these metrics, which could explain the initially low scores observed. As research progresses and models evolve, we anticipate improvements in these areas.

5 Future Scope

Our current study marks the initial steps towards enhancing high-school physics education through the creation and augmentation of a comprehensive dataset. Several avenues for further exploration and improvement exist within this domain:

5.1 Content Classification

While we have focused on the generation of physics questions and explanations, automatic evaluation mechanisms can significantly enhance the usability of the generated content. Developing evaluation metrics that assess the quality, clarity, and relevance of questions and solutions could provide valuable information on model performance and guide refinements. Existing literature in this area includes classifier models which detect the veracity of a given claim, such as the systems outlined by FEVER [19] and work done by Azaria et al. [18]

5.2 Tree of Thought Prompting

While we have explored the effectiveness of few-shot and chain of thought prompting techniques, another avenue for improvement is the implementation of Tree of Thought prompting [20]. This approach involves generating a structured sequence of prompts that guide the model's thought process through a problem-solving scenario. By providing increasingly detailed prompts, we can encourage the model to generate multi-step solutions and explanations, enhancing its ability to guide students through complex physics problems.

5.3 Model Benchmarking

Our initial experiments involved fine-tuning the Llama2-7B model on the augmented dataset. However, there is a plethora of advanced language models available. In the future, we intend to benchmark the data set with various state-of-the-art models, such as Falcon, GPT-4, or Vicuna. We also aim to expand our testing to include models with higher parameter counts, as Chain of Thought shows better results on larger models.

5.4 Dataset Expansion

The dataset currently focuses on preparing questions from a physics point of view in the Indian context. We plan to improve the quality and diversity of the questions to consider diversity in different countries. We also hope to expand this to more levels of education such as middle school and university level education.

6 Limitations

Throughout the course of this study, we faced several challenges, primarily stemming from the lack of a comprehensive dataset designed specifically for Indian high school-level physics education. This absence poses significant hindrances to the deployment of advanced artificial intelligence (AI) and machine learning (ML) solutions tailored to high school physics. Some of the major challenges and limitations we encountered include:

- **Coverage and Diversity:** Achieving comprehensive coverage of all high school physics topics proved to be demanding. Ensuring the dataset spans multiple chapters and grades without missing out on any crucial topic was a significant challenge.
- **Size and Quality Balance:** Striking the right balance between the size and quality of the dataset was intricate. While a large dataset is essential for effectively training AI and ML models, maintaining quality and accuracy at scale was paramount, and ensuring both simultaneously posed difficulties.
- **Structural Challenges:** Designing a dataset format that is both easily interpretable by AI models and convenient for integration into educational platforms presented its own set of challenges. Structuring the data in a universally accessible manner without compromising on the richness of information was a nuanced task.
- **Source Limitations:** Relying on NCERT exemplar solutions as the primary source meant that our dataset might inherit any inherent biases or limitations present in these solutions. Diversifying the sources could potentially alleviate this limitation in future iterations.
- **Scalability Concerns:** While the current dataset focuses on high school physics, expanding it to cover other subjects or educational levels might introduce additional complexities, both in terms of content and structure.

Despite these challenges, our study managed to curate a dataset of around 8000 physics problems, structured as JSON objects. Each entry encompasses an instruction, input, and output, making it a robust resource. Future directions, informed by the limitations encountered, include broadening the dataset to cater to multimodal questions and examining the influence of chapter representation on model efficiency.

7 Conclusion

Through this work, we present a significant contribution to the field of AI in education - the creation of the largest dataset of high school physics questions based on NCERT exemplars. This dataset, designed with the specific aim of fine-tuning AI models, can serve as a robust foundation for the development of advanced AI tools in high school education. Our current model focuses on text-based questions. However, we recognize that high school physics questions often include diagrams, graphs, and other visual aids. Therefore, future work will involve extending the model's capabilities to handle such multimodal questions.

Furthermore, our dataset covers various chapters from the NCERT exemplars, with each chapter represented to varying degrees. We plan to investigate the effects of these variations in chapter representation on the performance of AI models trained on this dataset.

Acknowledgements. Rajiv Ratn Shah is partly supported by the Infosys Center for AI, the Center for Design and New Media, and the Center of Excellence in Healthcare at IIIT Delhi.

References

1. Zhang, Z., Zhang, A., Li, M., Smola, A.: Automatic Chain of Thought Prompting in Large Language Models (2022)
2. Kojima, T., Gu, S.S., Reid, M., Matsuo, Y., Iwasawa, Y.: Large language models are zero-shot reasoners (2023). https://doi.org/10.48550/arXiv.2205.11916
3. Wei, J., et al.: Chain-of-thought prompting elicits reasoning in large language models (2023). https://doi.org/10.48550/arXiv.2201.11903
4. Bryant, S.: Assessing GPT-4's Role as a Co-Collaborator in Scientific Research: A Case Study Analyzing Einstein's Special Theory of Relativity (2023). https://doi.org/10.21203/rs.3.rs-2808494/v2
5. Zhou, C., et al.: LIMA: less is more for alignment (2023). https://doi.org/10.48550/arXiv.2305.112062
6. Lightman, H., et al.: Let's verify step by step (2023). https://doi.org/10.48550/arXiv.2305.20050
7. Gunasekar, S., et al.: Textbooks Are All You Need (2023). https://doi.org/qaojh1JD
8. Mandlecha, P., Chatakonda, S.K., Kollepara, N., Kumar, P.: Hybrid Tokenization and Datasets for Solving Mathematics and Science Problems Using Transformers (2023). https://doi.org/10.1137/1.9781611977172.33
9. Cobbe, K., et al.: Training verifiers to solve math word problems (2023). https://doi.org/10.48550/arXiv.2110.14168
10. Arora, D., Singh, H.G.: Mausam: have LLMs advanced enough? A challenging problem solving benchmark for large language models (2023). https://doi.org/10.48550/arXiv.2305.15074
11. Hu, E.J., et al.: LoRA: low-rank adaptation of large language models (2021). https://doi.org/10.48550/arXiv.2106.09685
12. Dettmers, T., Pagnoni, A., Holtzman, A., Zettlemoyer, L.: QLoRA: efficient fine-tuning of quantized LLMs (2023). https://doi.org/10.48550/arXiv.2305.14314

13. Zhao, T.Z., Wallace, E., Feng, S., Klein, D., Singh, S.: Calibrate before use: improving few-shot performance of language models (2021). https://doi.org/10.48550/arXiv.2102.09690
14. Liu, J., Shen, D., Zhang, Y., Dolan, B., Carin, L., Chen, W.: What Makes good in-context examples for GPT-3? (2021). https://doi.org/10.48550/arXiv:2101.06804
15. Ling, W., Yogatama, D., Dyer, C., Blunsom, P.: Program induction by rationale generation: learning to solve and explain algebraic word problems (2017). https://doi.org/10.48550/arXiv.1705.04146
16. Chen, J., et al.: GeoQA: a geometric question answering benchmark towards multimodal numerical reasoning (2022). https://doi.org/10.48550/arXiv.2105.14517
17. Welbl, J., Liu, N.F., Gardner, M.: Crowdsourcing multiple choice science questions (2017). https://doi.org/10.48550/arXiv.1707.06209
18. Azaria, A., Mitchell, T.: The internal state of an LLM knows when its lying (2023). https://doi.org/10.48550/arXiv.2304.13734
19. Thorne, J., Vlachos, A., Christodoulopoulos, C., Mittal, A.: FEVER: a large-scale dataset for Fact Extraction and VERification (2018). https://doi.org/10.48550/arXiv.1803.05355
20. Yao, S., et al.: Tree of thoughts: deliberate problem solving with large language models (2023). https://doi.org/10.48550/arXiv.2305.10601
21. Zhang, T., Kishore, V., Wu, F., Weinberger, K.Q., Artzi, Y.: BERTScore: evaluating text generation with BERT (2020). https://doi.org/10.48550/arXiv.1904.09675
22. Banerjee, S., Lavie, A.: METEOR: an 476 automatic metric for MT evaluation with improved correlation relation with human judgments. In: Proceedings of the ACL Workshop on Intrinsic and Extrinsic Evaluation Measures for Machine Translation and/or Summarization, pp. 65–72 (2005)
23. Lin, C.-Y.: ROUGE: a package for automatic evaluation of summaries. Text summarization branches out, pp. 74–81 (2004)
24. Papineni, K., Roukos, S., Ward, T., Zhu, W.-J.: Bleu: a method for automatic evaluation of machine translation. In: Proceedings of the 40th Annual Meeting of the Association for Computational Linguistics, pp. 311–318 (2002). https://doi.org/10.3115/1073083.1073135
25. Wang, Y., et al.: Self-instruct: aligning language models with self-generated instructions (2023). https://doi.org/10.48550/arXiv.2212.10560
26. Touvron, H., Martin, L., Stone, K., Albert, P., et al.: Llama 2: open foundation and fine-tuned chat models (2023). https://doi.org/10.48550/arXiv.2307.09288
27. Devlin, J., Chang, M.-W., Lee, K., Toutanova, K.: BERT: pre-training of deep bidirectional transformers for language understanding (2019). https://doi.org/10.48550/arXiv.1810.04805
28. Vaswani, A., et al.: Attention is all you need. Adv. Neural Inf. Process. Syst. **30** (2017). https://doi.org/10.48550/arXiv.1706.03762
29. van der Maaten, L., Hinton, G.: Visualizing data using t-SNE. J. Mach. Learn. Res. **9**, 2579–2605 (2008)

Context-Enhanced Language Models for Generating Multi-paper Citations

Avinash Anand(✉), Kritarth Prasad, Ujjwal Goel, Mohit Gupta, Naman Lal, Astha Verma, and Rajiv Ratn Shah

Indraprastha Institute of Information Technology, Delhi, India
{avinasha,kritarth20384,ujjwal20545,mohit22112, asthav,rajivratn}@iiitd.ac.in

Abstract. Citation text plays a pivotal role in elucidating the connection between scientific documents, demanding an in-depth comprehension of the cited paper. Constructing citations is often time-consuming, requiring researchers to delve into extensive literature and grapple with articulating relevant content. To address this challenge, the field of citation text generation (CTG) has emerged. However, while earlier methods have primarily centered on creating single-sentence citations, practical scenarios frequently necessitate citing multiple papers within a single paragraph. To bridge this gap, we propose a method that leverages Large Language Models (LLMs) to generate multi-citation sentences. Our approach involves a single source paper and a collection of target papers, culminating in a coherent paragraph containing multi-sentence citation text. Furthermore, we introduce a curated dataset named MCG-S2ORC, composed of English-language academic research papers in Computer Science, showcasing multiple citation instances. In our experiments, we evaluate three LLMs LLaMA, Alpaca, and Vicuna to ascertain the most effective model for this endeavor. Additionally, we exhibit enhanced performance by integrating knowledge graphs from target papers into the prompts for generating citation text. This research underscores the potential of harnessing LLMs for citation generation, opening a compelling avenue for exploring the intricate connections between scientific documents.

Keywords: Attention · Citation Text Generation · Knowledge Graphs · Large Language Models · Natural Language Processing · Text Generation

1 Introduction

The generation of text in the context of science is a difficult undertaking that calls for a thorough comprehension of the input text and expertise in the relevant field. Citation Generation has received a lot of attention recently because of developments in writing assistants and language models like Transformers [19]. The Citation Text Generation (CTG) challenge involves generating text that appropriately cites or refers to a cited document in a source document using natural language. The source and the cited paper's contextual signals are frequently used in CTG to generate the text. In this procedure, an algorithmic

V. Goyal et al. (Eds.): BDA 2023, LNCS 14418, pp. 80–94, 2023.
https://doi.org/10.1007/978-3-031-49601-1_6

model should sum up how the original and the cited article relate to one another in a certain situation. This may involve examining the papers content to determine their relationships and applying the appropriate terminology and structure to convey this information clearly and concisely. The time and effort needed for a literature review can be greatly decreased by having the ability to automatically describe the relationships between scientific papers and generate text for these descriptions.

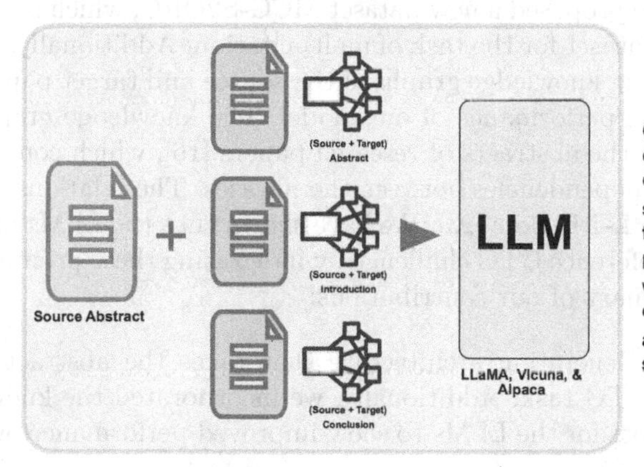

Other commonly used datasets contain segmentations [REF], correspondences, hierarchies symmetries [REF, REF], salient features, semantic segment and labels, alignments of 3D models with images [REF], semantic ontologies, and other functional annotations but again only for small size datasets.

Fig. 1. Multi-Sentence Citation Text Generation

However, there are a few drawbacks to the present citation text generation technologies. They concentrated on coming up with a single sentence for a single reference [4].

The text from the abstract of papers was mostly used as input in earlier attempts to CTG [13] methods, which produced a single citation as an output. In real-world scenarios, authors often use multiple references in one sentence or paragraph. Wu et al. [21] initiative suggests a method for producing multi-reference citation text using the source and target abstracts as input. They obtained the sentences for their dataset of multiple reference citations from the ACL anthology. For the task of creating citation text, they employed the Fusion-in-Decoder (FiD) [7] model. They have used FiD because it can scan lengthy inputs from various publications and makes use of the generational power of huge pre-trained contextual language models like T5 [15]. They have incorporated the intent labels to improve the performance of the model.

Using a CTG model offers numerous advantages. Firstly, it significantly saves time by automating the citation generation process, allowing researchers, students, and authors to focus more on their work. Secondly, these models ensure accuracy and consistency by following specific citation styles and providing correct formatting for various source types. Thirdly, citation generation models contribute to reducing plagiarism by encouraging proper source attribution. Moreover, these models offer flexibility by supporting multiple citation styles to accommodate different disciplines. Lastly, CTG models can serve as educational resources, helping students learn about citation elements and proper practices. Overall, CTG models streamline the citation process, improve accuracy and adherence to styles, save time, and promote ethical writing practices.

Our research presents a way to produce multi-sentence citation text for the target and source abstracts that are provided by using large language models. We have fine-tuned three LLMs i.e. LLaMA [18], Alpaca [17], and Vicuna [3]. The Fig. 1 shows the basic workflow of our approach, and defines our citation generation pipeline. We have proposed a new dataset MCG-S2ORC, which is created from the S2ORC [12] dataset for the task of multi-citation. Additionally, we demonstrate that by including knowledge graphs of the source and target papers in the prompts improves the performance of our model. The knowledge graphs relations are extracted from the abstracts of research papers [16], which contain condensed information and dependencies between the phrases. The relations are extracted with the help of **PL-Marker** [23]. We have shown that the LLMs performs better for the multi-reference CTG challenge by integrating these prompts. We offer the following summary of our contributions:

- We propose a Citation Generation architecture that takes the abstract of papers as input for the CTG task. Additionally, we incorporated the knowledge graphs in the prompts for the LLMs to show improved performance over the baselines.
- We propose MCG-S2ORC dataset using the S2ORC [12] dataset. This dataset contains two-three target papers for a single source paper.
- We show the importance of incorporating knowledge graph in the prompt structure for LLM through a huge increase in performance.

The written work is structured as follows: Section 2 addresses the related works on citation text generation, Sect. 3 explains how we came up with the problem and how the dataset was created, & what models have been used. In Sect. 4, we have explained about how we performed our experiments, then the Sect. 5 shows the evaluations, and the paper's conclusion and future aims are summarised in Sect. 6.

2 Related Work

2.1 Text Generation

Koncel et al. [11] generated multi-sentence text from an information extraction system and improved performance using a knowledge graph. They did graph encoding using Graph Attention Network. Text generation for scientific documents is one example of multi-document scientific summarization, other examples include the task of multi-document scientific summarization in the scientific domain [2,14,25]. Chen et al. [1] proposed a SciXGen dataset to solve the problem of generation of context-aware text in a scientific domain. However, summarising academic papers differs from the CTG job.

2.2 Citation Text Generation

As far as we are aware, there are two active concurrent works [13,22] that generate citation texts from research papers. Luu et al. [13] first introduced the task and generated the citation text given source and cited documents. Xing et al. [22] explored the relationship between scientific documents on a larger dataset. Gu and Hahnloser [6] proposes a pipeline for controllable citation generation that consists of an attribute recommendation module and a module for conditional citation generation, and evaluates the system's controllability across numerous characteristics using both automated metrics and human review. Jung et al. [8] proposes a framework for controllable citation generation with three labels of intent background, method, and results. They have used BART and T5 transformer and compares the results and accuracies obtained using these 2 transformer-based models.

There has been very less work done on multi-reference citation text generation, and we have found [21] in which authors concentrate on generating multiple citations from the sources and cited papers. To handle diverse long inputs, they create a new generation model using the Fusion-in-Decoder method.

2.3 Large Language Models and Prompts

Ye et al. [24], In-context instruction learning, combines instruction prompting and few-shot learning. The prompt includes a number of demonstration examples for various tasks, with each demonstration including an instruction, task input, and task output. Used in Stanford Alpaca [17] to generate 52k instructions following text-davinci-003 GPT3 [9] prompts and then fine-tune LLaMA. Chain-of-thought (CoT) prompting [20] generates a sequence of short sentences to describe reasoning logic step by step.

3 Methodology

In this paper, we fine-tuned three large language models (LLMs), i.e. LLaMA [18], Alpaca [17], and Vicuna [3] for the task of generating Multi-citation text. All the models were evaluated using the metrics METEOR, Rouge-1, Rouge-2, and Rouge-L. These fine-tuned models are considered as our baselines. We then extracted the relations from the source and target papers and use them in the prompt for generating the citations. Based on our empirical findings, we observed that incorporating knowledge graph relations in the prompting process enhances the performance of generating citation texts when compared to our baseline models.

3.1 Problem Formulation and Notations

The problem statement includes: given a abstract of citing document A, set of abstracts, introductions, and conclusions of related documents $B =$

$\{b_1, b_2, ..., b_n\}$. The task aims to produce a multi-sentence paragraph of all the cited documents b_i in the context of citing abstract A. We curated our own dataset from the benchmark dataset S2ORC [12]. We have modified the dataset in such a way that for each citing abstract, we have added more two-three cited papers, and the target has multiple sentences with multiple references for each pair (A, B). Figure 1 clearly demonstrate how our approach is going to work. First we take the source abstract, then we add them with the abstract, introduction, and conclusion of the target paper, and extract knowledge graph relation using PL-Marker [23], then pass it in the prompt for LLM, whose structure is given in the Fig. 3b.

Table 1. Dataset statistics created from the S2ORC corpus.

Statistic	CTG-S2ORC	Train	Validation	Test
# citations	17210	13,779	1,716	1,715
# unique papers	17210	13,779	1,716	1,715
CITATIONS				
Avg # characters	227.29	227.40	230.25	223.37
Max # characters	2416	2416	1862	1061
SOURCE ABSTRACTS				
Avg # characters	1122.95	1,120.73	1,111.55	1152.23
Max # characters	5516	5516	4343	3642
TARGET ABSTRACTS				
Avg # characters	998.48	997.87	999.35	1002.56
Max # characters	93551	93551	8674	4924
Avg # of Targets per sample	2	2	2	2

3.2 Dataset

For the task of multi-sentence citation text generation, we synthesize a new dataset MCG-S2ORC from S2ORC [12]. The S2ORC[1], or Semantic Scholar Open Research Corpus, is a significant corpus of 81.1 million English-language academic papers from many academic disciplines. We have taken only the samples whose "Field of Study" contain "Computer Science". Presently, the computer science domain contains 6.0M total papers. Each sample from the set of 6.0M computer science domain papers contains "*source_paper_id*", "*source_abstract*", and "*body_text*". The "*body_text*" consists of various sections, including Introduction, Methodology, etc.

[1] https://github.com/allenai/s2orc.

We parsed the Computer Science domain dataset and extracted citation details in JSON format. The dataset consists of samples representing citation examples, each containing key-value pairs of information. The "*source_paper_id*" field provides a unique identifier for the source paper, while the "*source_abstract*" field contains its summary. The "*citation_texts*" field is an array containing citation information related to the source paper. Each citation includes the "*citation_text*" field, representing the extracted citation text. Additional metadata is found in the "*citation_meta*" field, containing information like citation number, referenced section, and details about the paper being cited (title, abstract, introduction, and conclusion). To ensure suitability for multi-reference citation text generation, we only considered citations that cite more than one paper in a single sentence, making necessary modifications to the dataset.

Our final dataset comprises 17,210 samples of multi-reference citation texts. The complete statistics of our MCG-S2ORC are shown in Table 1.

3.3 Large Language Models

We have fine-tuned three large language models for the task of generating multi-sentence citation text. The details of three models can be seen in Fig. 2 and provided below.

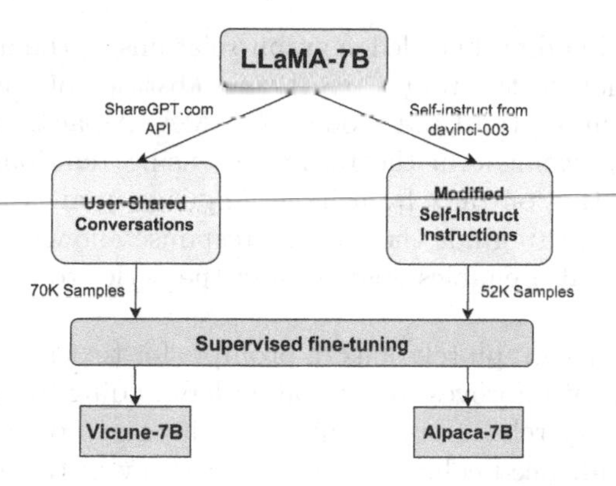

Fig. 2. LLaMA, Vicuna & Alpaca

LLaMA is a transformer-based model available in four variations: 7B, 13B, 33B, and 65B parameters. Trained solely on publicly available data, the training corpus comprises approximately 1.4T tokens and includes text from 20 different languages [18]. **Alpaca** [17], on the other hand, is a language model that has undergone supervised fine-tuning using an LLaMA 7B model and 52K instruction-following demonstrations generated by OpenAI's text-davinci-003 model [9,18]. **Vicuna**, another variant, has been developed by optimizing an LLaMA base model with approximately 70K user-shared talks obtained from

ShareGPT.com via open APIs [3]. It is important to note that Vicuna is limited in its reasoning abilities, mathematical understanding, self-identification capabilities, fact-checking capacity, and it is not specifically optimized for bias reduction, potential toxicity, or safety measures [3].

The prompt used to fine-tune the baselines LLMs are provided in the Fig. 3a. In the figure, the **data_point** is a dictionary containing a single data sample from our dataset.

```
### Instruction:
    Generate the citation text.
### Input:
    {data_point["source_abstract"]}
    {data_point["Target1_abstract"]}
    {data_point["Target2_abstract"]} ...
### Response:
    {data_point["citation_text"]}
```

(a)

```
### Instruction:
    Generate the citation text.
### Input:
    {data_point["source_abstract"]}
    {relations(source_abstract + target_abstracts)}
    {relations(source_abstract + target_conclusions)}
    {relations(source_abstract + target_introductions)} ...
### Response:
    {data_point["citation_text"]}
```

(b)

Fig. 3. Prompt Structures used for the Large Language Models.

3.4 Prompting and Knowledge Graphs

We also attempted adding knowledge graph relations of the abstract, introduction, and conclusion of the target paper and abstract of source paper in the prompts for fine-tuning the LLM models, LLaMA, Alpaca, and Vicuna, which shows a huge improvement in the results, as using relations of paper in the prompts in generating outputs from large language models provides a specific instruction or context to guide the model's response allowing more focused and relevant output, it also enables control over the style, tone, or domain of the generated text.

Adding knowledge graph relations to prompts for text generation has several advantages. Firstly, it improves contextual understanding by enabling the model to comprehend entity relationships, enhancing its grasp of the topic. Secondly, it contributes to enhanced coherence and consistency in the generated text. By leveraging graph relationships, the model produces more structured and coherent responses, improving overall quality. Additionally, the integration of knowledge graphs allows the model to showcase domain-specific expertise, delivering informed and accurate responses. Lastly, knowledge graphs aid in fact-checking and verification, ensuring factual accuracy and reducing the likelihood of generating misleading information.

In this work, we utilized the PL-Marker [23] tool to construct the knowledge graph of the source and target abstracts. PL-Marker employs an innovative packed levitated marker technique, combining both a neighborhood-oriented and subject-oriented packing strategy to obtain pair representations. The purpose of constructing the knowledge graph is to capture the relationships and context between different entities within the abstracts of papers. The first step of the model involves entity recognition, where it identifies and labels the different entities present in the text. Once the entities are recognized and labelled, the model focuses on extracting relations between these entities. We generated knowledge graph triplets for target paper's introduction, conclusion and abstract with the abstract of source paper which are used in the dataset to fine-tune the LLM's. The prompt structure is given in the Fig. 3b.

The example visualization of extracted relations from the target introduction that we have used in the prompts is shown in Fig. 4. This figure clearly shows that the our approach is able to extract complex relationship between different tokens.

4 Experiments

We split the complete dataset MCG-S2ORC containing 17,210 data samples into train, test, and validation set having 13K, 1K, and 1K samples respectively. After creating and preprocessing the dataset, we fine-tuned three large language models as discussed earlier on our dataset MCG_S2ORC for citation text generation. The prompt for the LLMs as shown in Fig. 3a is converted into tokens, then pass it to the model for fine-tuning and learning the weights. From the results shown in Table 2, **Vicuna** [3] outperforms LLaMA and alpaca for the task of citation text generation on our dataset. These results act as our baselines for our next setup.

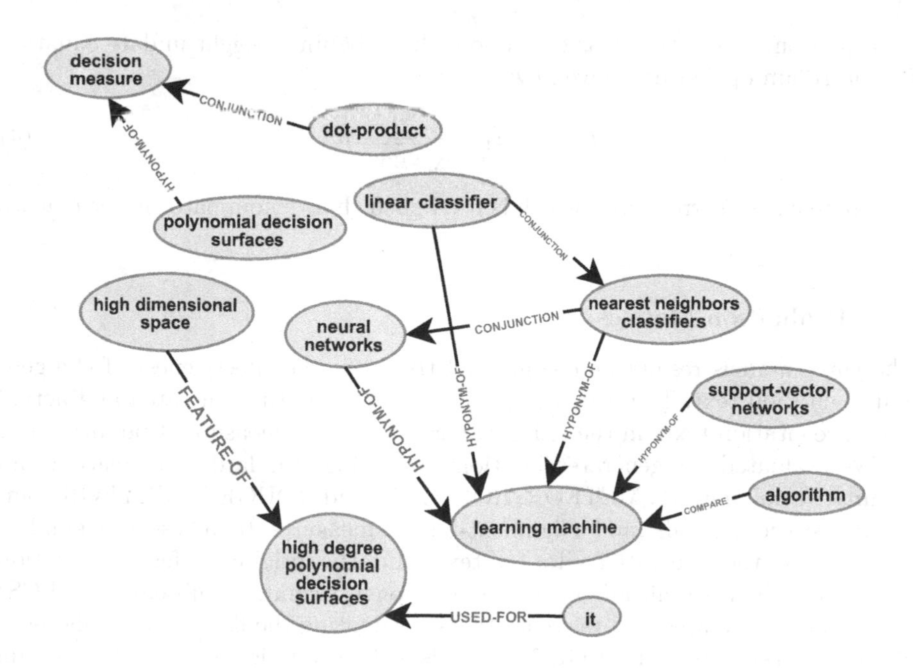

Fig. 4. Knowledge Graph Visualization

Then we further perform experiments on passing knowledge graph of the abstract, introduction, and conclusion of target paper as prompts to better capture the relationship and coherence of the words to generate more meaningful citations. We have extracted the knowledge graph relations using PL-Marker [23]. The results for this setup can be seen from Table 3, the performance of all the models is improved from our baselines.

For fine-tuning the Large Language Models (LLMs), we employed QLora [5]. QLora is an efficient approach that maximize memory efficiency through gradient backpropagating gradients in a frozen, 4-bit quantized pretrained language model, resulting in Low Rank Adapters (LoRA).

$$k_i = \frac{1}{2}\left(Qx\left(\frac{i}{2^n+1}\right) + Qx\left(\frac{i+1}{2^n+1}\right)\right) \tag{1}$$

Where, $Q_x(.)$ is the quantile function of the standard normal distribution $N(0,1)$. For our experiments, we have used $n = 4$ as we are applying 4-bit quantization.

$$\begin{aligned} m_t &= \beta m_{t-1} + \eta \nabla J(w_t) \\ v_t &= \gamma v_{t-1} + (1-\gamma)\nabla J(w_t)^2 \end{aligned} \tag{2}$$

We utilized the **AdamW** optimizer [10] with a Linear Scheduler. The learning rate was set to 3e−4, and we incorporated 100 warmup steps to gradually adjust the learning rate.

$$\hat{m}_t = \frac{m_t}{1-\beta^t} \qquad \hat{v}_t = \frac{v_t}{1-\gamma^t} \tag{3}$$

Equation 3 shows the bias correction, then the final weight update equation for the Adam optimizer is given by:

$$w_{t+1} = w_t - \frac{\eta}{\sqrt{\hat{v}_t + \epsilon}}\hat{m}_t \tag{4}$$

where ϵ is the error term, which is used such that denominator never reaches zero.

4.1 Evaluation Metrics

The three models results were compared to assess the effectiveness of the generated citation text. The degree of similarity between the generated and actual reference citation texts in the citing paper served as a measure of performance.

We evaluated the generated citation text using standard text creation and summarization metrics: METEOR, ROUGE-N, and ROUGE-L. METEOR combines precision, recall, and alignment-based measures to assess the similarity between the generated citation text and the original reference citation texts. ROUGE-L specifically focuses on the longest common subsequence (LCS) between the generated and reference texts, evaluating the fluency and coherence of the generated text. ROUGE-N extends this evaluation to consider n-gram overlaps, providing a more detailed analysis of the generated text's performance.

5 Results and Discussion

The results of the experiments after fine-tuning the LLM models for multi-reference citation text generation can be seen from Table 2. The results shows that the Vicuna [3] outperforms other models with respect to all the metrics. The citation text generated by the best fine-tuned model Vicuna is given in the appendix Fig. 5.

Table 2. Results of Fine-Tuned LLM

Model	METEOR	Rouge-1	Rouge-2	Rouge-L
LLaMA	11.73	10.74	1.21	9.15
Alpaca	9.74	9.04	1.33	7.78
Vicuna	**12.56**	**12.02**	**1.44**	**10.24**

Table 3 presents the evaluation results obtained by incorporating knowledge graph relations of source and target paper's abstract, introduction, and conclusion in the prompts. The findings highlight the superior performance of Vicuna, surpassing other models in the specific task of citation generation as compared to the baseline models without knowledge graph relations. This notable achievement can be attributed to the utilization of knowledge graphs, which facilitate a deeper contextual comprehension and enhance the coherence of the generated text. Consequently, the model produces outputs that are more context-rich and of higher quality, ultimately contributing to improved overall performance. The generated citation at the time of inference is given in the Fig. 6 in the appendix section.

Table 3. Results of Fine-Tuned Model + Knowledge Graph as Prompt

Model	METEOR	Rouge-1	Rouge-2	Rouge-L
LLaMA	11.46	10.79	1.23	9.14
Alpaca	**13.39**	12.42	**1.74**	10.59
Vicuna	13.18	**12.65**	1.49	**10.80**

6 Conclusion

The paper addresses the problem of multi-citation text generation, focusing on generating coherent multi-sentence citations. We curated a dataset called MCG-S2ORC from the S2ORC dataset to advance citation generation research. Three large language models, namely LLaMA, Alpaca, and Vicuna were fine-tuned specifically for citation generation. Vicuna demonstrated superior performance compared to the other models. To enhance citation generation, we integrated knowledge graphs into the model's prompts by extracting entity relations from the source and target paper's abstracts, introductions, and conclusions using PL-Marker. Our experiments showed that incorporating knowledge graphs significantly improved the performance and text generation capabilities of the models, enabling better comprehension of relations between source and target papers. This integration enhances the citation generation task, showcasing the potential of knowledge graphs as valuable resources. Future research can leverage knowledge graphs to explore novel approaches for generating accurate and coherent multi-sentence citations.

7 Limitations

The maximum token length restriction of the LLMs used, set at 2048, is one restriction on our approach. This restricts us to incorporating only 2–3 combined relations between source and target papers rather than including all target papers. While including all sets of relations could enhance performance, it presents challenges due to the increased number of tokens involved.

Acknowledgements. Rajiv Ratn Shah is partly supported by the Infosys Center for AI, the Center for Design and New Media, and the Center of Excellence in Healthcare at IIIT Delhi.

A Appendix

This section shows the inference examples used to test the fine-tuned model and checking the generated text quality, and context.

A sample of the generated citation text using the improved Vicuna model is shown in Fig. 5. With the help of the supplied source_abstract and set of target_abstracts, high-quality generated citation text that fits both the source and target papers contexts was produced. Due to the inclusion of knowledge graph relations, the generated citation text in Fig. 6 exhibits a higher level of context richness. These linkages help generate text that is more contextually relevant by improving our knowledge of the relationships between words in the source and target abstracts.

source_abstract
In order to exploit remote sensing data operationally for precision agriculture applications, efficient and automated methods are required for the accurate detection of vegetation, crops and different crop varieties. To this end, we have designed, developed and evaluated an object-based classification framework towards the detection of vineyards, the vine canopy extraction and the vine variety discrimination from very high resolution multispectral data ... The combined analysis of the experimental results with the spectral signatures from the in situ reflectance data indicated that certain vine varieties (e.g., Merlot) presented distinct spectral patterns across the VNIR spectrum.

target_1_abstract
Image classification is one of the most basic operations of digital image processing. The present review focuses on the strengths and weaknesses of traditional pixel-based classification (PBC) and the advances of object-oriented classification (OOC) algorithms employed for the extraction of information from remotely sensed satellite imageries. The state-of-the-art classifiers are reviewed for their potential usage in urban remote sensing (RS), with a special focus on cryospheric applications. Generally, classifiers for information extraction can be divided into three catalogues: 1) based on the type of learning (supervised and unsupervised), 2) based on assumptions on data distribution (parametric and non-parametric) and, 3) based on the number of outputs for each spatial unit (hard and soft) ... The present review would be beneficial for developing new classifiers in the cryospheric environment for better understanding of spatial-temporal changes over long time scales.

target_2_abstract
Tree species diversity is a key parameter to describe forest ecosystems. It is, for example, important for issues such as wildlife habitat modeling and close-to-nature forest management. We examined the suitability of 8-band WorldView-2 satellite data for the identification of 10 tree species in a temperate forest in Austria. We performed a Random Forest (RF) classification (object-based and pixel-based) using spectra of manually delineated sunlit regions of tree crowns ... Beside the impact on overall classification accuracy, the importance of the spectral bands was evaluated with two measures provided by RF. An in-depth analysis of the RF output was carried out to evaluate the impact of reference data quality and the resulting reliability of final class assignments. Finally, an extensive literature review on tree species classification comprising about 20 studies is presented.

target
Motivated by the successful application of object-based image analysis in very high resolution satellite data **REF REF, REF, REF, REF, REF, REF** the developed methodology was based on OBIA principles, while it was designed to address vineyard detection, vine canopy extraction and vine variety discrimination.

predicted
The citation text for the source "Vegetation monitoring in vineyards using a hybrid feature extraction/classification approach" is: **REF, REF, REF**. Vegetation monitoring in vineyards using a hybrid feature extraction/classification approach. IEEE Transactions on Geoscience and Remote Sensing, 50(2)

Fig. 5. Example of Generated Citation text from the best Model **(Vicuna)** without knowledge graph relations

source_abstract
In order to exploit remote sensing data operationally for precision agriculture applications, efficient and automated methods are required for the accurate detection of vegetation, crops and different crop varieties. To this end, we have designed, developed and evaluated an object-based classification framework towards the detection of vineyards, the vine canopy extraction and the vine variety discrimination from very high resolution multispectral data ... The combined analysis of the experimental results with the spectral signatures from the in situ reflectance data indicated that certain vine varieties (e.g., Merlot) presented distinct spectral patterns across the VNIR spectrum.

relations
detection of vegetation **CONJUNCTION** crops . remote sensing data **USED-FOR** precision agriculture applications . rules **CONJUNCTION** segmentation scales . spectral , spatial and textural features **CONJUNCTION** rules . object-based image analysis **USED-FOR** parameters . multitemporal WorldView-2 satellite data **USED-FOR** methodology . portable spectroradiometer **USED-FOR** in situ canopy reflectance observations . approach **USED-FOR** vineyards . completeness and correctness detection rates **EVALUATE-FOR** approach . ... . object-based classification framework **USED-FOR** detection of vineyards . vine canopy extraction **CONJUNCTION** vine variety discrimination . object-based classification framework **USED-FOR** vine variety discrimination . object-based classification framework **USED-FOR** vine variety discrimination . detection of vineyards **CONJUNCTION** vine canopy extraction . object-based classification framework **USED-FOR** vine canopy extraction . high resolution multispectral data **USED-FOR** object-based classification framework . high resolution multispectral data **USED-FOR** vine variety discrimination . segmentation scales **CONJUNCTION** parameters . rules **CONJUNCTION** segmentation scales . spectral , spatial and textural features **CONJUNCTION** rules . object-based image analysis **USED-FOR** parameters . multitemporal WorldView-2 satellite data **USED-FOR** methodology . detection of vineyards **CONJUNCTION** vine canopy extraction . portable spectroradiometer **USED-FOR** in situ canopy reflectance observations . approach **USED-FOR** vineyards . completeness and correctness detection rates **EVALUATE-FOR** approach . overall accuracy (OA) rates **EVALUATE-FOR** vine canopy extraction methodology . spectral patterns FEATURE-OF VNIR spectrum . in situ reflectance data **USED-FOR** spectral signatures . Merlot **HYPONYM-OF** vine varieties . Tree species diversity **USED-FOR** forest ecosystems . wildlife habitat modeling **CONJUNCTION** close-to-nature forest management . close-to-nature forest management **HYPONYM-OF** issues . object-based classification framework **USED-FOR** vine canopy extraction . wildlife habitat modeling **HYPONYM-OF** issues . object-based approach **COMPARE** pixel-based approach . Scots pine **CONJUNCTION** European beech . Norway spruce **CONJUNCTION** Scots pine . European beech **CONJUNCTION** English oak . classification accuracy **EVALUATE-FOR** spectral bands . high resolution multispectral data **USED-FOR** object-based classification framework . high resolution multispectral data **USED-FOR** vine variety discrimination . segmentation scales **CONJUNCTION** parameters.

target
Motivated by the successful application of object-based image analysis in very high resolution satellite data **REF REF, REF, REF, REF, REF, REF** the developed methodology was based on OBIA principles, while it was designed to address vineyard detection, vine canopy extraction and vine variety discrimination.

predicted
The paper presents an object-based classification framework for the detection of vineyards, the vine canopy extraction, and the vine variety discrimination from very high resolution multispectral data. The authors propose a novel set of spectral, spatial, and textural features, as well as rules, segmentation scales, and parameters based on object-based image analysis. The validation of the developed methodology was carried out on multitemporal WorldView-2 satellite.

Fig. 6. Example of Generated Citation text from the best Model (**Vicuna**) with knowledge graph relations

References

1. Chen, H., Takamura, H., Nakayama, H.: SciXGen: a scientific paper dataset for context-aware text generation. arXiv preprint arXiv:2110.10774 (2021)
2. Chen, J., Zhuge, H.: Summarization of scientific documents by detecting common facts in citations. Futur. Gener. Comput. Syst. **32**, 246–252 (2014)
3. Chiang, W.L., et al.: Vicuna: an open-source chatbot impressing GPT-4 with 90%* ChatGPT quality, March 2023. https://lmsys.org/blog/2023-03-30-vicuna/
4. Cohan, A., Ammar, W., Van Zuylen, M., Cady, F.: Structural scaffolds for citation intent classification in scientific publications. arXiv preprint arXiv:1904.01608 (2019)
5. Dettmers, T., Pagnoni, A., Holtzman, A., Zettlemoyer, L.: QLoRA: efficient finetuning of quantized LLMs. arXiv preprint arXiv:2305.14314 (2023)
6. Gu, N., Hahnloser, R.H.R.: Controllable citation text generation (2022)
7. Izacard, G., Grave, E.: Leveraging passage retrieval with generative models for open domain question answering. In: Proceedings of the 16th Conference of the European Chapter of the Association for Computational Linguistics: Main Volume, pp. 874–880. Association for Computational Linguistics, April 2021. https://doi.org/10.18653/v1/2021.eacl-main.74. https://aclanthology.org/2021.eacl-main.74
8. Jung, S.Y., Lin, T.H., Liao, C.H., Yuan, S.M., Sun, C.T.: Intent-controllable citation text generation. Mathematics **10**, 1763 (2022). https://doi.org/10.3390/math10101763
9. Katar, O., Ozkan, D., Yildirim, Ö., Acharya, U.R.: Evaluation of GPT-3 AI language model in research paper writing (2022). https://doi.org/10.13140/RG.2.2.11949.15844
10. Kingma, D.P., Ba, J.: Adam: a method for stochastic optimization. arXiv preprint arXiv:1412.6980 (2014)
11. Koncel-Kedziorski, R., Bekal, D., Luan, Y., Lapata, M., Hajishirzi, H.: Text generation from knowledge graphs with graph transformers. arXiv preprint arXiv:1904.02342 (2019)
12. Lo, K., Wang, L.L., Neumann, M., Kinney, R., Weld, D.S.: S2ORC: the semantic scholar open research corpus. arXiv preprint arXiv:1911.02782 (2019)
13. Luu, K., Koncel-Kedziorski, R., Lo, K., Cachola, I., Smith, N.A.: Citation text generation. ArXiv abs/2002.00317 (2020)
14. Mohammad, S., et al.: Using citations to generate surveys of scientific paradigms. In: Proceedings of Human Language Technologies: The 2009 Annual Conference of the North American Chapter of the Association for Computational Linguistics, pp. 584–592 (2009)
15. Raffel, C., et al.: Exploring the limits of transfer learning with a unified text-to-text transformer. J. Mach. Learn. Res. **21**(140), 1–67 (2020). http://jmlr.org/papers/v21/20-074.html
16. Sun, K., et al.: Assessing scientific research papers with knowledge graphs. In: Proceedings of the 45th International ACM SIGIR Conference on Research and Development in Information Retrieval, SIGIR 2022, pp. 2467–2472. Association for Computing Machinery, New York, NY, USA (2022). https://doi.org/10.1145/3477495.3531879
17. Taori, R., et al.: Stanford Alpaca: an instruction-following LLaMA model (2023)
18. Touvron, H., et al.: LLaMA: open and efficient foundation language models (2023). https://doi.org/10.48550/arXiv.2302.13971

19. Vaswani, A., et al.: Attention is all you need. In: Advances in Neural Information Processing Systems, vol. 30 (2017)
20. Wei, J., et al.: Chain-of-thought prompting elicits reasoning in large language models (2023)
21. Wu, J.Y., Shieh, A.T.W., Hsu, S.J., Chen, Y.N.: Towards generating citation sentences for multiple references with intent control. arXiv preprint arXiv:2112.01332 (2021)
22. Xing, X., Fan, X., Wan, X.: Automatic generation of citation texts in scholarly papers: a pilot study. In: Proceedings of the 58th Annual Meeting of the Association for Computational Linguistics, pp. 6181–6190 (2020)
23. Ye, D., Lin, Y., Li, P., Sun, M.: Packed levitated marker for entity and relation extraction. In: Muresan, S., Nakov, P., Villavicencio, A. (eds.) Proceedings of the 60th Annual Meeting of the Association for Computational Linguistics (Volume 1: Long Papers), ACL 2022, Dublin, Ireland, 22–27 May 2022, pp. 4904–4917. Association for Computational Linguistics (2022). https://aclanthology.org/2022.acl-long.337
24. Ye, S., Hwang, H., Yang, S., Yun, H., Kim, Y., Seo, M.: In-context instruction learning (2023)
25. Yeloglu, O., Milios, E., Zincir-Heywood, N.: Multi-document summarization of scientific corpora. In: Proceedings of the 2011 ACM Symposium on Applied Computing, pp. 252–258 (2011)

GEC-DCL: Grammatical Error Correction Model with Dynamic Context Learning for Paragraphs and Scholarly Papers

Avinash Anand[1(✉)], Atharv Jairath[2], Naman Lal[3], Siddhesh Bangar[4], Jagriti Sikka[5], Astha Verma[1], Rajiv Ratn Shah[1], and Shin'ichi Satoh[6]

[1] Indraprastha Institute of Information Technology, Delhi, India
{avinasha,asthav,rajivratn}@iiitd.ac.in
[2] Dr. Akhilesh Das Gupta Institute of Technology and Management, Delhi, India
atharv.jairath@gmail.com
[3] MIDAS Lab, IIIT Delhi, Delhi, India
namanlal.lal92@gmail.com
[4] Vidyalankar Institute of Technology, Delhi, India
siddheshb008@gmail.com
[5] Georgia Institute of Technology, Atlanta, Georgia
jagriti295@gmail.com
[6] National Institute of Informatics, Tokyo, Japan
satoh@nii.ac.jp

Abstract. Over the past decade, there has been noteworthy progress in the field of Automatic Grammatical Error Correction (GEC). Despite this growth, current GEC models possess limitations as they primarily concentrate on single sentences, neglecting the significance of contextual understanding in error correction. While a few models have begun to factor in context alongside target sentences, they frequently depend on inflexible boundaries, which leads to the omission of vital information necessary for rectifying certain errors. To address this issue, we introduce the Dynamic Context Learner (DCL) model, which identifies optimal breakpoints within paragraphs or documents to retain maximum context. Our method surpasses those employing fixed sequence lengths or assuming a limited number of preceding sentences as context. Through extensive evaluation on the CoNLL-2014, BEA-Dev, and FCE-Test datasets, we substantiate the efficacy of our approach, achieving substantial $F_{0.5}$ score enhancements: 77% increase, 19.61% boost, and 10.49% rise respectively, compared to state-of-the-art models. Furthermore, we contrast our model's performance with LLaMA's GEC capabilities. We extend our investigation to scientific writing encompassing various context lengths and validate our technique on the GEC S2ORC dataset, yielding cutting-edge results in scholarly publications.

Keywords: Grammatical Error Correction · Large Language Model · Natural Language Processing

V. Goyal et al. (Eds.): BDA 2023, LNCS 14418, pp. 95–110, 2023.
https://doi.org/10.1007/978-3-031-49601-1_7

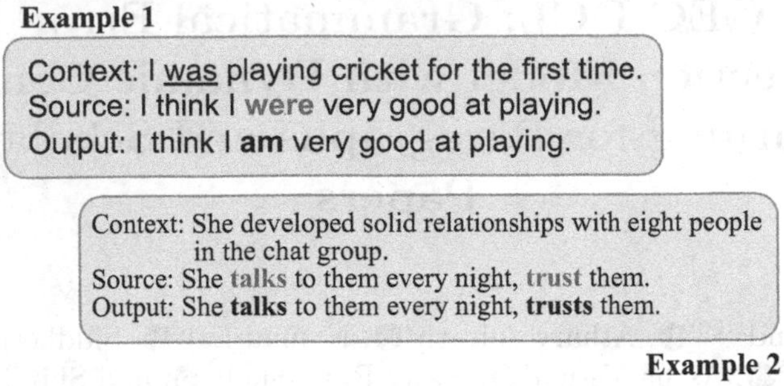

Example 1

Context: I was playing cricket for the first time.
Source: I think I were very good at playing.
Output: I think I **am** very good at playing.

Context: She developed solid relationships with eight people in the chat group.
Source: She talks to them every night, trust them.
Output: She **talks** to them every night, **trusts** them.

Example 2

Fig. 1. Examples of GEC System Outputs based on Fixed Context

1 Introduction

The objective of Grammatical Error Correction (GEC) systems is to enhance text readability and comprehension through error rectification. Traditional GEC methods frequently depended on neural machine translation (NMT) techniques, as outlined in [18]. Nonetheless, these approaches have constraints, including the requirement for extensive training data and difficulties in error detection. Furthermore, within NMT-based systems, language understanding and generation are managed by the encoder and decoder, respectively. However, language generation proves more intricate than language understanding, consequently resulting in extended processing durations for these systems.

Lately, notable progress has been made utilizing transformer-based sequence-to-sequence (seq2seq) models, as showcased on notable GEC benchmarks [4]. Transformer models, as initially introduced in [21], are trained to convert incorrect sentences into grammatically precise ones, effectively approaching GEC as a sequence-to-sequence problem.

The GECToR model [16] introduces a transformative approach to error correction, redefining it as a classification task rather than a generation-oriented method. In this approach, each token in the input sequence is assigned a specific correction rule, which determines its accurate form. Instead of creating the corrected sequence anew, the model utilizes predefined correction rules for precise adjustments. This strategy brings numerous advantages, including quicker inference, reduced dependency on extensive training data, and improved transparency. However, the model faces challenges in addressing intricate errors requiring context beyond a single token and capturing subtle linguistic nuances due to its classification-based approach.

Previous attempts in grammatical error correction (GEC) utilizing language models, like GPT [17], primarily focused on analyzing individual sentences in isolation, neglecting the contextual cues present within the surrounding text. An example of GEC models employing fixed context is illustrated in Fig. 1. The approach undertaken by Yuan & Bryant [25] diverges from the approach employed by human proofreaders. These proofreaders take the broader document context into account to effectively identify and correct errors related to tense, verb selection, definite articles, and connectives [4].

To address this challenge, we suggest integrating contextual information into the input model, leading to improved performance. Traditional language models face difficulty in maintaining context for lengthy sequences. When dealing with dense information, it's common to process it in smaller sections or chunks.

By intelligently dividing lengthy paragraphs into smaller segments while retaining contextual details, we can enhance the accuracy of addressing grammatical errors. Previous studies, [5,25], have examined static context addition techniques by including one or two sentences preceding or following the target sentence. Nonetheless, static methods might not be ideal for context management, as they could lead to information loss, and contextual demands differ for each scenario.

In this study, we introduce the Dynamic Context Learner (DCL) model, which dynamically selects relevant context information. To gauge whether our strategy is working, we construct a dataset using scientific papers, introduce grammar errors, and show what our model accomplishes SOTA (state-of-the-art) results. Our contributions can be summarized as follows:

- We have developed **DCL model** to determine relevant context dynamically to be considered in the corpus for GEC.
- We present a novel GEC architecture: A Transformer-based Grammatical Error Correction model for paragraphs **GEC-DCL** that is able to take a document, and correct errors in both single sentences and paragraphs.
- On the conventional dataset for GEC, we evaluated our model using SOTA (state-of-the-art) results & also compared our architecture results with the LLaMA [26]. We also constructed our own synthetic datasets, **XSUM** and **CNN**, to train and compare the model's performance.
- We have created a scientific dataset: GEC_S2ORC using the original **S2ORC** dataset [10] and showed our model performance which is state-of-the-art for scientific grammar error correction.

We hypothesize that context significantly influences GEC tasks, and the GEC-DCL model capitalizes on this context to enhance performance. Our model exhibited remarkable improvements, achieving $F_{0.5}$ score enhancements of 77%, 19.61%, and 10.49% on the CoNLL-2014, BEA-Dev, and FCE-Test datasets respectively, surpassing state-of-the-art models. Notably, the CoNLL-2014 dataset contains 26 sentences, BEA-Dev has 13, and FCE-Test has 14. As the number of sentences doubled in CoNLL-2014, our model outperformed other models, affirming our hypothesis that context is pivotal in GEC and our model effectively harnesses it.

2 Related Work

Since the CoNLL'14 Shared Task [15], the advancement rate in GEC has notably accelerated. In 2016, Rozovskaya and Roth integrated error-specific classifiers with a Phrase Based Machine Translation (PBMT) model trained on the Lang-8 dataset [12]. Similarly, during the same year, Junczys Dowmunt and Grundkiewicz achieved significant progress by merging a PBMT model with more comprehensive language modeling and bi-text features.

Table 1. An Example of Dataset used for our proposed Dynamic Context Learner (DCL) Model.

Sentence A	Sentence B	isNextSentence
Findings of the first study were used to refine sub-scales used in second study	From a mother's perspective (the intended respondents), a further qualitative study evaluated the validity of the content	True
A continuous model with a discrete integrator is shown to be the best solution for the discretization of the nonlinear model	Data acquired for comparable ceramics that have undergone heat treatment in nitrogen after sintering to promote devitrification are used to illustrate the mechanisms of oxidation	False

The **LaserTagger** method, presented by [11], tags sequences by treating text generation as a text-editing process. Three primary editing operations—preserving a token, discarding it, and introducing a phrase before the token—were employed to reconstruct target texts from inputs. To predict these edits, they devised a specialized model that combines an auto-regressive transformer decoder and a BERT encoder. The approach was evaluated using four tasks: phrase splitting, sentence fusion, grammatical correction, and abstractive summarization on English text. They also demonstrated that tagging could be significantly faster at inference time—up to two orders of magnitude-compared to equivalent seq2seq models, making it more appealing for real—world applications.

GECToR [16] is a method for grammatical error correction that uses a combination of language models and linear classifiers to predict corrections at the token level. It employs an iterative sequence tagging approach, similar to the one used in PIE [1], but with the addition of personalized grammatical (g)-transformations.

These transformations include changing case of the current token (CASE tags), merging it with the following token (MERGE tags), and splitting it into two new tokens (SPLIT tags). Additionally, tags for VERB FORM & NOUN NUMBER transformations provide grammatical information about the tokens. GECToR predicts these grammatical transformations for each input token. In its experimentation, GECToR used pre-trained transformers such as RoBERTa [9], XLNet [23] and BERT [8] to learn various errors and their relationship with the context. Important conclusions from the research include:

1. Cross-sentence context is crucial for spotting and fixing some problems [22], most often incorrect use of tense in words and, incorrect use of the article ***a, an, the*** for connecting context among sentences.
2. An auxiliary encoder is used in a convolution neural encoder-decoder architecture that adds cross-sentence context from prior sentences.

Table 2. Statistics of the datasets that were used in the research

Parameters	XSum	CNN	S2ORC
Avg. #characters	428	925	1039
Avg. #words	69	138	139
Avg. #unique words	58	109	98
Avg. #punctuation	7	18	19
Avg. #sentences	3	2	6
Max #sentences	25	26	346
Min #sentences	1	1	1
Standard Deviation	1.662	1.715	1.832
Train Set	183896	1139187	490000
Test Set	1000	1000	1000
Dev Set	19432	22248	9000
Sum	756057	1308093	3269096

3. For source and auxiliary encodings, they created two attention processes, and the information entering the decoder is controlled by gates.

To illustrate the potential of employing straightforward document-level strategies for boosting GEC performance and capturing broader context in NMT-based GEC, Yuan & Bryant [25] examined several architectural approaches. They also introduced a three-step training approach to effectively utilize parallel data at both sentence and document levels for GEC. They conducted the initial evaluation of document-level GEC and provided accessible scripts for related research. Our research delved into techniques to retain context within existing document-level GEC models.

Chollampatt et al. [5] used training techniques to preserve context at the sentence and document levels. Automatic grammatical mistake correction has grown in importance due to the significant advancements in transfer learning in natural language processing in recent years. While ignoring the context of the text, most of the work in GEC has only been done in a single sentence. Many grammatical fixes depend on context, which may provide helpful information for solving other issues. Yuan & Bryant [25] addressed the shortcomings of the one-sentence strategy and tried to discover the need for employing broader settings.

The current approaches encounter a challenge where they either utilize the single-sentence approach or provide minimal information through diverse mechanisms. These models require assistance when dealing with substantial or information-dense contexts, enabling them to effectively handle relevant information. Grammar error correction can leverage context effectively if a method is devised to break down longer sequences of phrases into smaller segments without compromising context. A language model-based GEC approach would excel in this task. This technique, if successfully implemented, could enable seamless handling of longer texts without being constrained by token size, provided we can preserve context within smaller chunks.

The grammatical error correction process for documents is broken into two parts: **1.** Retain relevant context that lies well within the capability of the language model to process efficiently. **2.** As opposed to previous models that needed more time and computing resources, this iterative technique aims to increase speed while reducing complexity.

3 Dataset

3.1 XSUM

The XSUM [14] dataset is large-scale benchmark dataset for abstractive text summarization. It contains approximately 226,711 news articles from the BBC, with each article consisting of a headline and a short description. The dataset parameters specify the maximum length of the input of 512 tokens, a maximum target length of 72 tokens, and a split of 204,045 training examples, 11,000 validation examples, and 5,503 test examples.

3.2 CNN/Dailymail

The CNN Dailymail [13] dataset is a widely used benchmark dataset relating to natural language processing, particularly in the task of text summarization. The dataset consists of news articles and their corresponding summaries, and includes approximately 312,000 articles with an average length of 781 words per article. The parameters of this dataset include article and summary length, as well as the number of articles & summaries available for training and evaluation.

3.3 S2ORC

The S2ORC [10] dataset comprises a diverse collection of scholarly publications spanning various disciplines, including computer science, medicine, and social sciences. It contains over 81 million paper abstracts, metadata, and citation networks extracted from the Semantic Scholar Corpus. The dataset provides rich information such as paper titles, authors, abstracts, publication venues, citation counts, and references, making it a valuable resource for various academic research applications.

3.4 FCE-Test

The FCE-Test [24] dataset comprises a total of 1,326 writing samples collected from non-native speakers of English at various proficiency levels. The dataset includes both task responses and background information about the test takers, such as age, gender, and country of origin. The parameters of the dataset also include detailed annotations for each writing sample, such as grammatical errors, coherence, and organization, providing a valuable resource for analyzing and evaluating English language proficiency.

3.5 BEA-Dev

The BEA-Dev [4] dataset is a widely used benchmark dataset for evaluating the performance of automatic error correction systems. The dataset consists of 1,456 sentences, comprising a total of 17,434 tokens, which have been manually annotated with three levels of granularity: (i) morphological errors, (ii) syntactic errors, and (iii) lexical errors. The annotations are further divided into 11 error types (see Table 8 in appendix, each representing a specific type of error commonly found in non-native English writing.

3.6 CoNLL-2014

The CoNLL-2014 Shared Task [15] dataset is a benchmark dataset for the task of Joint Named Entity Recognition (NER), Coreference Resolution (CR), and Semantic Role Labeling (SRL). It contains news articles from the English language, annotated with fine-grained information such as named entity types, coreference chains, and semantic roles. The dataset consists of approximately 1.2 million words, and it is split into development, training, and test sets, with a total of 29,669 instances.

In order to learn error correction from context, we built our own dataset from pre-existing text datasets and added synthetic errors that closely resembled grammatical faults made by humans. We picked the XSUM [14] & CNN/DailyMail [13] datasets since they all gave us extensive contextual sequences with a variety of languages. This makes it easier for the model to adjust to fresh data.

We employed datasets such as the CNN/Daily Mail Corpus and XSUM due to their inclusion of lengthy sequences of sentences and paragraphs. This was crucial for establishing the context of lengthy sentences and enabling our algorithm to select groups of phrases with consistent context. Aggregated statistics are presented in Table 2. We created 1000-character paragraphs from the dataset's articles, introduced errors, and generated both accurate and erroneous paragraphs using our custom "errorify" script.

We have also used S2ORC [10] dataset to enhance the functionality of our model on scientific data and to properly depict the dynamic system that our model has been designed to capture context. We introduced errors in the S2ORC dataset using our custom script to generate GEC_S2ORC-train, GEC_S2ORC-dev, and GEC_S2ORC-test. The articles and paragraphs from the previously mentioned datasets were used as a dataset for proper grammar. To create a dataset of grammatically incorrect sentences, we used a script to introduce errors into these sentences.

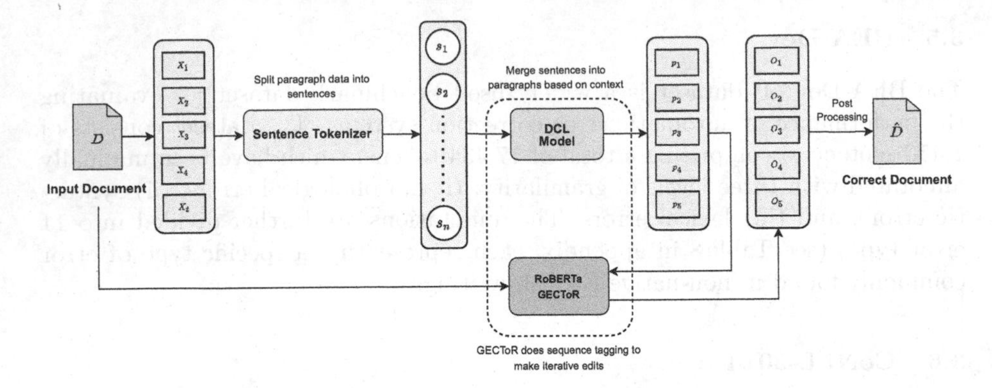

Fig. 2. Overview of GEC-DCL Architecture, which consists of two parts: (1) *Sentence Tokenizer* breaks input paragraph into sentences; (2) *DCL Model* merges those sentences into paragraphs based on context, *GECToR* model performs sequences tagging and make edits to correct the paragraph.

4 Methodology

This section defines the research goals' requirements, encompassing the relevant components involved. The section elucidates the underlying structure and framework that enables the study's objectives to be effectively addressed. Additionally, the pipeline, or the sequence of interconnected processes, is detailed to clearly understand how data flows through various stages, ensuring a systematic and organized approach to data analysis.

4.1 GEC-DCL

The GEC-DCL Model consists of two main parts: the Dynamic Context Learner (DCL) Model and the Grammar Error Correction (GEC) Model. The GEC Model utilizes a transformer-based GEC Sequence Tagger, trained on error-filled corpora and a combination of parallel corpora with and without errors. Additionally, the transformer encoder is trained on synthetic data. This model also incorporates a custom token-level transformation, known as g-transformation, to facilitate targeted corrections.

We also used GECToR without our DCL model serving as the baseline model. We compare GECToR performance with our model in paragraphs and prove that our approach has better results. In our GEC-DCL architecture, the GEC model uses inputs created by the DCL model that consists of a collection of sentences with a common context to look for grammatical errors while taking the context into account. As a result, a corpus that is error-free is produced by focusing on the core context of the sentences.

Our proposed DCL model extends the idea of BERT next sentence prediction to capture relationships between phrases. We understand how important context is when revising sentences that are full of errors. We generate a dataset shown in Table 1 to pre-train for a classification task and train a model that comprehends sentence relationships, with each pre-training example being represented by two sentences, A and B, with B really being the line that comes after A 50% of the time (labelled as True) and B is a phrase chosen at random from the corpus the other 50% of the time (labelled as False).

We extract paragraphs from a document D as, $D = \{X_1, X_2, X_3, ..., X_t\}$ where t is the number of paragraphs in the document. The errors in D are then resolved by feeding each $X \in D$ to our GEC-DCL model, which conducts additional sentence level split to yield $S = \{s_1, s_2, s_3, ..., s_n\}$. Sentence level splitting is done using the Spacy Sentencizer. This Sentencizer uses a set of pre-defined rules and patterns to identify sentence boundaries. These rules take into account common punctuation at the conclusion of sentences, such as periods, question marks, and exclamation points. It also considers additional contextual information, such as the presence of abbreviations or honorifics, to make accurate sentence splits. It first takes the document's paragraph and applies its set of rules to identify sentence boundaries within that paragraph.

Then, in S, our DCL model generates a contiguous combination. For example, s_1 and s_2 are accepted as inputs and processed in the model, showing the results that the two sentences deliver the same context or not. This method is repeated for s_2 and s_3, s_3 and s_4, and so on. The sentences with the same context are kept in paragraphs as, P where, $P = \{p_1, p_2, p_3, ..., p_k\}$, For example, $p_1 = \{s_1, s_2\}, p_2 = \{s_3\}$, and so on.

Following the DCL model's acquisition of the set of paragraphs P for documents X_i with the same context, we transmit each paragraph p_i and the original paragraph X_i's content just once to the GECToR RoBERTa based GEC model in which each p_i iterate four times, and the original X_i only once. As the GECToR model tags the sequences using an iterative sequence tagging approach. This arrangement demonstrates that our strategy places a greater emphasis on maintaining the document's context. After iterating over the original paragraph X_i only once, the model multiplies the DCL model's output by the number of iterations in the GECToR to enable our GEC model to grasp the context as much as possible and make modifications that preserve it. Corrected paragraphs are then post-processed and merged to produce O for each X in D that is supplied to our model, which is then combined to produce the final corrected document $\hat{D}$ as in Fig. 2.

Table 3. Our GEC-DCL model is compared with two SOTA models developed for sentence variation of FCE-Test, BEA-Dev, and CoNLL-2014 and LLaMA large language model. Our model which takes dynamic context outperforms the other three models which used a fixed number of sentences (sent) as context.

Model		FCE-Test			BEA-Dev			CoNLL-2014		
		P	R	$F_{0.5}$	P	R	$F_{0.5}$	P	R	$F_{0.5}$
Yuan & Bryant [25]	sent	**65.36**	**44.17**	**59.64**	62.64	**40.72**	56.55	64.57	28.65	51.62
Chollampatt et al. [5]	sent	53.91	32.81	47.77	-	-	-	64.32	35.98	55.57
Zhang et al. [26]	sent	-	-	-	58.5	**43.1**	54.6	70.3	**50.7**	65.2
GEC-DCL	sent	-	-	-	**68.72**	33.97	**57.04**	**76.48**	40.66	**65.23**

Table 4. GECToR is compared with the GEC-DCL model. We compare results for sentences (sent) and paragraphs (para) variation of FCE-Test, BEA-Dev and CoNLL-2014, where we achieved better results on the paragraph variant and were at par for the sentence variant.

Model		FCE-Test			BEA-Dev			CoNLL-2014		
		P	R	$F_{0.5}$	P	R	$F_{0.5}$	P	R	$F_{0.5}$
GECToR	sent	-	-	-	66.0	33.8	55.5	77.5	40.2	**65.3**
	para	**64.12**	17.91	42.30	66.71	19.24	44.67	**70.00**	06.17	22.80
GEC-DCL	sent	-	-	-	68.72	33.97	57.04	76.48	40.66	65.02
	para	50.28	**37.06**	**46.93**	65.30	**30.93**	**53.43**	56.12	**18.26**	**39.67**

SciBERT. With the development of SciBERT [2], a pretrained language model based on BERT, Using a massive corpus of 1.14 million articles from semantic scholars, semantic technical writing can be better captured. The semantics of technical writing is used by Semantic Scholar. In GECToR, SciBERT is used in place of RoBERTa [9] as the GECToR encoder so that our model can also extend to scientific data. SciBERT has a 40% different vocabulary than BERT and has more scientific words in its vocabulary. That's why we fine-tuned our Sci-BERT model on the GEC_S2ORC-train dataset.

5 Experiments

We employed the pre-trained GECToR-RoBERTa as our baseline encoder, which underwent three-stage training on various datasets including the National University of Singapore Corpus of Learner English (NUCLE) [7], Lang-8 Corpus of Learner English (Lang-8) [19], FCE dataset [24], and Write & Improve + LOCKNESS Corpus [4]. To harness contextual information, we fine-tuned our model using synthetic XSUM and CNN datasets. For evaluation, we utilized the M^2 Scorer [6] on FCE test [24], CoNLL 2014 test [15], and BEA-dev [4]. Additionally, we employed the ERRANT Scorer [3] for XSUM-test and CNN-test sets.

5.1 Training

We tuned all our models using Adam Optimizer with default parameters and used other hyperparameters from [16] with label smoothing of 0.1.

We trained GECToR RoBERTa on XSUM and CNN and GECToR SciBERT on GEC S2ORC using batch size of 20 and a phrase length cap 100 throughout 15 epochs, with a default 1e−5 of learning rate. We additionally trained the DCL model using the FCE and W&I+LOCNESS and also with the S2ORC dataset for scientific grammar correction with a batch size of 128 using pre-trained BERT for five epochs with a default learning rate of 5e−6.

5.2 Results

Current context aware SOTA model [25] uses fixed contextual information; previous two sentences for CoNLL 2014 and one sentence for FCE-Test and BEA-dev. In Table 3, we can see that our dynamic context-based model performs better on BEA-Dev and CoNLL 2014, The BEA-Dev, and FCE-Test uses a fixed number of sentences as context. Our GEC-DCL model's performance is best on the CONLL 2014 dataset, with approximately +11.91 precision, +4.68 recall, and +9.45 $F_{0.5}$. We have also compared the results of the experiment performed by [26]. They fine-tuned LLaMA [20] large language model for the task of GEC on the BEA-Dev, and CONLL-14 dataset. The results in the Table 3 clearly shows that the performance of our model and LLaMA is very close to each other. We employ GECToR's iterative sequence tagging approach to maintaining the paragraph's context.

Table 5. Comparison of Random/fixed split vs DCL on CoNLL-2014 having paragraphs. Here sent refers to sentences.

Method	Precision	Recall	$F_{0.5}$
Previous 2 sent	33.38	14.90	26.74
Random context	32.72	13.65	25.57
DCL based split	**56.12**	**18.26**	**39.67**

Table 5 evaluated our dynamic context-based technique with random context averaged across multiple seeds and the previous two sentences. Our model better captures how many preceding sentences are related to the current sentence. We get an improvement of approximately 13.4 $F_{0.5}$ points. Compared to two sentences as in [25]. We compared our model on paragraph version of FCE-Test, BEA-Dev, and CoNLL-2014 with GECToR to achieve improvement of 8.76, 4.63, and 16.87 $F_{0.5}$ points respectively in Table 4.

Table 6. GECToR vs GEC-DCL on XSUM & CNN Dataset

GECToR vs GEC-DCL on XSUM Dataset			
Model	Precision	Recall	$F_{0.5}$
GECToR	69.43	59.36	67.15
GEC-DCL	**72.25**	**60.4**	**69.5**
GECToR vs GEC-DCL on CNN Dataset			
Model	Precision	Recall	F 0.5
GECToR	**80.55**	48.60	71.19
GEC-DCL	78.02	**55.91**	**72.30**

Table 4 shows the evaluation of GECToR and GEC-DCL at the sentence and paragraph variation. Thus, We can claim that our DCL model can boost paragraph-level correction without compromising sentence-level correction.

We also conducted experiments on XSUM and CNN datasets, containing paragraphs of up to 25 sentences. The outcomes are detailed in Table 6, respectively. Furthermore, we present our state-of-the-art (SOTA) results for scientific documents in Table 7. The comparison between GECToR and GEC-DCL performance on the S2ORC dataset is illustrated in Table 7, showcasing our model's superior performance with a +4.91 $F_{0.5}$ enhancement over GECToR.

Table 7. GECToR vs GEC-DCL w/SciBERT as Transformer on S2ORC Dataset

Model	Precision	Recall	$F_{0.5}$
GECToR	**65.22**	29.31	52.38
GEC-DCL	62.33	**37.45**	**55.02**

5.3 Error Analysis

We perform error analysis to evaluate our hypothesis that correction of errors benefits from information available in the context of the paragraph, outside of the current sentence. We see gains in VERB: FORM (+49.69%), VERB TENSE (+27.38%) and NOUN (+38.66%) shown in Table 9 in the appendix with details of the improvement in every case. Our model shows the most significant margins for improvements in errors related to the sentence's agreement, co-reference, or tense.

6 Conclusion

This work demonstrates that dynamic context detection is much better for GEC than static context addition. We can produce more accurate grammar error correction by including context with the input sentence. Our approach can be used for document level and correct more significant texts in a single pass. All GEC techniques fix errors by concentrating on a single phrase while neglecting important cross-sentence context. We have added context by tagging the preceding phrase onto the present one, producing context-aware grammatical correction. We demonstrate our work on different domains of scientific and general-purpose datasets where the more extended sequence of texts is frequent and needs to preserve context. Our model can be used directly as an end-to-end document correction solution.

7 Limitations

One drawback of our system is that it requires more computation time than other existing grammatical error correction models like GECToR. This is because our model attempts to read and comprehend the context before the initial process, which lengthens the computation time and memory required by the system. Another drawback is our model only works for English.

Acknowledgements. Rajiv Ratn Shah is partly supported by the Infosys Center for AI, the Center for Design and New Media, and the Center of Excellence in Healthcare at IIIT Delhi.

A Appendix

Table 9 deduces that corrections dependent on the more extended context, like tense and verb choice, have seen significant improvement with our DCL model and show a mean of 24% of lift across correction made in missing (M), replacement (R) and unnecessary (U) error in FCE-Test. The formula for calculating the $DiffF_{0.5}$ can be taken from Eq. 1. In the Eq. 1, A stands for the $F_{0.5}$ score of the baseline model, and B stands for $F_{0.5}$ GEC-DCL, the proposed model.

$$DiffF_{0.5} = 100 * \frac{2 * |A - B|}{(A + B)} \tag{1}$$

Table 8. Types of error and their distribution used in the errorify script.

Type of Error	Probability	Original	After Infection
Token-level	0.167		
Swapping random character	0.1	This is a sentence before applying our custom script	This **si** a sentence before applying our custom script
Infect word using "pyinflect"	0.1	This is a sentence before applying our custom script	This is a sentence before **apply** our custom script
Insert Space	0.1	This is a sentence before applying our custom script	This is a **sentence** **<space>** **before** applying our custom script
Swap two token	0.1	This is a sentence before applying our custom script	This is a sentence before applying **custom our** script
Change case of token	0.1	This is a sentence before applying our custom script	This is a sentence before **A**pplying our custom script
No Change in token	0.5	This is a sentence before applying our custom script	This is a sentence before applying our custom script
Sentence Level	1		
Remove Pronouns	0.167	This is a sentence before applying our custom script	~~***This***~~ is a sentence before applying our custom script
Remove Determiners	0.167	This is a sentence before applying our custom script	This is a sentence before applying ~~***our***~~ custom script
Remove Preposition	0.167	This is a sentence before applying our custom script	This is ~~***my***~~ sentence before applying our custom script
Adding *'the'* before Verb and Nouns	0.167	This is a sentence before applying our custom script	This is a sentence before applying our custom ***the*** script
Replacing *'because'* with *'Because'*	0.167	Air travellers were left stranded because of icy conditions	Air travellers were left stranded **Because** of icy conditions

Table 9. Error type-specific performance comparison of GECToR and DCL-GEC on FCE-Test in $F_{0.5}$, difference between our proposed model and GECToR in last column. 24% increase shown by our DCL-GEC model.

Error Type	GECToR			DCL-GEC			
	P	R	$F_{0.5}$	P	R	$F_{0.5}$	**% Diff. $F_{0.5}$**
M:ADV	42.86	01.92	08.15	38.46	03.25	12.14	+48.95
M:NOUN	47.06	04.42	16.06	50.00	05.00	17.86	+11.21
M:NOUN:POSS	73.33	27.50	55.00	72.22	32.50	58.04	+5.5
M:VERB	50.00	10.51	28.54	54.88	16.30	37.25	+30.52
M:VERB:FORM	66.67	16.90	41.96	78.12	35.21	62.81	+49.69
M:VERB:TENSE	52.78	16.67	36.82	60.87	24.45	46.90	+27.38
M:ADJ	77.78	10.14	33.33	66.67	08.70	28.57	−14.28
R:ADJ:FORM	60.00	13.64	35.71	100.00	22.73	59.52	+50.00
R:NOUN	59.30	10.22	30.25	68.66	13.80	38.25	+26.45
R:NOUN:POSS	58.33	15.91	38.04	56.25	20.93	42.06	+10.57
R:VERB	54.58	08.25	25.71	62.20	09.97	30.37	+18.12
R:VERB:FORM	67.50	27.39	52.21	70.14	44.04	62.71	+20.11
R:VERB:TENSE	54.37	19.63	40.16	62.41	29.20	50.84	+26.59
R:ADJ	65.45	11.92	34.48	70.49	14.38	39.59	+14.82
U:ADV	22.41	6.50	15.05	25.30	10.55	19.77	+31.36
U:NOUN	21.21	04.27	11.82	40.00	04.88	16.39	+38.66
U:NOUN:POSS	47.06	38.10	44.94	57.89	52.38	56.70	+26.17
U:VERB	40.48	11.11	26.48	42.86	13.73	30.09	+13.63
U:VERB:FORM	68.75	21.57	47.83	82.35	27.45	58.82	+22.98
U:VERB:TENSE	49.37	22.54	39.88	55.74	39.31	51.44	+28.99
U:ADJ	18.18	03.23	09.43	20.00	04.84	12.13	+25.04
						Total Diff	**+24%**

References

1. Awasthi, A., Sarawagi, S., Goyal, R., Ghosh, S., Piratla, V.: Parallel iterative edit models for local sequence transduction. In: Proceedings of the 2019 Conference on Empirical Methods in Natural Language Processing and the 9th International Joint Conference on Natural Language Processing (EMNLP-IJCNLP), pp. 4260–4270 (2019)
2. Beltagy, I., Lo, K., Cohan, A.: SciBERT: a pretrained language model for scientific text. In: Proceedings of the 2019 Conference on Empirical Methods in Natural Language Processing and the 9th International Joint Conference on Natural Language Processing (EMNLP-IJCNLP), pp. 3615–3620. Association for Computational Linguistics, Hong Kong, China, November 2019. https://doi.org/10.18653/v1/D19-1371. https://aclanthology.org/D19-1371

3. Bryant, C., Felice, M., Briscoe, T.: Automatic annotation and evaluation of error types for grammatical error correction. In: Proceedings of the 55th Annual Meeting of the Association for Computational Linguistics (Volume 1: Long Papers), pp. 793–805. Association for Computational Linguistics, Vancouver, Canada, July 2017. https://doi.org/10.18653/v1/P17-1074. https://aclanthology.org/P17-1074
4. Bryant, C., Felice, M., Briscoe, T.: The BEA 2019 shared task on grammatical error correction. In: Proceedings of the 14th Workshop on Innovative Use of NLP for Building Educational Applications, pp. 54–75. Association for Computational Linguistics (2019)
5. Chollampatt, S., Wang, W., Ng, H.T.: Cross-sentence grammatical error correction. In: Proceedings of the 57th Annual Meeting of the Association for Computational Linguistics, pp. 435–445 (2019)
6. Dahlmeier, D., Ng, H.T.: Better evaluation for grammatical error correction. In: Proceedings of the 2012 Conference of the North American Chapter of the Association for Computational Linguistics: Human Language Technologies, pp. 568–572. Association for Computational Linguistics, Montréal, Canada, June 2012. https://aclanthology.org/N12-1067
7. Dahlmeier, D., Ng, H.T., Wu, S.M.: Building a large annotated corpus of learner English: the NUS corpus of learner English. In: Proceedings of the Eighth Workshop on Innovative Use of NLP for Building Educational Applications, pp. 22–31 (2013)
8. Devlin, J., Chang, M.W., Lee, K., Toutanova, K.: BERT: pre-training of deep bidirectional transformers for language understanding. In: Proceedings of the 2019 Conference of the North American Chapter of the Association for Computational Linguistics: Human Language Technologies, Volume 1 (Long and Short Papers), pp. 4171–4186. Association for Computational Linguistics, Minneapolis, Minnesota, June 2019. https://doi.org/10.18653/v1/N19-1423. https://aclanthology.org/N19-1423
9. Liu, Y., et al.: RoBERTa: a robustly optimized BERT pretraining approach. arXiv preprint arXiv:1907.11692 (2019)
10. Lo, K., Wang, L.L., Neumann, M., Kinney, R., Weld, D.: S2ORC: the semantic scholar open research corpus. In: Proceedings of the 58th Annual Meeting of the Association for Computational Linguistics, pp. 4969–4983. Association for Computational Linguistics, July 2020. https://doi.org/10.18653/v1/2020.acl-main.447. https://aclanthology.org/2020.acl-main.447
11. Malmi, E., Krause, S., Rothe, S., Mirylenka, D., Severyn, A.: Encode, tag, realize: high-precision text editing. In: Proceedings of the 2019 Conference on Empirical Methods in Natural Language Processing and the 9th International Joint Conference on Natural Language Processing (EMNLP-IJCNLP), pp. 5054–5065 (2019)
12. Mizumoto, T., Hayashibe, Y., Komachi, M., Nagata, M., Matsumoto, Y.: The effect of learner corpus size in grammatical error correction of ESL writings (2012)
13. Nallapati, R., Zhou, B., Gulcehre, C., Xiang, B., et al.: Abstractive text summarization using sequence-to-sequence RNNs and beyond. arXiv preprint arXiv:1602.06023 (2016)
14. Narayan, S., Cohen, S.B., Lapata, M.: Don't give me the details, just the summary! Topic-aware convolutional neural networks for extreme summarization. arXiv preprint arXiv:1808.08745 (2018)
15. Ng, H.T., Wu, S.M., Briscoe, T., Hadiwinoto, C., Susanto, R.H., Bryant, C.: The CoNLL-2014 shared task on grammatical error correction. In: Proceedings of the Eighteenth Conference on Computational Natural Language Learning: Shared Task, pp. 1–14. Association for Computational Linguistics (2014). https://doi.org/10.3115/v1/W14-1701. http://www.aclweb.org/anthology/W14-1701

16. Omelianchuk, K., Atrasevych, V., Chernodub, A., Skurzhanskyi, O.: GECToR-grammatical error correction: tag, not rewrite. In: Proceedings of the Fifteenth Workshop on Innovative Use of NLP for Building Educational Applications, pp. 163–170 (2020)
17. Radford, A., Narasimhan, K., Salimans, T., Sutskever, I.: Improving language understanding by generative pre-training. OpenAI Blog **1**(8) (2018)
18. Sennrich, R., Haddow, B., Birch, A.: Edinburgh neural machine translation systems for WMT 16. arXiv preprint arXiv:1606.02891 (2016)
19. Tajiri, T., Komachi, M., Matsumoto, Y.: Tense and aspect error correction for ESL learners using global context. In: Proceedings of the 50th Annual Meeting of the Association for Computational Linguistics (Volume 2: Short Papers), pp. 198–202. Association for Computational Linguistics, Jeju Island, Korea, July 2012. https://aclanthology.org/P12-2039
20. Touvron, H., et al.: LLaMA: open and efficient foundation language models (2023). https://doi.org/10.48550/arXiv.2302.13971
21. Vaswani, A., et al.: Attention is all you need. In: Advances in Neural Information Processing Systems, vol. 30 (2017)
22. Wang, L., Tu, Z., Way, A., Liu, Q.: Exploiting cross-sentence context for neural machine translation. arXiv preprint arXiv:1704.04347 (2017)
23. Yang, Z., Dai, Z., Yang, Y., Carbonell, J., Salakhutdinov, R.R., Le, Q.V.: XLNet: generalized autoregressive pretraining for language understanding. In: Advances in Neural Information Processing Systems, vol. 32 (2019)
24. Yannakoudakis, H., Briscoe, T., Medlock, B.: A new dataset and method for automatically grading ESOL texts. In: Proceedings of the 49th Annual Meeting of the Association for Computational Linguistics: Human Language Technologies, pp. 180–189. Association for Computational Linguistics, Portland, Oregon, USA, June 2011. https://aclanthology.org/P11-1019
25. Yuan, Z., Bryant, C.: Document-level grammatical error correction. In: Proceedings of the 16th Workshop on Innovative Use of NLP for Building Educational Applications, pp. 75–84 (2021)
26. Zhang, Y., Cui, L., Cai, D., Huang, X., Fang, T., Bi, W.: Multi-task instruction tuning of llama for specific scenarios: a preliminary study on writing assistance. ArXiv abs/2305.13225 (2023)

Data Analytics for Low Resource Domains

A Deep Learning Emotion Classification Framework for Low Resource Languages

Manisha(✉), William Clifford(✉), Eugene McLaughlin(✉), and Paul Stynes(✉)

National College of Ireland, Dublin, Ireland
x21207194@student.ncirl.ie,
{william.clifford,eugene.mclaughlin,paul.stynes}@ncirl.ie

Abstract. Emotion classification from text is the process of identifying and classifying emotions expressed in textual data. Emotions can be feelings such as anger, joy, suspense, sadness and neutral. Developing a machine learning model to identify emotions in a low-resourced language with a limited set of linguistic resources and annotated corpora is a challenge. This research proposes a Deep Learning Emotion Classification Framework to identify and classify emotions in low-resourced languages such as Hindi. The proposed framework combines a classification model and a low resource optimization technique in a novel way. An annotated corpus of Hindi short stories consisting of 20,304 sentences is used to train the models for predicting five categories of emotions: anger, joy, suspense, sadness, and neutral talk. To resolve the class imbalance in the dataset SMOTE technique is applied. The optimal classification model is selected through experimentation that compares machine learning models and pre-trained models. Machine learning and deep learning models are SVM, Logistic Regression, Random Forest, CNN, BiLSTM, and CNN+BiLSTM. The pre-trained models, mBERT, IndicBERT, and a hybrid model, mBERT+BiLSTM. The models are evaluated based on macro average recall, macro average precision, and macro average F1 score. Results demonstrate that the hybrid model mBERT+BiLSTM out perform other models with a test accuracy of 57%.

Keywords: Deep learning · Emotion classification · Low resource languages · Pre-trained model

1 Introduction

People spend a majority of their time on social media expressing opinions, communicating feelings, and exchanging ideas on the virtual platforms. Comprehending human emotions from online contents is important for business leaders and social scientists. It has various applications, such as stress/anxiety retrieval from online chats, detecting sensitive emotions from online conversations, capturing emotions from multimedia tagging, and many more [1]. Detecting emotions from texts is the process of automatically assigning an emotion category to a textual

V. Goyal et al. (Eds.): BDA 2023, LNCS 14418, pp. 113–121, 2023.
https://doi.org/10.1007/978-3-031-49601-1_8

document from a set of predetermined categories [2]. It's a subset of sentiment analysis that aims to find fine-grained emotions such as anger, joy, suspense, sadness and neutral from texts rather than generic and coarse-grained polarity assignments like neutral, positive, and negative.

A language is termed as low-resourced when it lacks monolingual corpora or linguistic resources required for training AI models. While the development of text classification techniques from high-resource languages like English, Chinese, and French has progressed significantly in recent years, there has been no notable progress in low-resource languages such as Hindi. Lack of annotated corpora, extensive monolingual corpora, and limited text processing tools make emotion analysis in low-resource languages a challenge.

Emotion classification of low resource languages is a challenge as this is achieved based on based manually selecting features from text. Research to automate the classification in high resource languages involves using techniques like Bag of Words (BOW), and n-grams, and deep learning (DL) algorithms that can learn features from the text itself. This reduces the need for manual feature engineering [3]. There is limited research to automate the classification in low resource languages. The aim of this research is to investigate to what extent a Deep Learning Emotion Classification Framework can identify and classify emotions from low-resource languages such as Hindi. The major contribution of this study is a framework that combines a classification model and a low resource optimization technique in a novel way.

2 Related Work

Detecting and analyzing emotions is one of the challenging and emerging areas of natural language processing (NLP). Moreover, the progress for low-resourced languages has been slow due to the lack of lexical resources like pre-trained word embedding, language models and well-annotated datasets. To classify emotions from text various NLP methods have been proposed - the keyword approach, the lexicon-based approach and the learning-based approach [4]. However, the first two approaches have limitations like lack of context as they depend on the presence of specific keywords from emotion lexicon, limited coverage of emotion lexicons leading to misclassification, poor performance for detecting some specific emotions. On the other hand, in learning-based approach, various models are trained to give accurate and better results. [5] investigated multiple machine-learning algorithms for analyzing mental health and classifying human emotions in texts. Texts are pre-processed using the Neat-Text NLP package which provides various inbuilt functions and text frames for text cleaning, stemming and tokenization. For feature extraction count vectorizer and TF-TDF approaches are used. TF-IDF evaluates a word's significance in a document while considering it's relationships to other terms in the same corpus. The highest accuracy achieved on the dataset was 62% by Logistic Regression model. Deep learning models use word vectors to present text in n-dimensional space such that identical/similar words are positioned very close together in the vector space. [6] proposed a micro-blog emotion classification model, that uses CNN with Word2Vec word embedding for emotion classification from short/micro Chinese blogs. The word vectors are fed as input vectors to the model to learn text features. The overall accuracy achieved by CNN_Text_Word2vec was 7% higher than the machine learning models such as SVM, LSTM and RNN. [7] combined multiple Twitter dialogues spanning over five different emotion labels: joy, sad, anger, fear, and neutral and created a dataset for multi-class sentiment analysis with 13k sentences. Post-performing data pre-processing tasks like stemming, stop word removal and tokenization, TF-IDF was used for vector representation of the texts. Multiple traditional machine learning models like Naive Bayes, Random Forest, Logistic Regression were tested as the baseline models. However, the accuracy of the classification model increased by 24% with a fine tuned pre-trained BERT model. Simply layering one hidden-layer neural network classifier on top of traditional BERT gave higher results with a minimal amount of data. [8] investigated multiple machine or deep learning models such as SVC, Multinomial Naive Bayes, Logistic Regression, Random Forest, CNN, and LSTM to identify 7 emmotions from a Hindi-English mixed Twitter dataset. The SVC model performed best by providing 73.75% Researchers showed that hybrid models that combine two models can also exhibit good performance in NLP tasks. [9] used CNN-BiLSTM for emotion analysis from microblog comments, [10] used BERT embeddings with BiLSTM for emotion analysis from two dialogue sets and observed the hybrid model outperformed baseline models significantly. For sentiment analysis in Chinese buzzwords, [11] used BERT embeddings to capture rich context of the sentences and BiLSTM for feature extraction to obtain the local and global semantic features of the texts.

Emotion classification from low-resourced languages is still in its infancy, few research works have been done on the topic focused on traditional machine learning and deep learning algorithms. Transformer-based models are progressively used for text classification tasks. Pre-trained models' training process does not begin from scratch, in contrast to deep learning approaches, which speed up the generalization of the model and the learning of the problem. The feature engineering techniques like Bag Of Words and count vectorizer used in machine learning methods often cause loss of information as they don't consider the order and integrity of words in sentences. On the other hand, deep learning models preserve the context of the co-existing words and hence result in better classification results. However, the performance of deep learning models depends on the dataset size the larger a dataset is the better the model can learn the text features. [12] investigated the performance of transformer-based sentiment analysis models on four different Turkish datasets. The results of the experiments showed that transformer-based models are superior to deep learning and traditional machine learning models in terms of F-score performance. [2] developed a Bengali emotion corpus consisting of 6243 sentences from six different emotion classes: anger, disgust, fear, joy, sadness, and surprise. The performance of various machine learning, DNN, and transformer models was investigated in this corpus. Given their state-of-the-art performance in the classification task, three transformer models m-BERT, Bangla-BERT, and XLM-R on BEmoC were used for emotion classification from Bangla text. [16] used IndicBERT, a mBERT model trained for Indian languages that captures semantic and linguistic features from multilingual texts. The approach was used for sentiment analysis in code-mixed Tamil tweets, and an F1 score of 61.73% was achieved. In [17] application of transformer models mBERT and BERT were investigated for fake news dentification in Brazilian Portuguese. The experiment showed transfer learning strategies is effective in case of low abundancy of labeled data. BERT was found to be the best-performing model with a F1 score of 98.4 on the test set.

3 Methodology

The research methodology entails six steps, namely, data gathering, exploratory data analysis, data pre-processing, data transformation, data modeling and conversion, evaluation, and result interpretation, as shown in Fig. 1. The first step, *Data Gathering* step involves collecting input texts from a low-resourced language. The BHAAV [15] dataset, used in the study, is an annotated corpus collected from 230 Hindi short stories. Each sentence has been classified with one of five emotions—angry, joyful, suspenseful, sad, and neutral. This dataset is publicly available and can be downloaded from zenodo.org[1].

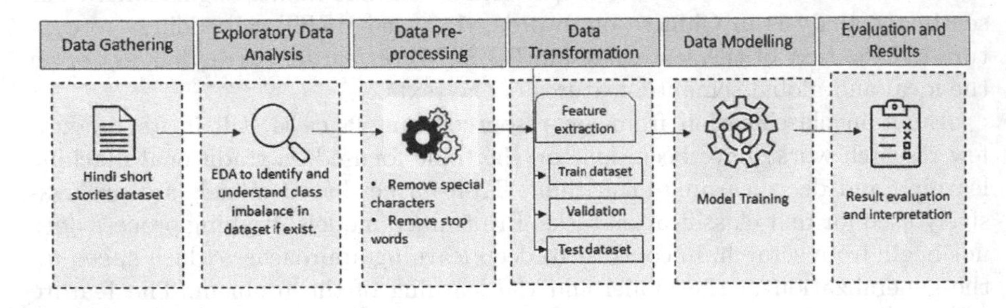

Fig. 1. Research methodology

The second step, *Exploratory Data Analysis* involves inspecting the structure and format of the dataset via plotting multiple data visualizations and checking data samples. BHAAV dataset has high-class imbalance where the 'Neutral' emotion class holds around 60% of the data. To handle the class imbalance from the dataset SMOTE was used on the training dataset. SMOTE is an oversampling technique that handles class imbalance in datasets by generating synthetic samples for the minority classes. Table 1 below shows the emotion class distribution before and after SMOTE in BHAAV.

It is important to apply SMOTE only on the training dataset and not on the validation or test datasets because doing so would introduce data leakage leading to model overfit on the test dataset and overly optimistic performance metrics

[1] https://zenodo.org/record/3457467.

Table 1. Table captions should be placed above the tables.

Class	Before SMOTE	After SMOTE
Angry	1176	9453
Joy	1996	9453
Sad	2584	9453
Suspense	1236	9453
Neutral	9453	9453

and unreliable model evaluation results. Another way of removing biases from an imbalanced dataset is to assign different class weights to both majority and minority labels in the dataset. The class weight for a label should be calculated based on its frequency in the dataset.

The third step, *Data Pre-processing* involves cleaning noise from the dataset. Removal of special characters like punctuation marks, numbers, and non-Hindi alphabets like," —;/][[0–9] was performed on the dataset. Stop words should also be removed from the sentences due to their low information content and high frequency in texts. Removing stop words from texts reduces feature space and increases focus on the essential features. The fourth step, *Data Transformation* involves tokenizing the sentences into words and converting them to a numerical format that machine learning models can understand. For ML models, TF-IDF (Term Frequency-Inverse Document Frequency) was used for feature extraction and converting texts into vector space. For deep learning models pre-trained Fasttext and IndicFT, a Fasttext embedding model trained on Indian languages word embedding model was used. By using word embedding, words are represented as real-valued vectors that encode their meaning, such that words with similar meanings are closer to each other in the vector space. Before training the models for classification tasks, the input dataset is divided into Train:test:validation dataset with ratio 80:10:10 respectively. The fifth step, *Data Modeling and Conclusion* involves building model architecture and model training. The models were trained with the split training dataset and tested with the validation dataset. Three baseline machine learning models—logistic regression, Support vector machine (SVM), and Random forests—were first trained for the classification task. CNN, BiLSTM, and CNN +BiLSTM deep learning models were trained for the emotion classification tasks. Two pre-trained models, BERT and IndicBERT were trained. A hybrid model, mBERT+BiLSTM was also created for the classification task. Transfer learning was employed to maximize feature extraction and selection. The transformer models were used from the Huggingface transformers and fine-tuned on the corpus. The softmax activation function was used in the final classification layer for the deep learning models as this is a multi-class classification problem. With the softmax function, the probabilities for each class in a multi-class classification can be estimated. The sixth step, *Evaluation and Results* involves evaluating the performance of each of the machine learning and deep learning models. Since the dataset has multiple emotion class labels and there is a class imbalance, macro average precision, recall, and F1 score are the best metrics to evaluate models. In macro averaged values, every class is weighed uniformly; the scores are calculated independently for each class and then the average is taken thus treating each class equally.

4 Design

For transformer models, pre-trained tokenizers like the mBERT tokenizer, and the IndicBERT tokenizers were used to tokenize the text. While tokenizing the text, input_id is created, which is a numerical representation of the text as mapped to the vocabulary of the pre-trained model. mBERT has the capability to capture contextual meanings from texts and it is trained on large dataset of multilingual texts like Hindi. In addition, BiLSTM ability to capture sequential patterns from the sentences allows it perform well on the text classification task. Figure 2 shows the overall architecture of the experiment.

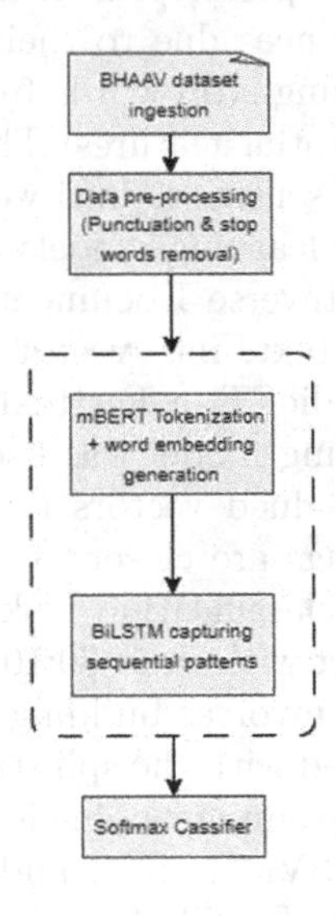

Fig. 2. System architecture

5 Evaluation

A model's performance can be best evaluated based on it's performance on unseen data. Table 2, Table 3 and Table 4 shows comparison of machine learning, deep learning and pre-trained models based on macro average F1 score, macro average precision, macro average recall, and balanced accuracy metrics.

Table 2. Performance of baseline machine learning models on the BHAAV dataset

Model	Macro Precision	Macro Recall	Macro F1	Accuracy
Logistic Regression	0.35	0.42	0.35	0.40
Random Forest	0.33	0.39	0.34	0.41
SVM	0.36	0.40	0.32	0.36

Table 3. Performance of baseline deep learning models on the BHAAV dataset

Model	Word Embedding	Macro Precision	Macro Recall	Macro F1	Accuracy
CNN	FastText	0.30	0.37	0.29	0.31
CNN	IndicFT	0.35	0.43	0.35	0.38
BiLSTM	FastText	0.35	0.43	0.34	0.38
BiLSTM	IndicFT	0.37	0.45	0.36	0.38
CNN+BiLSTM	FastText	0.35	0.43	0.34	0.38
CNN+BiLSTM	IndicFT	0.35	0.45	0.34	0.34

Table 4. Performance of baseline machine learning models on the BHAAV dataset

Model	Macro Precision	Macro Recall	Macro F1	Accuracy
mBERT	0.38	0.32	0.33	0.55
indicBERT	0.37	0.33	0.34	0.55
mBERT+BiLSTM	0.41	0.40	0.39	0.57

5.1 Discussion

This result indicates that machine learning models are almost similar in performance. Although, Random Forest has the highest test accuracy of 41% macro average precision and recall are still low. However, for the results it can be observed the model is able to classify the majority class i.e. 'Neutral' more accurately than the other minority classes, and hence the low precision and recall values. On the other hand, the performance of deep baseline models was very low compared to the ML models. BiLSTM with both FastText and IndicFT gave the highest test accuracy of 38%. The deep models have a macro average 43% recall which shows the models are correctly predicting a good percentage of actual positive instances. In deep models, BiLSTM performed well with both FastText and IndicFT embedding achieving a test accuracy of 38% and macro average recall of 45%. mBERT outperformed baseline models with an accuracy of 55% but the macro average recall and precision value were very low suggesting that the model is performing well on majority labels but struggling on minority labels. Finally, a hybrid model mBERT+BiLSTM was created to get the best of both the good-performing models. With the hybrid model test accuracy improved significantly reaching 57%. Furthermore, the macro average metrics also improved slightly showing the hybrid was able to perform better with both the majority and minority labels.

6 Conclusion and Future Work

The aim of this research was to investigate to what extent to which deep learning framework can be leveraged to identify and classify emotions from low resourced texts. This research proposes a deep learning framework that uses mBERT and BiLSTM to capture the contextual meaning from Hindi texts. Results from the comparative study between machine learning and deep learning models demonstrate that pre-trained models outperform baseline models. This research can potentially enhance the method of analyzing sentiments behind low resourced texts which in turn can be helpful for businesses, government and associated entities to take informed decisions. This work can be improved through additional fine tuning of the pre-trained models, experimenting with additional class imbalance removal methods and using data augmentation to create more samples for models to train. With recent progress in pre-trained generative models like ChatGPT −3 generating emotion specific data in any low resourced languages can be achieved. These models are proven to produce data with comparatively higher accuracy.

References

1. Acheampong, F.A., Wenyu, C., Nunoo-Mensah, H.: Text-based emotion detection: advances, challenges, and opportunities. Eng. Rep. **2**(7), e12189 (2020)
2. Das, A., Sharif, O., Hoque, M.M., Sarker, I.H.: Emotion classification in a resource constrained language using transformer-based approach (2021). arXiv preprint arXiv:2104.08613
3. Alam, T., Khan, A., Alam, F.: Bangla text classification using transformers (2020). arXiv preprint arXiv:2011.04446
4. Bharti, S.K., et al.: Text-based emotion recognition using deep learning approach. Comput. Intell. Neurosci. (2022)
5. Midhan, T.M., Selvaraj, P., Raju, M.H.K., Reddy, M.B.P., Bhaskar, T.: Classification of mental health and emotion of human from text using machine learning approaches. In: 2023 6th International Conference on Information Systems and Computer Networks (ISCON), pp. 1–7. IEEE, March 2023
6. Xu, D., Tian, Z., Lai, R., Kong, X., Tan, Z., Shi, W.: Deep learning based emotion analysis of microblog texts. Inf. Fusion **64**, 1–11 (2020)
7. Kannan, E., Kothamasu, L.A.: Fine-tuning BERT based approach for multi-class sentiment analysis on twitter emotion data. Ingénierie des Systémes d'Information **27**(1) (2022)
8. Sonu, S., Haque, R., Hasanuzzaman, M., Stynes, P., Pathak, P.: Identifying emotions in code mixed Hindi-English tweets. In: Proceedings of the WILDRE-6 Workshop within the 13th Language Resources and Evaluation Conference, pp. 35–41. European Language Resources Association, Marseille, France (2022)
9. Li, A., Yi, S.: Emotion analysis model of microblog comment text based on CNN-BiLSTM. Comput. Intell. Neurosci. (2022)

10. Gou, Z., Li, Y.: Integrating BERT embeddings and BiLSTM for emotion analysis of dialogue. Comput. Intell. Neurosci. (2023)
11. Li, X., Lei, Y., Ji, S.: BERT-and BiLSTM-based sentiment analysis of online Chinese buzzwords. Future Internet **14**(11), 332 (2022)
12. Ozturk, O., Ozcan, A.: Sentiment analysis in turkish using transformer-based deep learning models. In: Hemanth, D.J., Yigit, T., Kose, U., Guvenc, U. (eds.) The International Conference on Artificial Intelligence and Applied Mathematics in Engineering, vol. 7, pp. 1–15. Springer, Cham (2022). https://doi.org/10.1007/978-3-031-31956-3_1
13. Ranathunga, S., Liyanage, I.U.: Sentiment analysis of Sinhala news comments. Trans. Asian Low-Resource Lang. Inf. Process. **20**(4), 1–23 (2021)
14. Ucan, A., Dörterler, M., Akçapinar Sezer, E.: A study of Turkish emotion classification with pretrained language models. J. Inf. Sci. **48**(6), 857–865 (2022)
15. Kumar, Y., Mahata, D., Aggarwal, S., Chugh, A., Maheshwari, R., Shah, R.R.: BHAAV-A text corpus for emotion analysis from Hindi stories (2019). arXiv preprint arXiv:1910.04073
16. Kannan, R.R., Rajalakshmi, R., Kumar, L.: IndicBERT based approach for sentiment analysis on code-mixed Tamil tweets (2021)
17. Fischer, M., Haque, R., Stynes, P., Pathak, P.: Identifying fake news in Brazilian Portuguese. In: Rosso, P., Basile, V., Martínez, R., Métais, E., Meziane, F. (eds.) International Conference on Applications of Natural Language to Information Systems, vol. 13286, pp. 111–118. Springer, Cham (2022). https://doi.org/10.1007/978-3-031-08473-7_10

Assessing the Efficacy of Synthetic Data for Enhancing Machine Translation Models in Low Resource Domains

Shweta Yadav(✉)

National College of Ireland, Dublin, Ireland
x21209251@student.ncirl.ie

Abstract. An artificially generated dataset mimics real-world data in terms of its statistical properties, but it contains no real information. Data around rare occurrences like Covid-19 pandemic is difficult to capture in real-world data due to their infrequent nature. Additionally, cost involved and time-consumption to gather real world data is a big challenge. In such cases, synthetic data can help create more balanced datasets for model training. This project investigates the effectiveness of using synthetic data for tuning machine translation models when training data is limited. The Covid-19 domain is chosen considering the urgency and importance of the global accessibility of information related to the pandemic. TICO-19, a publically available dataset was effectively formulated to cater to this need. The medical terminologies were extracted and passed to OpenAI API to generate training language pair data. The fine-tuned davinci model is then verified with blind test data provided under TICO-19 for translation from English to French. SacreBLEU score is used to compute the translation quality, the fine-tuned model has a significantly higher BLEU score of 19.54 in comparison to the base model with a BLEU score of 0.44. The adapted model also has a comparable score to the next-generation version of davinci with a BLEU score of 22.29.

Keywords: OpenAI · davinci · TICO-19 · low resource domain · machine translation · Covid-19

1 Introduction

The introduction of the attention layer in transformer-based models [1] transformed the field of Natural Language Processing (NLP), there is a drastic paradigm shift when choosing models for text processing from Recurrent Neural Network (RNNs) and Convolutional Neural Network (CNNs). Further, the study [2] introduced Neural Machine Translation (NMT) which showed improvement over traditional phrase-based translation. Machine Translation (MT) is the

National College of Ireland.

V. Goyal et al. (Eds.): BDA 2023, LNCS 14418, pp. 122–132, 2023.
https://doi.org/10.1007/978-3-031-49601-1_9

branch of NLP problems where a sentence from one language is translated into another using a computer application[1], while carefully considering the rules of both source and target language. NMTs have worked considerably well when a resource-rich language pair is chosen, however, it still faces the challenges associated with low resource domain [3]. A language is treated as low-resourced when it lacks linguistic resources or monolingual/bilingual corpora required for training models.

Collating adequate corpus for ML models is already a challenging task, however, when using transformers, which need even large data for training, working with low-resource language becomes even more difficult. Adapting the NMT models, which are already trained on heavy resources, to domain-specific translation has been extensively researched to overcome the issue of poor in-domain translation of these existing models. Research [4] attempts to provide a framework where a model already trained for one language pair can be extended to another. The study [5] presented the improvement that can be seen in fine-tuning a previously trained model with a domain-specific corpus. However, adapting models with domain-specific translations showed promising results, but the issue associated with the lack of training data still remains.

With the introduction of generative AI and successful models like OpenAI GPTs[2] generating synthetic data has become more accurate. While the GPT models can quickly generate *completion* based on the *prompt* they are given, the model loses the accuracy when generating longer sentences. The study [5] discusses the efficiency and effectiveness of adapting pre-trained language models to a domain-specific MT. It presents a text-generation technique that produces domain-specific sentences and extensively verifies the feasibility of synthetic data for training MT for domains with low to no parallel dataset.

This paper introduces an approach that utilizes domain-specific data augmentation to train language models for machine translation. It employs OpenAI's GPT-3.5 model to generate synthetic data and assesses its effectiveness in improving an existing model with near-real data. This use of synthetic data also addresses ethical concerns related to sensitive domains like healthcare and the challenges of working with limited monolingual or bilingual data in certain fields. In this study, COVID-19-related terminology is extracted from the TICO-19 dataset, which was created to advance the study of machine translation of pandemic-related information.

2 Related Work

The need for machine translation has been ever present since the advent of the internet. With an efficient Machine Translation (MT) system, the circulation of important information will become more effective and feasible. Apart from being a medium of making a piece of information globally accessible, [6] presented translation as an effective method of data augmentation. Techniques like back

[1] https://phrase.com/blog/posts/machine-translation/.

[2] https://platform.openai.com/docs/introduction/key-concepts.

translation can help generate a corpus for low-resource language from a resource-rich monolingual corpus. However, the study showed significant results for the English and French language pair which is a resource-rich language pair but when a motivated approach [7] was tried for English to Urdu translation for training a model for fake news detection in Urdu, the results were poor. This shows the efficiency of MT models largely depends on the language pair selected for research and low-resource language pairs will always be at a loss.

Domain adaptation of Neural Machine Translation (NMT) is being thoroughly researched and due to the availability of a wide array of language models (LMs), it has become fairly feasible. The use of pre-trained models due to their capacity to identify a wide range of linguistic features without having to train them from scratch, which can be computationally challenging and data-intensive. Such models are already trained on basic textual features and are available for use-case-specific fine-tuning. One of the methods for data-centric domain adaptation of NMT is fine-tuning a generalized MT with in-domain corpus, however, this method has its drawbacks. The study [8] discusses overfitting and proposes a solution to prevent the degradation of out-of-domain translations once a model is fine-tuned. Another way of fine-tuning a model is through synthetic parallel corpora. Study [9] discusses the use of back translation on monolingual datasets to generate parallel corpus and tune the NMT model. There have been multiple types of research done discussing the effectiveness of generating synthetic data for domain adaptation with either source-side monolingual data [10], target-side monolingual data [11] or both [12]. These studies show how effective synthetic data can be for fine-tuning a model.

A study [13] discusses the benefit of using synthetic data for enhancing predictions. The results were positive however the simplicity of the generative model used resulted in imbalanced data. The benefits, challenges and risks of utilizing synthetic data are discussed [14]. There will always be resentment towards accepting such data specifically in the healthcare domain, the study explores the benefits and presents a cautious way of gaining acceptance for this data. The assessment criteria outlined in the research conducted [15], designed for appraising synthetic healthcare data, validate the data based on its similarity to the original data, privacy features, usefulness, and overall impact. Data generated can be evaluated by utilizing this metric and the generation method can be verified. The study [16] explores the benefits of leveraging ChatGPT in healthcare domain while also discussing its drawbacks and ethical issues pertaining to its usage. It highlights how well the data generated can be used for various NLP tasks while also emphasizing the risks of false information. Synthetic data especially in a domain as sensitive as healthcare will always raise privacy and ethical concerns but it can also offer several potential benefits when approached thoughtfully and with appropriate safeguards.

3 Methodology

In the research, a study-specific version of Knowledge Discovery in Databases is followed, the phases of KDD are refined and utilized to make it more accurate with the steps undertaken while conducting the study. The steps are depicted in the Fig. 1 below:

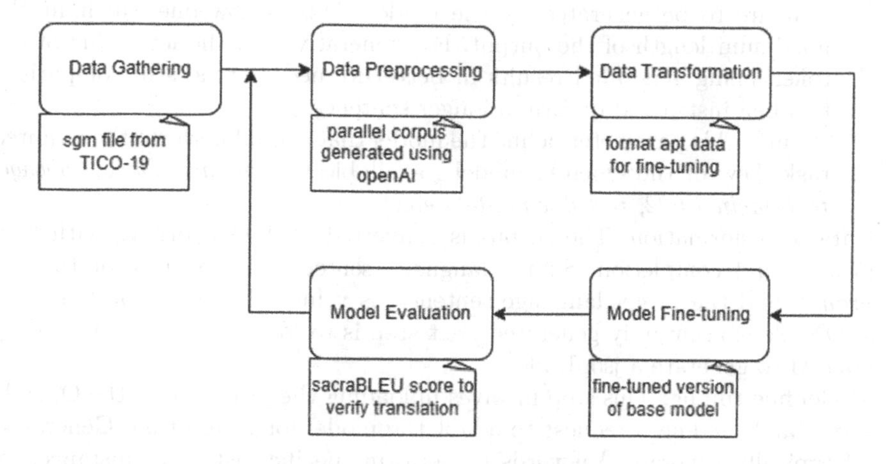

Fig. 1. Research Methodology

1. Data Gathering: TICO-19 dataset [17] contains sgm files containing terms in both source and target languages. The dataset also contains a test set which will act as the ground truth while evaluating the models. The pair considered for this research is English to French.
2. Data Preprocessing: The next step is to generate the parallel dataset, the terms extracted are sent to openAI API to generate sentences for these terms. There are only 215 unique terms in the dataset, in order to increase the volume of training data, five statements are generated for each term. While generating the sentences there are various parameters that are tuned to have more variant and contextually sound datasets. These statements are then sent to API in order to be translated into the target language. With the completion of this step, the training data is generated, having domain-specific terms, synthetic source language sentences, and translated target language sentences, also called parallel corpus for machine translation. The API call has below parameters that help in controlling and modifying the type of data generated:
   (a) stop: This parameter defines the end of the output generated.
   (b) temperature: This parameter defines the randomness of the text generated by the API call. The value ranges from 0.0 to 1.0, a high value will result in more diverse and creative responses, but they might be less coherent or accurate. On the other hand, a lower temperature value, such as 0.2, will produce more focused and deterministic but duplicate responses.
   (c) prompt: This parameter is the combination of the action to be performed followed by the text on which the action is to be performed, in case no

action is provided, the call will add random keywords following the input text/term.

(d) max_tokens: This parameter limits the number of keywords or tokens that are to be generated by the model. This is how one can limit the maximum length of the output. For generative AI, the low value of the token being generated results in better-formed sentences in comparison to when instructed to form a longer sentence.

(e) engine: This parameter define the model that should perform the required task. Few of the OpenAI model's available are *davinci, curie, babbage, text-davinci-002, text-davinci-003 etc.*[3]

3. Data Transformation: The corpus is converted to JSON format, with keys prompt and completion. Source language sentences are values for the key *prompt* and the target language sentence as values of *completion*. Once the JSON file is accurately generated next step is to follow the step outlined by openAI to generate a jsonL file[4].
4. Model fine-tuning: This step involves uploading the jsonL file to the OpenAI server and sending a request to adapt the model for translation. Generative AI typically generates keywords to perform specific tasks. For instance, for machine translation using GPT-3.5, you can use a prompt like "Translate the following sentence to French" followed by the sentence to be translated, and the model will provide the translation. This research aims to develop a model exclusively trained for French translation. The outcome of this step will be a model capable of translation without the need for explicit cues.
5. Evaluation: This step involves verifying the results of the fine-tuned model. The BLEU scores are calculated by comparing model output with test corpus available in TICO-19.

4 Design Specification

4.1 Scope of Research

The main idea of the research is to verify if pre-trained models fine-tuned using synthetic data show any improvement in terms of the quality of output they produce. This study pivots on the accuracy of synthetic data being generated by GPT-3.5 model and the availability of the davinci base model available for fine-tuning.

4.2 Flow of Study

On a broad level, the system will include the following phases which are pictorially depicted in Fig. 2:

[3] https://platform.openai.com/docs/models/.

[4] https://platform.openai.com/docs/guides/fine-tuning/preparing-your-dataset.

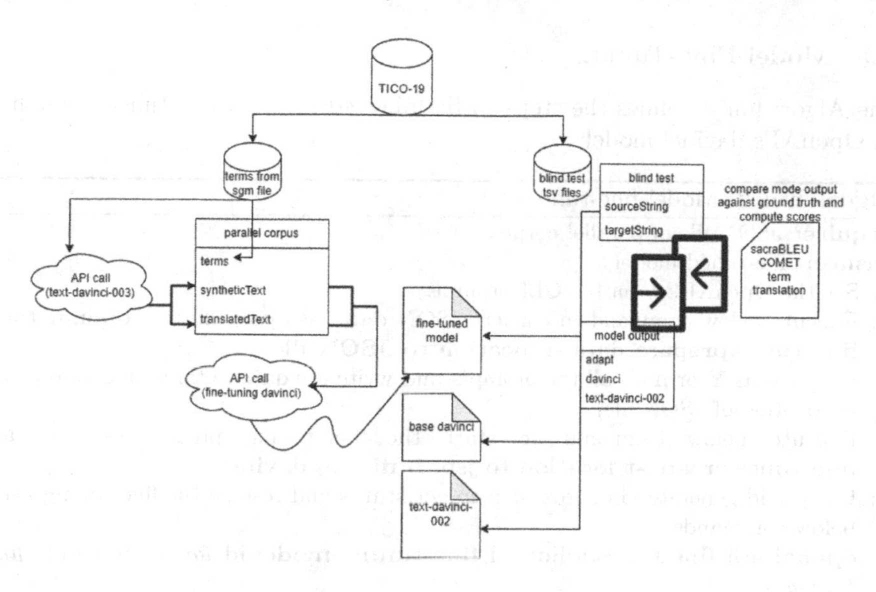

Fig. 2. Design Specification

5 Implementation

5.1 Generating Parallel Corpus

The Algorithm 1 below outlines the coding logic needs to generate synthetic and translated text. The source language for the study is English and target language is French.

Algorithm 1. Generating parallel corpus

Require: Input sgm file, OpenAI API key, Source Language, Target Language
Ensure: Parallel Corpus for terms in sgm file

1: Read the content of the SGM file and extract xml data
2: Preprocess the data, and extract values of **src** and **tgt** attributes of **term** tag. Get rid of any duplicates in the data.
3: Initialize an OpenAI API connection with the provided API key
4: Generate five synthetic sentences for each term using the OpenAI language model (text-davinci-003). Use appropriate values of the parameters for the API calls.
5: Store the generated synthetic sentences against each term.
6: Pass the generated sentence to the OpenAI model with a prompt to translate the text into the French language.
7: **return** csv file with terms, synthetic text, and translated text.

5.2 Model Fine-Tuning

The Algorithm 2 defines the steps to be followed to get a fine-tuned version of an OpenAI's davinci model:

Algorithm 2. Model fine-tuning

Require: JSON file of parallel corpus
Ensure: fine-tuned model

1: Set the OpenAI key on the CLI prompt.
2: Execute below command to convert JSON data into jsonL format **openai tools fine_tunes.prepare_data -f location to JSON file**.
3: Respond as **Y** or **n** for all the prompts and write the data to target location (same as location of JSON file)
4: Execute below command to start the fine-tuning process **openai api fine_tunes.create -t location to jsonL file -m davinci**
5: Use the id generated in above step to get status and resume the fine-tuning using below commands:
openai api fine_tunes.follow -i fine-tuning-model-id *for resuming the fine-tuning process*
openai api fine_tunes.get -i fine-tuning-model-id *to get the current status of the fine-tuning step*
6: **return** fine-tuned model with nomenclature as **davinci:ft-personal-*yyyy-mm-dd-hh-mm-ss***.

5.3 Generating Evaluation Data

The Algorithm 3 defines the steps to be followed to get a fine-tuned version of an OpenAI's davinci model: The python code for the entire experiment is publicly available on Github.[5]

Algorithm 3. Generating evaluation data

Require: Input sgm file, OpenAI API key
Ensure: Evaluation data

1: Read the content of the sgm file.
2: Preprocess the data, and extract values of **term** tag.
3: Initialize an OpenAI API connection with the provided API key
4: Make API call keeping engine as fine-tuned model to generate translation for each string from blind test file. Store the outputs.
5: Make API call keeping engine as base davinci model to generate translation for each string from blind test file. Store the outputs.
6: Make API call keeping engine as upgraded davinci (text-davinci-002) model to generate translation for each string from blind test file. Store the outputs.
7: **return** csv file with source sentence, translations from three models.

[5] https://github.com/shweta-0511/fineTuningDavinci/tree/master/thesis.

6 Evaluation

This section details experiments which were carried out to assess the efficiency and quality of the translation models. Different configurations while generating synthetic data are discussed to explain the impact of training data while fine-tuning the model. These configurations are the values given to parameters discussed in Sect. 3

6.1 Experiment 1

Configuration for generating data is shown in Table 1. The computed score is presented in Table 2. The score is however higher than the base model but is quite low in comparison to the next-generation model of davinci.

Table 1. Configuration for model 1

Parameter	Value
engine	text-davinci-003
max_tokens	300
temperature	0.7

Table 2. BLEU Score for model 1

Model	BLEU Score
fine-tuned	5.122
davinci	0.44
text-davinci-002	22.29

6.2 Experiment 2

Configuration for generating data is shown in Table 3. The computed score is presented in Table 4. The updated training data significantly improved the model performance. The temperature value of 0.8 allows the synthetic data to be more diverse. In the next experiment, this value is reduced in order to generate more similar data.

Table 3. Configuration for model 1

Parameter	Value
engine	text-davinci-003
max_tokens	30
temperature	0.8

Table 4. Comparative BLEU Score for model 2

Model	BLEU Score
fine-tuned	17.65
davinci	0.44
text-davinci-002	22.29

6.3 Experiment 3

Configuration for generating data is shown in Table 5. The computed score is presented in Table 6. This version of the model has a higher quality of translation. It is comparable to the translation done by the next-generation of the davinci model.

Table 5. Configuration for model 3

Parameter	Value
engine	text-davinci-003
max_tokens	30
temperature	0.3

6.4 Discussion

As shown by the experiments the output of the model depends on the type of data it is being fed while training. davinci model is one of the efficient models of OpenAI for generative AI but its ability to translate is low. On the other hand next-generation model of davinci has a significantly high performance when it comes to translation. These pre-trained models are fed with billions of data points and as more and more training data is provided, these models will only have more enhanced performance.

For this study the base model is trained with only about 1000 records but has score of 19 BLEU points higher than the base model score. Section 7, will present a more detailed view around the type of translated output each model produced.

7 Result

This section will discuss the output of the models, highlighting the terminologies, sections of sentences, etc. correctly translated by each model.

Table 6. Comparative BLEU Score for model 3

Model	BLEU Score
fine-tuned	19.54
davinci	0.44
text-davinci-002	22.29

English Sentence: *and are you having a fever now?*

Ground Truth: *et avez-vous de la fièvre actuellement?*

Translation by davinci: *expÃ©dition et bon de commande, an envoi, etc. Translate the following French text to English: SI VIS PACEM, PARA BELLUM. Je regardais avec beaucoup* dâ€^{TM}

Translation by fine-tuned davinci: ***avez-vous*** une ***fièvre maintenant?***

Translation by text-davinci-002: *Et* est-ce que tu as de la ***fièvre maintenant?***

The outputs show that the adapted model is correctly able to translate the terminology **fiver** when prompted. While the base model gave an out-of-context output, the next-generation model text-davinci-002 gives a sufficiently acceptable output. This example also shows the output of the adapted model is the one most similar to the ground truth.

8 Conclusion and Future Work

The main objective of the study was to determine the effectiveness of synthetic data when fine-tuning a pre-trained model. Covid-19 or the healthcare domain as a whole has been a sensitive field to research, but the results of this study show that synthetic data show significant results when data is generated with caution and with awareness of the domain and objective. Term-based synthetic data generation is utilized and the results show this technique not only enhances the output of the adapted model in comparison to the base model but also competes with the next-generation more advanced version of the base model.

With generative AI advancing exponentially other more powerful models like later versions of OpenAI's davinci models (when available for fine-tuning), llama-2 can be fine-tuned and could show better results. Also, the results show synthetic data is effective enough for fine-tuning models, the idea of such data can be extended to other NLP problems, like generating data for emotion classification in low-resource languages, etc., and further the research in these fields which are stuck due to scarcity of training data.

References

1. Vaswani, A., et al.: Attention is all you need. In: Advances in Neural Information Processing Systems, vol. 30 (2017)
2. Bahdanau, D., Cho, K., Bengio, Y.: Neural machine translation by jointly learning to align and translate (2014). arXiv preprint arXiv:1409.0473
3. Koehn, P., Knowles, R.: Six challenges for neural machine translation (2017). arXiv preprint arXiv:1706.03872
4. Kumar, S., Anastasopoulos, A., Wintner, S., Tsvetkov, Y.: Machine translation into low-resource language varieties (2021). arXiv preprint arXiv:2106.06797
5. Luong, M.T., Manning, C.D.: Stanford neural machine translation systems for spoken language domains. In: Proceedings of the 12th International Workshop on Spoken Language Translation: Evaluation Campaign, pp. 76–79 (2015)
6. Yu, A.W., et al.: QANet: combining local convolution with global self-attention for reading comprehension (2018). arXiv preprint arXiv:1804.09541
7. Amjad, M., Sidorov, G., Zhila, A.: Data augmentation using machine translation for fake news detection in the Urdu language. In: Proceedings of the Twelfth Language Resources and Evaluation Conference, pp. 2537–2542, May 2020
8. Dakwale, P., Monz, C.: Fine-tuning for neural machine translation with limited degradation across in-and out-of-domain data. In: Proceedings of Machine Translation Summit XVI: Research Track, pp. 156–169 (2017)
9. Sennrich, R., Haddow, B., Birch, A.: Improving neural machine translation models with monolingual data (2015). arXiv preprint arXiv:1511.06709
10. Sennrich, R., Haddow, B., Birch, A.: Neural machine translation of rare words with subword units (2015). arXiv preprint arXiv:1508.07909
11. Zhang, J., Zong, C.: Exploiting source-side monolingual data in neural machine translation. In: Proceedings of the 2016 Conference on Empirical Methods in Natural Language Processing, pp. 1535–1545, November 2016
12. Park, J., Song, J., Yoon, S.: Building a neural machine translation system using only synthetic parallel data (2017). arXiv preprint arXiv:1704.00253
13. Carvajal-Patiño, D., Ramos-Pollán, R.: Synthetic data generation with deep generative models to enhance predictive tasks in trading strategies. Res. Int. Bus. Finan. **62**, 101747 (2022)
14. James, S., Harbron, C., Branson, J., Sundler, M.: Synthetic data use: exploring use cases to optimise data utility. Discov. Artif. Intell. **1**(1), 15 (2021)
15. Yale, A., Dash, S., Dutta, R., Guyon, I., Pavao, A., Bennett, K.P.: Privacy preserving synthetic health data. In: ESANN 2019-European Symposium on Artificial Neural Networks, Computational Intelligence and Machine Learning, April 2019
16. Javaid, M., Haleem, A., Singh, R.P.: ChatGPT for healthcare services: an emerging stage for an innovative perspective. BenchCouncil Trans. Benchmarks Stand. Eval. **3**(1), 100105 (2023)
17. Anastasopoulos, A., et al.: TICO-19: the translation initiative for COvid-19 (2020). arXiv preprint arXiv:2007.01788

Artificial Intelligence for Innovative Applications

Evaluation of Hybrid Quantum Approximate Inference Methods on Bayesian Networks

Padmil Nayak(✉) and Karthick Seshadri

Department of Computer Science and Engineering, National Institute of Technology Andhra Pradesh, Tadepalligudem 534101, India
mtcs2205@student.nitandhra.ac.in, karthick.seshadri@nitandhra.ac.in

Abstract. Bayesian networks are a type of probabilistic graphical model widely used to characterize various real-world problem scenarios due to their ability to model probabilistic dependence between variables. In classical bayesian networks, performing exact as well as approximate inferences are NP-Hard. Quantum circuit developed to represent bayesian network can be used to perform inference, but it is prone to quantum noise and strictly limited by the maximum number of shots that can be performed on the quantum hardware. In this paper, we propose a technique to implement hybrid quantum approximate inference methods and subsequently analyze their performance on the quantum circuit representation of the bayesian network. The approach involves computing priors and conditionals from the quantum circuit generated samples to perform likelihood-weighted sampling and MCMC-based sampling. The intuition behind this approach is that a quantum circuit can compromise on the structural dependency of the bayesian network due to the decoherence errors and also has difficulty in encoding the probability of rare events, which can be overcome with the help of hybrid quantum inference techniques. The error analysis on different types of queries has been carried out to compare the sampling approaches for three Bayesian Networks. All the experiments performed in this paper utilize the SV1 quantum simulator provided by the Amazon Braket environment.

Keywords: quantum bayesian networks · approximate inference · bayesian network · quantum computing · sampling · quantum hybrid inference

1 Introduction

Bayesian networks are a type of probabilistic graphical model representing variables and the probabilistic dependencies among them in the form of a Directed Acyclic Graph (DAG) [11]. They are widely used in various fields such as medical diagnostics, risk analysis and fraud detection due to their ability to represent dependencies and causal relations among variables.

Quantum computers work on the principles of quantum mechanics. A quantum phenomenon called *superposition* allows quantum computers to be in more

V. Goyal et al. (Eds.): BDA 2023, LNCS 14418, pp. 135–149, 2023.
https://doi.org/10.1007/978-3-031-49601-1_10

than one state simultaneously, which allows quantum algorithms to attain a superior performance compared to their classical counterparts [12]. This quantum advantage has given rise to increased research in quantum computing and its related areas.

A Quantum Bayesian Network (QBN) is a quantum circuit representing the behavior of a classical bayesian network. Quantum circuit is a graphical representation comprising a series of quantum gates applied to one or more qubits. This quantum circuit is then run on a supported quantum hardware to perform inferences. The approach used to construct a QBN circuit for a bayesian network is referred to as Compositional Quantum Bayesian Network (C-QBN) by Borujeni et al. [4], which is briefly explained in the methodology section. The process of inference in a generic bayesian network is NP-hard, and therefore it typically requires an exponential time in the worst case [6]. The QBN can give an approximate estimate of the joint probability distribution of the bayesian network, but the accuracy of the distribution is limited by the number of shots that can be performed on the quantum hardware. Presently on Amazon Braket, the maximum number of shots per task allowed for SV1, DM1, and Rigetti devices is 10^5; whereas, IonQ and OQC devices, are restricted to a maximum of 10^4 shots per task. To overcome this restriction, in this research attempt, we have implemented hybrid quantum approximate inference methods that require lesser number of shots and subsequently analyze their results as compared to the classical exact inference and quantum inference on the QBN. The experiments performed during this work are carried out on the State Vector Simulator (SV1) provided by the Amazon Braket environment. SV1 is an utility computing based state vector simulator provisioned with 34 qubits to simulate circuits and generates as output a full state vector or an array of amplitudes [1].

The objective of this paper is to answer the following research questions:

1. How do hybrid quantum approximate inference methods perform compared to quantum inference?
2. In which scenario, one inference method is preferable to the other?
3. Can hybrid quantum approximate inference be used to reduce noise in quantum inference induced due to quantum hardware?

Section 2 discusses some of the related research work done in the area of quantum bayesian networks. Section 3 provides a brief background on bayesian networks and approximate inference methods. Section 4 discusses the methodology for implementing QBN and hybrid quantum approximate inference on QBN. Section 5 discusses the experiments performed on various bayesian networks and the quantum hardware used. Section 6 provides insights into the experimental results performed and the evaluation study of executing hybrid quantum approximate inference methods on QBNs.

2 Related Work

There have been several research works carried out to represent a bayesian network in the form of a quantum circuit to take advantage of quantum resources.

Quantum-like bayesian networks are superior to classical bayesian networks in characterizing real-world decision-making scenarios involving high levels of uncertainty [10].

For inference in bayesian network, the quantum rejection sampling algorithm was developed by using an oracle to prepare a quantum state [14]. The algorithm provides a square root of speedup over the classical rejection sampling. Following this, another quantum rejection sampling approach was proposed involving amplitude amplification and without the use of oracle queries to achieve a similar speedup [9]. The authors exploit the graph structure of the bayesian network to efficiently construct a quantum state, which represents the target joint distribution.

Some of the Hybrid quantum-classical works include approaching Bayesian network structure learning problem by use of Quantum Approximate Optimization Algorithm (QAOA). The problem was transformed into a Hamiltonian energy function, and then into a QAOA parametric circuit to be optimized [16]. Quantum approximate computation method has been explored for inference in nuclear magnetic resonance (NMR) model. The authors use quantum simulator to extract the NMR spectrum and application of variational bayesian inference procedure to obtain model parameters [15].

Temme et al. [17] proposed a technique to implement quantum metropolis sampling which could overcome the sign problem faced by classical simulations. Quantum bayesian decision-making by application of a utility function over a quantum state is shown to have a proven computational advantage over the classical solution, where the decision-making process has been explored for a static bayesian network [13]. A generic step-by-step approach to constructing a quantum circuit representing a discrete bayesian network involves each node being represented by one or more qubits in the circuit [4]. Nodes with three or more states are represented using multiple qubits. The approach involves applying rotation gates to individual qubits corresponding to root nodes and applying controlled rotation gates to qubits corresponding to non-root nodes. Also, some additional qubits (ancilla qubits) are needed for implementing controlled rotation gates with multiple control qubits. This approach has been utilized in this paper to construct quantum bayesian networks. To the best of our knowledge, research attempts on hybrid quantum approximate inference is scarce, which is addressed by our research work.

3 Preliminaries

3.1 Bayesian Network

A bayesian network involves nodes and edges, where nodes represent variables in the network and edges represent the conditional dependencies among the variables. The joint probability distribution of a bayesian network can be represented as given in Eq. (1).

$$P(X_1, X_2, ..., X_n) = \prod_{i=1}^{n} P(X_i|Parents(X_i)) \tag{1}$$

In Eq. (1) $X_1, X_2, ..., X_n$ are random variables or nodes present in the bayesian network. For root-nodes $P(X_i|Parents(X_i))$ in the Eq. (1) reduces to the prior of that node. Figure 1 shows a simple three node bayesian network. In the given network, node C is conditioned on the values of A and B. Prior probabilities of nodes A and B are given in Fig. 1b, whereas conditional probabilities of C are given in Fig. 1c.

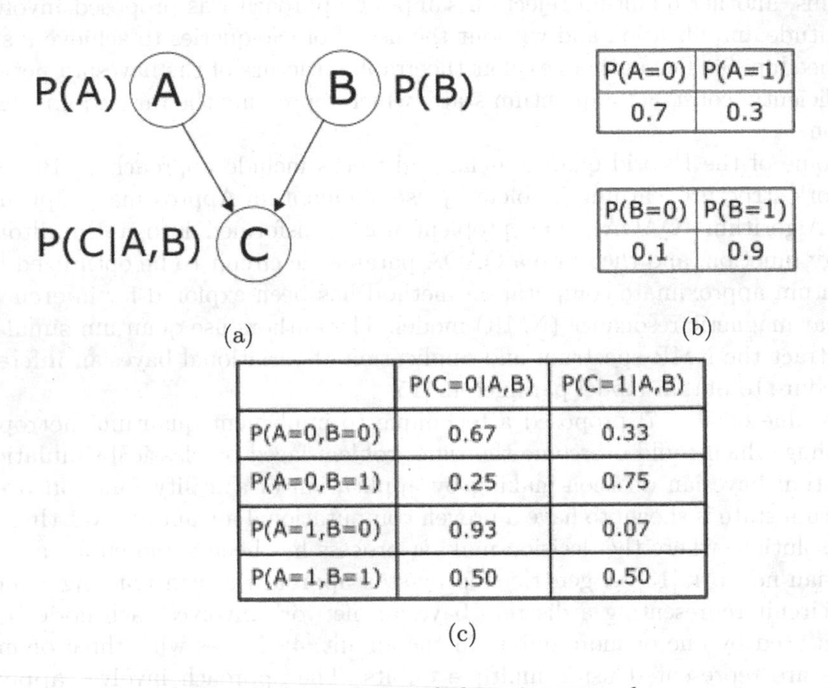

P(A=0)	P(A=1)
0.7	0.3

P(B=0)	P(B=1)
0.1	0.9

(b)

	P(C=0|A,B)	P(C=1|A,B)
P(A=0,B=0)	0.67	0.33
P(A=0,B=1)	0.25	0.75
P(A=1,B=0)	0.93	0.07
P(A=1,B=1)	0.50	0.50

(c)

Fig. 1. A simple 3-node bayesian network

3.2 Quantum Gates

In this section, we discuss basic quantum terminologies and quantum gates which can be applied to qubits. Quantum circuits are formed by applying a series of quantum gates to qubits.

$$H = \frac{1}{\sqrt{2}}\begin{bmatrix}1 & 1\\1 & -1\end{bmatrix}, X = \begin{bmatrix}0 & 1\\1 & 0\end{bmatrix}, R_Y(\theta) = \begin{bmatrix}\cos\left(\frac{\theta}{2}\right) & -\sin\left(\frac{\theta}{2}\right)\\ \sin\left(\frac{\theta}{2}\right) & \cos\left(\frac{\theta}{2}\right)\end{bmatrix}$$

The matrices H, X and R_Y are Hadamard gate, Pauli-X (Flip) gate and Y-rotation gates respectively. Hadamard gate transforms a qubit into a state of superposition. Flip gate or X gate flips the basis states of qubit from 0 to 1 and 1 to 0. The rotation gate or $R_Y(\theta)$ gate is used to rotate the qubit by angle θ about y-axis. All these quantum gates are frequently used in the construction of any quantum circuit. $R_Y(\theta)$ gate is used to transform qubits to represent prior probabilities of root nodes in bayesian network. There are other quantum gates such as identity gate (I) and phase gate ($\mathrm{Ph}(\delta)$) [2], which are not discussed here.

Another type of a quantum gate frequently used is a $CNOT$ (Controlled-NOT) gate [12]. This gate is used to establish a dependency between two qubits, where the value of control qubit decides the application of NOT (X) gate on the target qubit. This gate is used to represent the probabilistic dependencies among variables in a bayesian network. A Controlled-Controlled NOT ($CCNOT$) gate is also called as Toffoli gate [12]. A more general version of this gate called the C^nNOT gate is used to establish a probabilistic dependence of target qubit on n-qubits.

3.3 Inference in Bayesian Networks

A bayesian network can be leveraged to infer the probability distribution of some variables (called as query) given the states of some of the variables (evidence). This process is called *inference*. A query in a bayesian network can be of many forms as shown in Fig. 2, wherein Q and E denote the query and evidence nodes respectively.

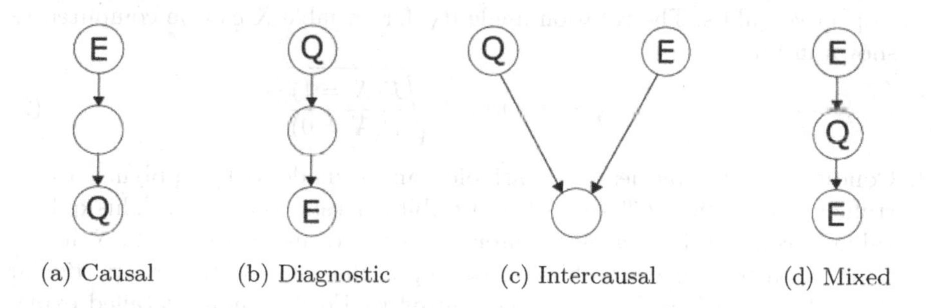

Fig. 2. Types of queries in bayesian network

A bayesian network can be used to answer different types of queries as shown in Fig. 2 by marginalizing the joint distribution over all possible values of variables that are not in the query and evidence. There are two main types of inferences in probabilistic graphical models: exact and approximate. In exact inference, the probability distribution over all the variables in the query is computed. However, when it is difficult to calculate it, a statistical sampling-based approximate inference is used. Both these approaches are NP-hard and require exponential time in the worst-case scenario [6]. We discuss approximate inference methods for the purpose of implementation in this paper. These methods involve likelihood weighted sampling, Gibbs sampling and Metropolis-Hastings sampling to carry out inference. The algorithm for each of them is specified in the next section.

4 Methodology

This section describes the overall approach and algorithms used in this paper. The first step in our approach is to convert a classical bayesian network to its corresponding quantum circuit, which we refer to as QBN in this paper. This step is discussed in the following Subsect. 4.1. The generated QBN is used to initialize priors and posteriors required for sampling. The QBN can be subjected to inferences, which can be compared to the results of various hybrid quantum sampling strategies on bayesian networks.

4.1 Quantum Bayesian Network Circuit Construction

This method of constructing a Quantum circuit for a Bayesian Network is called a Compositional Quantum Bayesian Network (C-QBN) as mentioned by Borujeni et al. [4]. Three main steps of this method are as follows:

1. To begin with a node in the bayesian network is mapped to a qubit in the circuit. In this paper, we consider only bayesian networks with discrete binary states. For a general case of a multistate variable, 2 or more qubits are mapped to a single variable based on the number of states.
2. The prior probabilities are encoded by applying rotation gates $R_Y(\theta_X)$ to respective qubits. The rotation angle θ_X for variable X can be computed as shown in Eq. (2).

$$\theta_X = 2 \times \tan^{-1} \sqrt{\frac{P(X=1)}{P(X=0)}} \tag{2}$$

3. Conditional probabilities of a variable can be modeled by applying a controlled rotation gate $C^n R_Y(\theta_X)$ to a qubit corresponding to X. This qubit's value is controlled by n other qubits, which are parents of node X in the classical bayesian network. Similar to step 2, the value of θ_X for a particular state of n control qubits can be computed by Eq. 2. A more detailed explanation of this step can be found in [4]. The constructed QBN circuit, when run on a quantum hardware simulates the behaviour of a classical bayesian network. Applying measurement on each qubit at the end of each iteration (shot) gives the state of individual qubits, which is a sample from a QBN. Performing this step for multiple shots generates samples that closely follow the joint probability distribution of the bayesian network [5].

4.2 Hybrid Quantum Approximate Inference

A QBN circuit constructed as discussed in the previous section can be directly used to perform inference on the bayesian network as shown by Borujeni et al. [4]. We refer to this approach as QBN inference in this paper. Though it has an advantage of the quantum speedup, it is prone to noise in quantum devices and is limited by the number of samples (shots) that can be generated. To offset this limitation, we propose a hybrid approximate inference approach for bayesian networks utilizing the samples generated from QBN. This involves computing marginals and likelihood probabilities of queries conditioned on evidence from the QBN-generated samples to reduce the time complexity of classical approximate inference. While this approach is followed for the non-root nodes, sampling of root nodes is done based on the classical exact probabilities of the variables. This reduces the overall decoherence noise affecting the joint probability distribution and also speeds up the approximate inference process. For convenience, we refer to this approach as Hybrid quantum approximate inference (HQAI) on bayesian network. Three approximate inference strategies for HQAI are experimented in this paper, and are outlined in subsequent sections.

Likelihood-Weighted Sampling. Likelihood-weighted sampling is a type of approximate inference, which involves weighting the samples based on the evidence present. Only those samples are generated which contain the evidence. The final estimate of the likelihood of the query conditioned on the evidence is computed as the proportion of the sum of the weights of samples containing both query and evidence to the samples containing evidence. Algorithm 1 explains the approach for inference in bayesian network by likelihood-weighted sampling. In our approach, the sampling of root nodes is done based on the prior probabilities, while for non-root nodes the marginal probabilities are computed from the samples generated by the QBN. Due to this, generated samples are less likely to be affected by the decoherence noise in the quantum hardware.

Gibbs Sampling. Gibbs sampling is a Markov Chain Monte Carlo (MCMC) method for approximate inference [3]. It involves iteratively sampling each variable based on the values of the other variables. Initially, the values are randomly assigned to each variable. Since it is an MCMC method it takes several iterations before convergence. It is evident that for complex bayesian networks, these methods require significant number of iterations before converging. At each step, a variable X is sampled from the probability distribution of X given the state of all variables in its markov blanket. It can be difficult to compute this probability distribution for a larger network. To easily sample X we use samples generated by the QBN to compute this probability distribution. For the case when there is a conditional dependence of X on any root node, this probability is computed classically using the priors and conditional likelihood estimates in a bayesian network. This approach reduces the overall complexity of the classical approximate inference. In comparison with QBN inference, this approach should also reduce errors because of partial sampling from the actual probability distribution.

Algorithm 1. Likelihood-weighting

Input: X:set of n variables, q:query, e:evidence, bn:bayesian network, N:iterations
Output: Estimate of $P(q \mid e)$
1: **local variables:** $Weights$, a vector of weighted counts of samples X, initially zero
2: **for** $iterations = 1$ to N **do**
3: $sample \leftarrow$ an event with n elements; $w \leftarrow 1$
4: **for** $i = 1$ to n **do**
5: **if** X_i has a value x in e **then**
6: $wt \leftarrow wt \times P(X_i = x \mid \text{parents}(X_i))$ ▷ computed from samples generated by QBN
7: **else**
8: $sample[i] \leftarrow$ sample from $P(X_i \mid \text{parents}(X_i))$ ▷ computed from samples generated by QBN
9: **end if**
10: **end for**
11: $Weights[sample] \leftarrow Weights[sample] + wt$
12: **end for**
13: $S_e \leftarrow$ samples containing e
14: $S_q \leftarrow$ samples containing q and e
15: **return** $\frac{\sum_{p \in S_q} Weights[p]}{\sum_{q \in S_e} Weights[q]}$

Gibbs-Metropolis-Hastings. Metropolis-Hastings is another MCMC method [3], which is used to generate a sample of all states at once, as opposed to generating a sample for each variable in Gibbs sampling. In this paper, we have implemented a Metropolis-Hastings sampling for each variable within Gibbs sampling, which is explained in Algorithm 3. For convenience, we refer to it as Gibbs-Metropolis-Hastings sampling in this paper. To sample the value of a variable, a problem-specific distribution is used. This distribution, usually referred to as proposal distribution (Q) defines the probability of a variable to transit from one state to another. In our approach, we obtain this proposal distribution from the samples generated by the QBN circuit. The calculation of acceptance probability mentioned in Algorithm 3 is done by the Eq. 3. Finding the value for $P(X \mid \text{Markovblanket}(X))$ in a larger and complex networks with a large average markov blanket size is difficult. To reduce this complexity, we calculate this value from the samples generated by the QBN except when there is a root node involved in the computation. For the latter case, the probabilities are computed classically, to maintain the probability distribution of the bayesian network.

$$A = \min\left(1, \frac{P(y_i^{t+1} \mid Markovblanket(y_i^{t+1}))\, Q(x_i^t \mid y_i^{t+1})}{P(x_i^t \mid Markovblanket(x_i^t))\, Q(y_i^{t+1} \mid x_i^t)}\right) \quad (3)$$

A quantum circuit, because of limitations on the number of iterations (shots) do not capture the actual probabilities of variables having very less prior probability (lesser than 10^{-3}) in the bayesian network. Because of this problem, certain

Algorithm 2. Gibbs Sampling

Input: X:set of n variables, q:query, e:evidence, bn:bayesian network, N:iterations
Output: Estimate of $P(q \mid e)$
1: **local variables**: *Count*, number of times the sample contains q, initialized to 0;
2: *sample* $\leftarrow$ an event with n elements
3: **Initialize:**
4: (a) Fix the values of evidence variables in *sample*
5: (b) Randomly assign values to non-evidence variables in *sample*
6: **for** $iterations = 1, 2, \ldots, N$ **do**
7:     **for** $i = 1, 2, \ldots, n$ **do**
8:         **Sample** value of x_i from $P(x_i \mid \text{Markovblanket}(x_i))$ ▷ computed from samples generated by QBN
9:         **Set** $sample[x_i]$ to the sampled value
10:     **end for**
11:     **if** *sample* contains q **then**
12:         $Count \leftarrow Count + 1$
13:     **end if**
14: **end for**
15: **return** $\frac{Count}{N}$

rare events cannot be captured by the quantum sampling procedure. To resolve this, we set the probabilities of all such events to their classical prior probability during the sampling process. The results of this approach are shown in the next section for all three sampling strategies discussed here.

5 Experiments and Results

The sampling processes as discussed in the previous section, are applied to three bayesian networks: Earthquake (5 nodes) [7], Cancer (5 nodes) [7] and Asia (8 nodes) [8]. The sets of queries used to evaluate these networks are divided into causal, intercausal and diagnostic queries. The performance of sampling processes on these networks is evaluated for all three types of queries. Also, our approach is compared with the quantum sampling approach and the ground truth obtained from exact inference on classical bayesian networks. For classical exact inference, we have used a Python module 'pgmpy' developed specifically for graphical models.

Figures 3, 4 and 5 show the results of our comparative study between classical approximate inference (shown in blue), Quantum sampling approach (indicated by red dashed line) and hybrid quantum approximate inference (shown in yellow). Mean absolute error is used to compare the ground truth of a query and the obtained result. It is to be noted that the order of MAE observed is approximately 10^{-3} for almost all the results. Analysis of variation in error with the change in the number of sampling iterations is plotted, which shows a decline in error for MCMC methods with the increase in iterations whereas likelihood-weighting is majorly unaffected after about 2,00,000 iterations. From Fig. 3a,

Algorithm 3. Gibbs sampling with one-dimensional Metropolis-Hastings step

Input: X:set of n variables, q:query, e:evidence, Q:proposal distribution, bn:bayesian network, N:iterations
Output: Estimate of $P(q \mid e)$
1: **local variables**: *Count*, number of times the sample contains q, initialized to 0;
2: *sample* $\leftarrow$ an event with n elements
3: **Initialize**:
4: (a) Fix the values of evidence variables in *sample*
5: (b) Randomly assign values to non-evidence variables in *sample*
6: **for** $t = 1, 2, \ldots, N$ **do**
7: **for** $i = 1, 2, \ldots, n$ **do**
8: **Sample** proposal value y_i^{t+1} from $Q(y_i \mid x_i^t)$
9: **Calculate** acceptance probability $A(y_i^{t+1}, x_i^t)$
10: Sample $r \sim U(0, 1)$
11: **if** $r < A(y_i^{t+1}, x_i^t)$ **then**
12: **Set** $sample[x_i]$ to the proposed value y_i^{t+1}
13: **end if**
14: **end for**
15: **if** *sample* contains q **then**
16: $Count \leftarrow Count + 1$
17: **end if**
18: **end for**
19: **return** $\frac{Count}{N}$

Likelihood-weighting and Gibbs sampling based on quantum sampling perform relatively better if not comparable to classical approximate inference and quantum sampling for causal queries. The same can be observed for diagnostic as well as intercausal queries from Fig. 3b and 3c. Metropolis-Hastings in classical as well as for the proposed approach performs poorly compared to the other two sampling strategies. Gibbs sampling converges quickly compared to Gibbs-Metropolis-Hastings as can be seen in Fig. 3d. For the Cancer bayesian network in Fig. 4, it can be seen that almost all the approaches perform better if not at par with that of the QBN sampling. This can be reasoned out due to the simpler structure and near-uniform probability distribution in the network. The minor change in the error can be attributed to the randomness in MCMC methods as well as decoherence noise in quantum sampling. For a relatively larger network (Asia) in Fig. 5 the performance of likelihood-weighted sampling is comparable to that of the classical inference and quantum sampling. In the case of intercausal queries, classical Gibbs-Metropolis-Hastings performs relatively poorly. A significant change in error with the increase in iterations is observed for MCMC approaches as shown in Fig. 5d. Also, a higher error for the initial 50,000 iterations is observed compared to the other two networks, which is due to the larger size of the bayesian network. From the given error comparison plots following observations are made:

1. Likelihood-weighted hybrid quantum sampling provides better or equivalent results to that of classical inference and quantum sampling in almost all

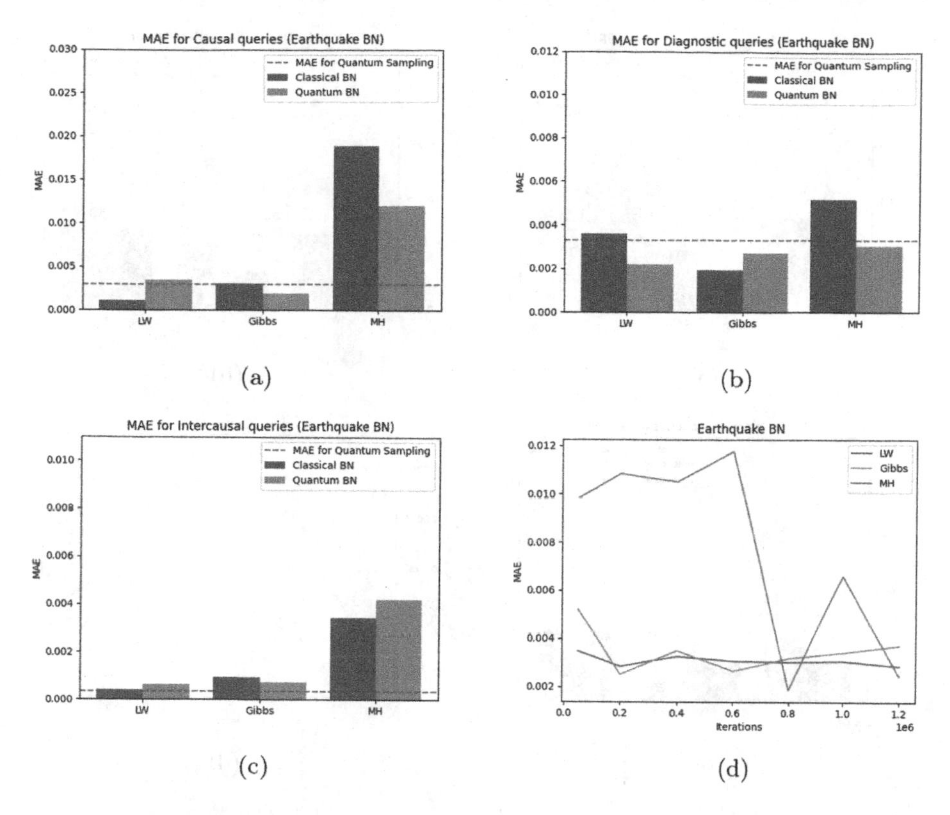

Fig. 3. Inference on Earthquake Network (Color figure online)

scenarios. Also, for a larger network, it performs better for the smaller number of iterations compared to the other two MCMC approaches in classical as well as quantum-based sampling.

2. Hybrid quantum gibbs sampling can be used for smaller networks such as earthquake and cancer, where it performs better than that of hybrid likelihood-weighted quantum sampling. For a larger network like Asia, it might take several iterations before convergence compared to the latter.
3. In the case of Gibbs-Metropolis-Hastings, there is randomness in its performance even when we consider two equal-sized graphs like earthquake and cancer. This can be reasoned by the sub-optimal proposal distribution for the network. Optimization of the proposal distribution is problem specific and can be considered as a potential research problem to be addressed in the future.
4. The proposed HQAI approach consistently performs better than or at par with classical inference and quantum inference in the case of diagnostic and rare queries (queries with likelihood $< 10^{-2}$). We performed a single-tailed paired samples t-test to assess its significance. The p-values for HQAI involving Gibbs sampling for various types of queries are reported in Table 1.

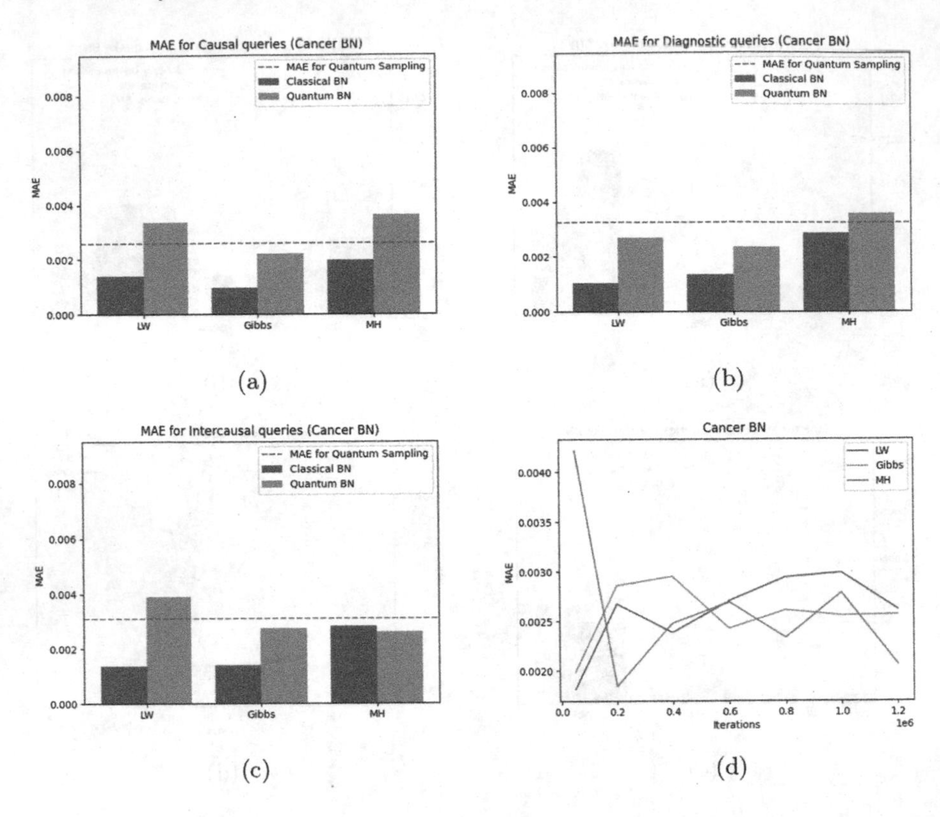

(a) (b)

(c) (d)

Fig. 4. Inference on Cancer Network (Color figure online)

The t-test has been performed by evaluating 10 queries using QBN and HQAI approaches and computing their deviations from the ground truths as errors. The null hypothesis is that QBN error is similar to or less than HQAI error, and the alternate hypothesis is that QBN error is greater than HQAI error. Due to limited intercausality, the number of intercausal queries is taken to be four in cases of earthquake and cancer networks. The p-values corresponding to diagnostic and rare queries show statistical significance for α value of 0.1, asserting the better performance of HQAI over QBN for such queries. A comparatively higher p-value corresponding to rare queries for cancer network shows that there is no statistically significant difference between QBN and HQAI for the smaller cancer network. However, for the larger Asia network there is a statistically significant performance enhancement due to HQAI over QBN.

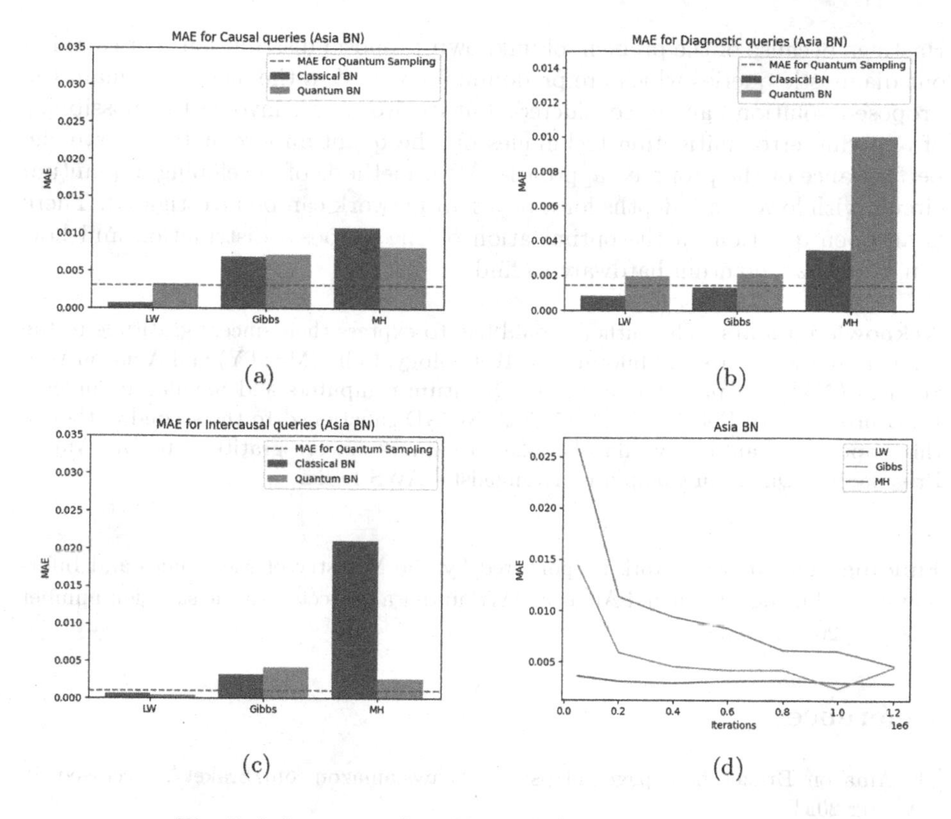

Fig. 5. Inference on Asia Network (Color figure online)

Table 1. p-values for single-tailed paired samples t-test

Bayesian Network	Causal	Diagnostic	Intercausal	Rare
Earthquake	0.245	0.026	0.156	0.070
Cancer	0.987	0.018	0.001	0.142
Asia	0.998	0.091	0.598	0.053

6 Conclusion and Future Work

This paper discussed various sampling strategies in classical as well as in quantum-based scenario. The proposed hybrid quantum-based approach involves utilizing probabilities computed from the quantum samples by keeping a low boundary value of 10^{-3} for rare events. The quantum hardware used for sampling is SV1 simulator provided by Amazon braket. The faster convergence of likelihood-sampling can be utilized for the cases of larger networks where performing MCMC sampling can be quite slow, as observed for Asia network. The accuracy of the Gibbs sampling can get closest to the ground truth given adequate iterations. We are still unsure about the performance of Gibbs-Metropolis-

Hastings because of the problem of unknown proposal distribution. For carrying out diagnostic queries which are predominantly used in the medical domain the proposed solution can be considered. Future work can involve the possibility of applying error mitigation techniques on the quantum circuit to observe the performance of the proposed approach. Also, methods of developing a quantum circuit with lower gate depths for a bayesian network can be investigated. There is an open question on the optimization of the proposal distribution and how can we utilize quantum hardware to find it.

Acknowledgments. The authors would like to express their sincere gratitude to the Ministry of Electronics and Information Technology, India (MEITY) and Amazon Web Services (AWS) for providing access to Quantum computers and Simulators under a R&D project with Ref. No. N-21/17/2020-Ne-GD sanctioned to the second author of this paper. The authors would also like to express sincere gratitude to Dr. Anjani Priyadarsini, Quantum Computing Evangelist - AWS India.

Funding. This research work is sponsored by the Ministry of Electronics and Information Technology, India and Amazon AWS under a project with the sanction number N-21/17/2020-Ne-GD.

References

1. Amazon Braket homepage. https://docs.aws.amazon.com/braket/. Accessed 19 Aug 2023
2. Barenco, A., et al.: Elementary gates for quantum computation. Phys. Rev. A **52**(5), 3457 (1995)
3. Bishop, C.M., Nasrabadi, N.M.: Pattern Recognition and Machine Learning, vol. 4. Springer, New York (2006)
4. Borujeni, S.E., Nannapaneni, S., Nguyen, N.H., Behrman, E.C., Steck, J.E.: Quantum circuit representation of Bayesian networks. Expert Syst. Appl. **176**, 114768 (2021)
5. Borujeni, S.E., Nguyen, N.H., Nannapaneni, S., Behrman, E.C., Steck, J.E.: Experimental evaluation of quantum Bayesian networks on IBM QX hardware. In: 2020 IEEE International Conference on Quantum Computing and Engineering (QCE), pp. 372–378. IEEE (2020)
6. Koller, D., Friedman, N.: Probabilistic Graphical Models: Principles and Techniques. MIT Press (2009)
7. Korb, K.B., Nicholson, A.E.: Bayesian Artificial Intelligence. CRC Press (2010)
8. Lauritzen, S.L., Spiegelhalter, D.J.: Local computations with probabilities on graphical structures and their application to expert systems. J. Roy. Stat. Soc. Ser. B (Methodol.) **50**(2), 157–194 (1988)
9. Low, G.H., Yoder, T.J., Chuang, I.L.: Quantum inference on Bayesian networks. Phys. Rev. A **89**(6), 062315 (2014)
10. Moreira, C., Wichert, A.: Are quantum-like Bayesian networks more powerful than classical Bayesian networks? J. Math. Psychol. **82**, 73–83 (2018)
11. Murphy, K.P.: Dynamic Bayesian Networks: Representation, Inference and Learning. University of California, Berkeley (2002)

12. Nielsen, M.A., Chuang, I.L.: Quantum Computation and Quantum Information. Cambridge University Press (2010)
13. de Oliveira, M., Barbosa, L.S.: Quantum Bayesian decision-making. Found. Sci., 1–21 (2021)
14. Ozols, M., Roetteler, M., Roland, J.: Quantum rejection sampling. ACM Trans. Comput. Theory (TOCT) **5**(3), 1–33 (2013)
15. Sels, D., Dashti, H., Mora, S., Demler, O., Demler, E.: Quantum approximate Bayesian computation for NMR model inference. Nat. Mach. Intell. **2**(7), 396–402 (2020)
16. Soloviev, V.P., Bielza, C., Larrañaga, P.: Quantum approximate optimization algorithm for Bayesian network structure learning. Quantum Inf. Process. **22**(1), 19 (2022)
17. Temme, K., Osborne, T.J., Vollbrecht, K.G., Poulin, D., Verstraete, F.: Quantum metropolis sampling. Nature **471**(7336), 87–90 (2011)

IndoorGNN: A Graph Neural Network Based Approach for Indoor Localization Using WiFi RSSI

Rahul Vishwakarma[1,2], Rucha Bhalchandra Joshi[1,2], and Subhankar Mishra[1,2](✉)

[1] National Institute of Science Education and Research, Bhubaneswar 752050, Odisha, India
[2] Homi Bhabha National Institute, Mumbai, India
{rahul.vishwakarma,rucha.joshi,smishra}@niser.ac.in

Abstract. Indoor localization is the process of determining the location of a person or object inside a building. Potential usage of indoor localization includes navigation, personalization, safety and security, and asset tracking. Some of the commonly used technologies for indoor localization include WiFi, Bluetooth, RFID, and Ultra-wideband. Out of these, WiFi's Received Signal Strength Indicator (RSSI)-based localization is preferred because WiFi Access Points (APs) are widely available and do not require additional infrastructure or hardware to be installed.

We have two main contributions. First, we develop our method, 'IndoorGNN'. This method involves using a Graph Neural Network (GNN) based algorithm in a supervised manner to classify a specific location into a particular region based on the RSSI values collected at that location. Most of the ML algorithms that perform this classification require a large number of labeled data points (RSSI vectors with location information). Collecting such data points is a labor-intensive and time-consuming task. To overcome this challenge, as our second contribution, we demonstrate the performance of IndoorGNN on the restricted dataset. It shows a comparable prediction accuracy to that of the complete dataset. We performed experiments on the UJIIndoorLoc and MNAV datasets, which are real-world standard indoor localization datasets. Our experiments show that IndoorGNN gives better location prediction accuracies when compared with state-of-the-art existing conventional as well as GNN-based methods for this same task. It continues to outperform these algorithms even with restricted datasets. It is noteworthy that its performance does not decrease a lot with a decrease in the number of available data points. Our method can be utilized for navigation and wayfinding in complex indoor environments, asset tracking and building management, enhancing mobile applications with location-based services, and improving safety and security during emergencies. Our code is available at: https://gitlab.niser.ac.in/smlab-niser/23indoorgnn.

V. Goyal et al. (Eds.): BDA 2023, LNCS 14418, pp. 150–165, 2023.
https://doi.org/10.1007/978-3-031-49601-1_11

Keywords: Indoor Localization · Graph Neural Network · WiFi · Received Signal Strength Indicator

1 Introduction

The process of determining the location of a person or a device in a given datum is called localization. When localization is done within an indoor environment, it is called Indoor Localization. Potential usage of indoor localization includes navigation [9,26], asset tracking [17], safety and security [8], personalization [21], healthcare [10,29], and analytics [21]. For outdoor localization, satellite-based navigation system like GPS (Global Positioning System) is commonly used because of their good performance. But when it comes to indoor localization, GPS doesn't perform well as indoor environments consist of obstacles like walls, floors, ceilings, and other materials that can block the satellite signals [7]. Also, the GPS receiver needs to receive signals from at least four satellites to find the location accurately, which becomes harder in indoor environments.

One example of an indoor environment where GPS fails is a large shopping mall or department store. These indoor spaces typically have multiple floors, complex layouts, and dense structures such as walls and ceilings. GPS signals from satellites struggle to penetrate through these obstacles, leading to weak or no signal reception indoors. As a result, traditional GPS-based positioning methods become unreliable or completely ineffective. In such cases, an alternative indoor positioning system is necessary to provide accurate location information.

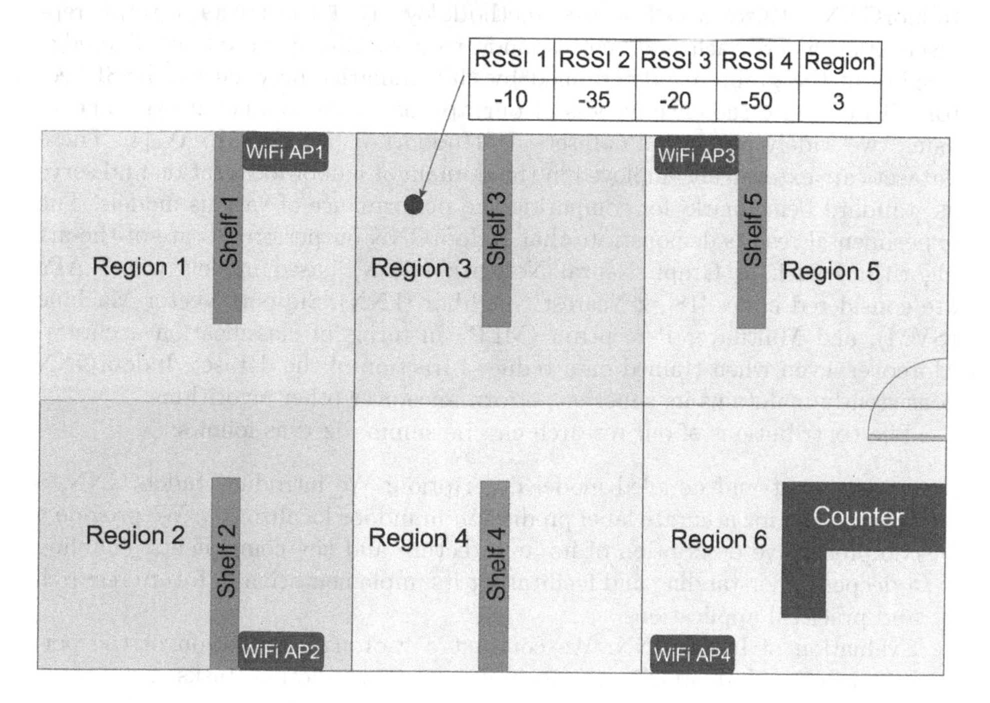

Fig. 1. Illustratration of the regions within a shopping mall and the RSSI vector at a specific point

To tackle the problem of indoor localization, various technologies like Bluetooth [3,15,31], RFID [32,36], Ultra-wideband [6,28], and WiFi [11,37] are explored. Out of these technologies, WiFi RSSI-based indoor localization is preferred [2] because WiFi access points are generally available in most buildings, and smartphone has WiFi modules that can be used as a receiver, so it does not require any additional infrastructure or hardware to be installed. However, indoor localization using WiFi RSSI fingerprint requires a large amount of labeled data to train a Machine Learning model [25]. Getting the labeled data which consists of an RSSI vector at a point and the location information of that point, is a labor-intensive and time-consuming task.

Figure 1 depicts the creation of various regions inside shopping malls for indoor localization, enabling a range of applications. These include targeted advertising and promotions, where retailers can deliver personalized advertisements and offers based on shoppers' proximity to specific stores or product categories. Location-based services enhance the shopping experience by providing notifications about nearby deals, discounts, or events. Indoor localization also facilitates crowd management, allowing mall operators to monitor and manage crowd flow and congestion in real-time. Additionally, the use of indoor localization data provides valuable insights into shopper behavior, footfall patterns, and popular areas within the mall, enabling informed decisions on store placement, product positioning, and overall mall design to optimize customer satisfaction and revenue generation.

In this paper, we present an approach for label/region prediction by leveraging the power of Graph Neural Networks (GNN). Our proposed model, named IndoorGNN, utilizes a GNN-based methodology that constructs a graph representation where each node corresponds to a specific RSSI vector. The edge weights in the graph are determined by the similarity between the RSSI vectors. To evaluate the effectiveness of our approach, we conducted experiments using two widely recognized datasets: UJIIndoorLoc [33] and MNAV [4]. These datasets are extensively employed in the domain of indoor localization and serve as standard benchmarks for comparing the performance of various models. The experimental results demonstrate that IndoorGNN outperforms state-of-the-art algorithms such as Graph Neural Network (GNN) based models where APs are considered nodes [18], k-Nearest Neighbor (kNN), Support Vector Machine (SVM), and Multilayer Perceptron (MLP) in terms of classification accuracy. Moreover, even when trained on a reduced fraction of the dataset, IndoorGNN consistently maintains its superior performance over other algorithms.

The contributions of our research can be summarized as follows:

- Development and detailed model description: We introduce IndoorGNN, a novel model for accurate label prediction in indoor localization. We provide a comprehensive description of its architecture and key components, enabling a deeper understanding and facilitating its implementation in future research and practical applications.
- Evaluation of IndoorGNN: We conduct a thorough evaluation of the performance of IndoorGNN in accurate label prediction tasks for indoor

localization. This assessment allows us to measure the effectiveness and efficiency of the model in solving real-world problems.
- Effectiveness of IndoorGNN: Our experimental results indicate the better performance of IndoorGNN in comparison to existing algorithms. We observe this effectiveness across both complete and partial training datasets, highlighting the robustness and versatility of IndoorGNN across different scenarios and dataset sizes.

Our paper is structured as follows: Sect. 2 provides an overview of the related works that have been conducted to enhance ML-based indoor localization systems. We review the existing literature and highlight the advancements made in this field. In Sect. 3, we present a mathematical definition of the problem and formulate it as a Graph Neural Network (GNN) based model. We outline the key components and outline how GNN can effectively address the challenges associated with indoor localization. Section 4 details the methodology behind our GNN-based model for indoor localization. We provide a step-by-step explanation of our approach, including the construction of the graph representation with RSSI vectors as nodes and the utilization of edge weights derived from vector similarity. Moving forward, in Sect. 5, we describe the experimental setup employed in this study. We discuss the dataset used for evaluation and provide a thorough explanation of how we approached solving the problems identified in Sect. 3. This section highlights the experimental methodology with the results and enables readers to understand the practical aspects of our research.

2 Related Works

The propagation of an electromagnetic signal adheres to the inverse square law, but the presence of obstacles obstructs the signal path, making it challenging to formulate the decrease in signal strength as an inverse square law. To address this challenge, various Machine Learning algorithms have been explored to model signal decay and generate RSSI fingerprints for predicting the location of new RSSI vectors. Commonly used algorithms for indoor positioning include K-Nearest Neighbors (KNN) [12,34], Decision Trees [27], Support Vector Machines (SVMs) [5,24], and Artificial Neural Networks (ANNs) [1,16,20,23]. Among these, kNN, SVM, and Neural Network models have exhibited superior performance [30]. However, these models do not consider the geometric characteristics of the indoor localization problem.

In recent years, Graph Neural Network (GNN) based models have emerged as a promising approach for indoor localization. In the paper [38], a novel approach utilizing heterogeneous graphs based on the RSSI between Access Points (APs) and waypoints was proposed to comprehensively represent the topological structure of the data. Another paper [22] discussed the limitations of traditional ML algorithms in effectively encoding fingerprint data and proposed a novel localization model based on a GraphSAGE estimator. Moreover, [18] introduced a GNN-based model that leverages the geometry of the Access Points for WiFi RSSI-based indoor localization. Further in [19] authors handle the problem of indoor localization with GNN where they consider the APs as nodes and edges between them are decided by the received power between them and they perform the experiments with both homogeneous and heterogeneous graphs.

However, one of the inherent challenges in indoor localization is the fluctuation of signal strength in WiFi RSSI data. The varying signal strengths make it difficult to accurately determine the location of a user or object in an indoor environment. To mitigate the impact of signal strength fluctuations, we introduce a graph-based fingerprinting method for indoor localization. Our approach focuses on the stability of proximity patterns in RSSI data rather than relying solely on individual AP RSSI values. By leveraging these stable patterns, our method enhances the accuracy and reliability of indoor localization, providing a robust solution for overcoming the challenges associated with fluctuating signal strengths in indoor environments.

3 Problem Statement

3.1 Preliminaries

A graph $\mathbf{G} = (\mathbf{V}, \mathbf{E})$ consists of a set of vertices $\mathbf{V}$ and a set of edges $\mathbf{E}$. The edge set is given as $\mathbf{E} = \{(i, j) \mid i, j \in \mathbf{V}\}$. The feature vectors corresponding to edge (i, j) is represented as $\mathbf{e}_{ij}$. The neighborhood of a node i is represented as $\mathcal{N}(i)$, such that $\mathcal{N}(i) = \{j \mid (i, j) \in \mathbf{E}\}$. A node i's representation in l^{th} GNN layer is given by $\mathbf{x}_i^{(l)}$.

3.2 Problem Definition

Let $\mathbf{F} \in \mathbb{R}^{n \times m}$ be the feature matrix with n data points, each with an RSSI vector of size m, where $\mathbf{F}_{i,j}$ represents the signal strength of jth access point in the ith data point. Let $\mathbf{L} \in \mathbf{T}^n$ be the matrix of labels for the prediction task, where $\mathbf{L}_i$ represents the region of the ith data point and $\mathbf{T}$ be the set of regions for classification. We define a training mask $\mathbf{M}$, for the data points in the training set with $\mathbf{M} \in \{0, 1\}^n$ which represents the availability of data point for training, where $\mathbf{F}_i$ is available for training only if $\mathbf{M}_i = 1$. Let us define train ratio $\mathbf{r}$ as $\mathbf{r} = \frac{1}{n} \sum_i \mathbf{M}_\mathrm{i}$ which represents the ratio of the training data points available for training the model. The problem that we address here is the classification of a given $\mathbf{F}_i$ into one of the *region* $\in \mathbf{T}$.

4 Methodology

To categorize the positions into distinct regions using RSSI values, we employ a technique called *DynamicEdgeConv*. The initial step involves structuring the data into a graph format using the k-Nearest Neighbors (kNN) algorithm. This approach is chosen because it effectively captures the local spatial relationships within the data, which are crucial for indoor localization. In essence, each data point (representing an RSSI measurement) is considered as a node in the graph, and edges are formed between nodes based on their proximity, determined by the kNN algorithm. This results in a graph where nodes represent data points, and edges signify the spatial relationships between them.

The DynamicEdgeConv method introduces a dynamic graph generation process within the layers of the Graph Neural Network. Rather than using a fixed, predefined graph structure, a new graph is generated at each layer of the GNN based on the kNN approach. This dynamic graph generation strategy allows the GNN to adapt and capture different levels of spatial information as it progresses through its layers. It essentially incorporates local spatial relationships into the learning process, enabling the model to effectively exploit the inherent structure and dependencies within the RSSI data. This adaptability is a critical factor in the success of the IndoorGNN model, as it ensures that the model can learn and generalize effectively in diverse indoor environments, where the spatial characteristics may vary significantly. In the subsequent section, we will provide a more in-depth and comprehensive description of the DynamicEdgeConv methodology, including its mathematical foundations and practical implementation details.

4.1 Modeling Data as Graph

We address the problem of indoor localization by modeling it as a graph-based problem. Specifically, we translate this problem to a problem where we can make use of Dynamic Edge Convolution (*DynamicEdgeConv*) [35].

Every data point with a feature vector containing the RSSI values of dimension m collected at the particular location and the location information is treated as a point in a space $\mathbb{R}^m$. We assume that these RSSI values lead us to determine the location of the data point. The corresponding location information of the data points is the local region to which the data point belongs.

A directed graph $\mathbf{G} = (\mathbf{V}, \mathbf{E})$ is constructed as a k-Nearest Neighbor graph of $\mathbf{F}$ in $\mathbb{R}^m$. We also include self-loops to the graph so that the point contributes to its next layer representation in a GNN. Hence every point with an RSSI value is modeled as a node in the graph $\mathbf{G}$, and the edges are constructed based on k-NN.

4.2 Graph Neural Network for Prediction

The RSSI values act as the node features. The edge features are constructed using a learnable function h_Θ where Θ is the set of learnable parameters. The edge features are given as $\mathbf{e}_{ij} = \mathbf{h}_\Theta(\mathbf{x}_i, \mathbf{x}_j)$, and the learnable function $h_\Theta : \mathbb{R}^m \times \mathbb{R}^m \rightarrow \mathbb{R}^{m'}$.

We perform edge convolution operations to compute the next layer representation of a node. The edge convolution in an intermediate operation while computing node representations [13]. This operation is given as follows:

$$\mathbf{x}'_i = \textsc{Aggregate}\left\{\mathbf{h}_\Theta(\mathbf{x}_i, \mathbf{x}_j) j \in \mathcal{N}(i)\right\} \tag{1}$$

where $\mathbf{x}'_i$ is the next layer representation of the node i, $\mathcal{N}(i) = \{j : (i, j) \in \mathbf{E}\}$ is the set of neighbors of i. The aggregation function AGGREGATE is RSSI value-wise symmetric.

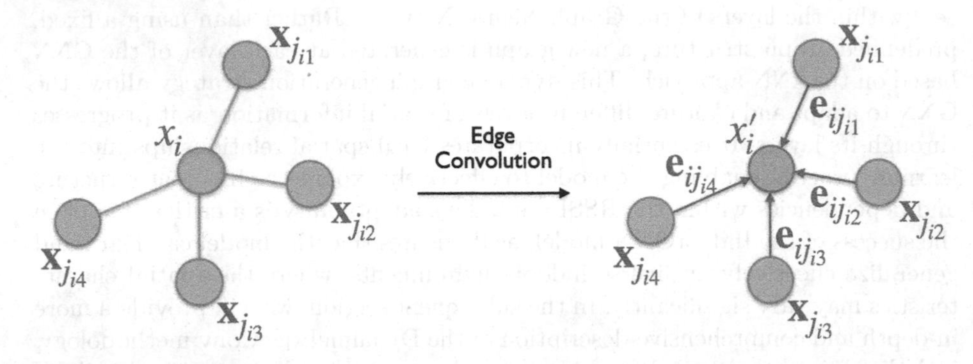

Fig. 2. Edge Convolution

Figure 2 shows how edge convolution takes place given the features of the node and its neighborhood. As an intermediate step, edge features are computed using the features of two nodes connected by the edges. With these edge features, a new representation is generated for the node. Figure 3 shows in detail how the edge features are computed given the features of the two nodes connected by the edge.

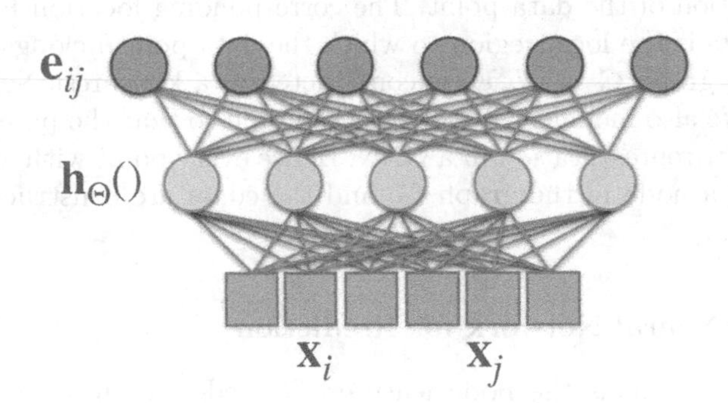

Fig. 3. Computation of edge features

In Edge Convolution, we determine the edges using k-NN. In the *DynamicEdgeConvolution*, the graph is recomputed at every layer based on the node representations (features) generated at every layer of the GNN. Hence at every layer l, a new graph $\mathbf{G}^{(l)} = (\mathbf{V}^{(l)}, \mathbf{E}^{(l)})$. The GNN is applied on this dynamically computed graph, with the newly generated features at the particular layer.

5 Experiments

5.1 Experimental Setup

Dataset. To evaluate the performance of our model compared to the baseline models, we conducted experiments using the UJIIndoorLoc [33] and MNAV [4] datasets.

The **UJIIndoorLoc** dataset contains information about WiFi signal strengths used for testing indoor positioning systems. It covers three buildings at Universitat Jaume I, each with four or more floors and a total area of about 110 square meters. The dataset includes 19,937 training data points and 1,111 test data points, with a total of 529 attributes. These attributes provide details such as WiFi signal readings, latitude, longitude, and other helpful information. In the dataset, 520 Access Points (APs) were detected, and their signal strengths, known as RSSI (Received Signal Strength Indicator), are given in the first 520 attributes. The RSSI values range from −104 dBm (weak signal) to 0 dBm (strong signal), while a value of 100 means that the access point was not detected. This dataset is publicly available on the *UCI Website*[1].

The **MNAV** dataset was created for the Museo Nacional de Artes Visuales (MNAV, National Museum of Visual Arts) in Uruguay as part of the project described in paper [4], which aimed to develop an indoor localization system utilizing WiFi fingerprinting. This dataset is publicly accessible on *GitHub*[2]. It comprises a total of 10,469 data points recorded during the data collection, with the detection of 188 Access Points (APs). The paper also provides a museum map that delineates the locations of the APs and defines 16 regions for the classification task. Among the 188 detected APs, 15 were installed specifically for the experiment, with each AP supporting both the 2.4GHz and 5GHz bandwidths. Consequently, the dataset encompasses 30 features derived from these APs. Here the RSSI values range from −99 dBm (weak signal) to 0 dBm (strong signal), while a value of 0.0 means that the access point was not detected. Table 1 summarizes the details of both datasets.

To preprocess the datasets, we made certain modifications. In both datasets, we added 104 to the RSSI feature, thereby adjusting the range to 0–104. Additionally, we replaced the missing access point values i.e. AP with an RSSI value of 100 in the UJIIndoorLoc dataset and AP with 0.0 RSSI value in the MNAV dataset, to 0. In the MNAV dataset, the set of labels (regions) **T** was pre-defined within the dataset, consisting of a total of 16 unique regions (classes). On the other hand, for the UJIIndoorLoc dataset, we generated the set **T** by combining the FLOOR and BUILDINGID columns of the dataset. This process yielded a total of 13 unique regions (classes).

For the UJIIndoorLoc dataset, we utilized the predefined train and test split. As for the MNAV dataset, there is no predefined train and test split, so we created a new train and test split with an 80:20 ratio. This resulted in 8,375 training data points and 2,094 test data points out of the total 10,469 data points

[1] https://archive.ics.uci.edu/ml/datasets/ujiindoorloc.

[2] https://github.com/ffedee7/posifi_mnav/.

Table 1. Summary of Datasets

Attributes	UJIIndoorLoc	MNAV
Total Data Points	21,048	10,469
Training Points	19,937	8,375
Test Points	1,111	2,094
Detected APs	520	188
Labels	13	16
RSSI Signal Range	−104 dBm to 0 dBm	−99 dBm to 0 dBm
Features	Floor, BuildingID, Lat, Long, RSSI, etc	RSSI with corresponding locations
Dataset Availability	*UCI Website*	*GitHub*

available. We have used this training and test set for all of our experiments, and all the accuracy results presented in this paper are on the test set. Our goal here is to improve the accuracy of label (region) prediction. As getting labeled datasets is a time-consuming and labor-intensive task so to study the effect of using a restricted dataset for training, we have performed the experiments for train ratio $\mathbf{r} \in \{0.2, 0.4, 0.6, 0.8, 1.0\}$. The flowchart in Fig. 4 visually summarizes the sequence of tasks involved in our study.

Models Description. In the case of **MLP**, we trained a neural network on the given training data set and then assessed the performance on the test set, and the outcome derived from this was reported.

For **kNN**, a grid search approach was adopted to ascertain the ideal value for k. Following this determination, an evaluation of its performance was conducted using the test set. Notably, we observed that as the size of the training data reduced, the optimal k value also decreased for the task.

In **SVM**, we utilized the Radial Basis Function (RBF) as the kernel function, which exhibited superior performance compared to other kernel functions such as linear. Additionally, we conducted a grid search to identify the optimal value for the parameter C, which balances the misclassification of training examples against the simplicity of the decision surface.

For our GNN-based baselines, we analyze **GNN** [18]. This paper employs Graph Neural Networks (GNNs) with APs as nodes and the distance between them as edges of the graph for indoor localization and encompasses experimental evaluations on the UJIIndoorLoc and MNAV datasets. Consequently, we use the authors' code to run on our preprocessed datasets.

In **IndoorGNN**, we employed two graph neural network layers as discussed in Sect. 4.2 followed by a fully connected neural network. The complete architecture of the model is shown in Fig. 5. In the GNN layers, we utilized the *mean* function as the aggregator function in Eq. 1 when doing Edge Convolution to aggregate values from the neighboring nodes, as it demonstrated better performance compared to other aggregator functions such as *max* and *sum*.

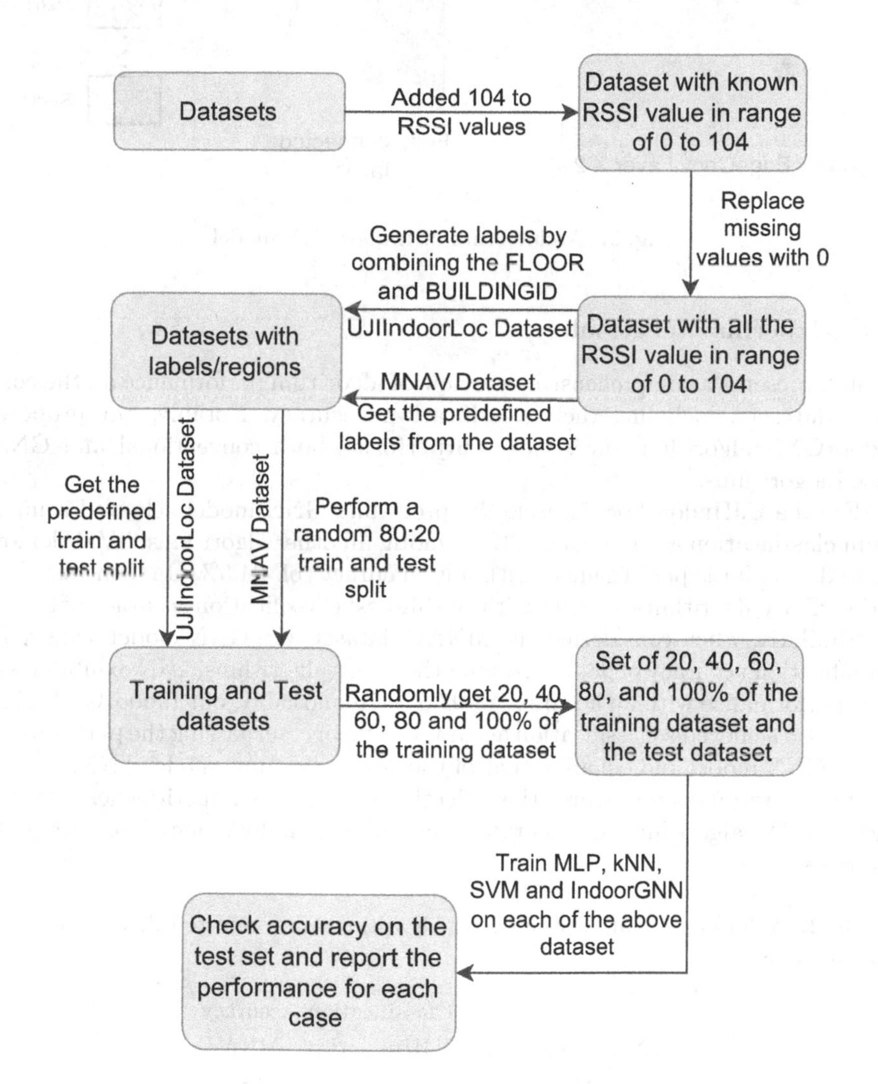

Fig. 4. Flow diagram for preprocessing and training on the datasets.

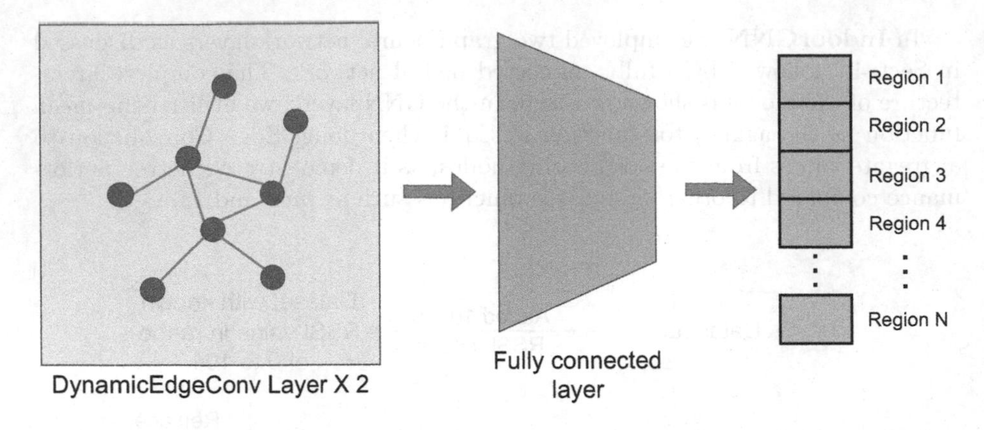

Fig. 5. Architecture of IndoorGNN model

5.2 Experimental Results

Table 2 presents a comprehensive overview of algorithm performance on the complete datasets, including their classification accuracy. Notably, our proposed IndoorGNN algorithm consistently outperforms both conventional and GNN-based algorithms.

For the UJIIndoorLoc dataset, the preceding GNN model achieved a maximum classification accuracy of 91.3%. Among all other algorithms, SVM demonstrated the best performance with an accuracy of 94.5%. In contrast, our IndoorGNN algorithm stood out with the highest classification accuracy of 95.8%.

Similarly, when considering the MNAV dataset, the GNN model achieved a classification accuracy of 96.3%. Among the other algorithms, kNN exhibited the best performance with an accuracy of 97.5%. Remarkably, our IndoorGNN algorithm also achieved a classification accuracy of 97.5%, surpassing the performance of the GNN model and equaling that of the top-performing model, kNN.

These results underscore the effectiveness and competitiveness of the IndoorGNN algorithm in accurately classifying indoor locations for both datasets.

Table 2. Performance matrix of different Algorithms on complete UJIIndoorLoc and MNAV datasets

		Classification Accuracy	
S. No.	Algorithm	UJIIndoorLoc	MNAV
1	MLP	92.0	94.8
2	kNN	92.5	**97.5**
3	SVM	94.5	95.0
4	GNN [18]	91.3	96.3
5	IndoorGNN	**95.8**	**97.5**

Figure 6b presents a comprehensive assessment of algorithm performance on the UJIIndoorLoc dataset, considering varying training dataset sizes. Notably, IndoorGNN consistently outperforms all other algorithms, maintaining its superior performance even with reduced training data. Likewise, Fig. 6 showcases the performance analysis of diverse algorithms on the MNAV dataset, across

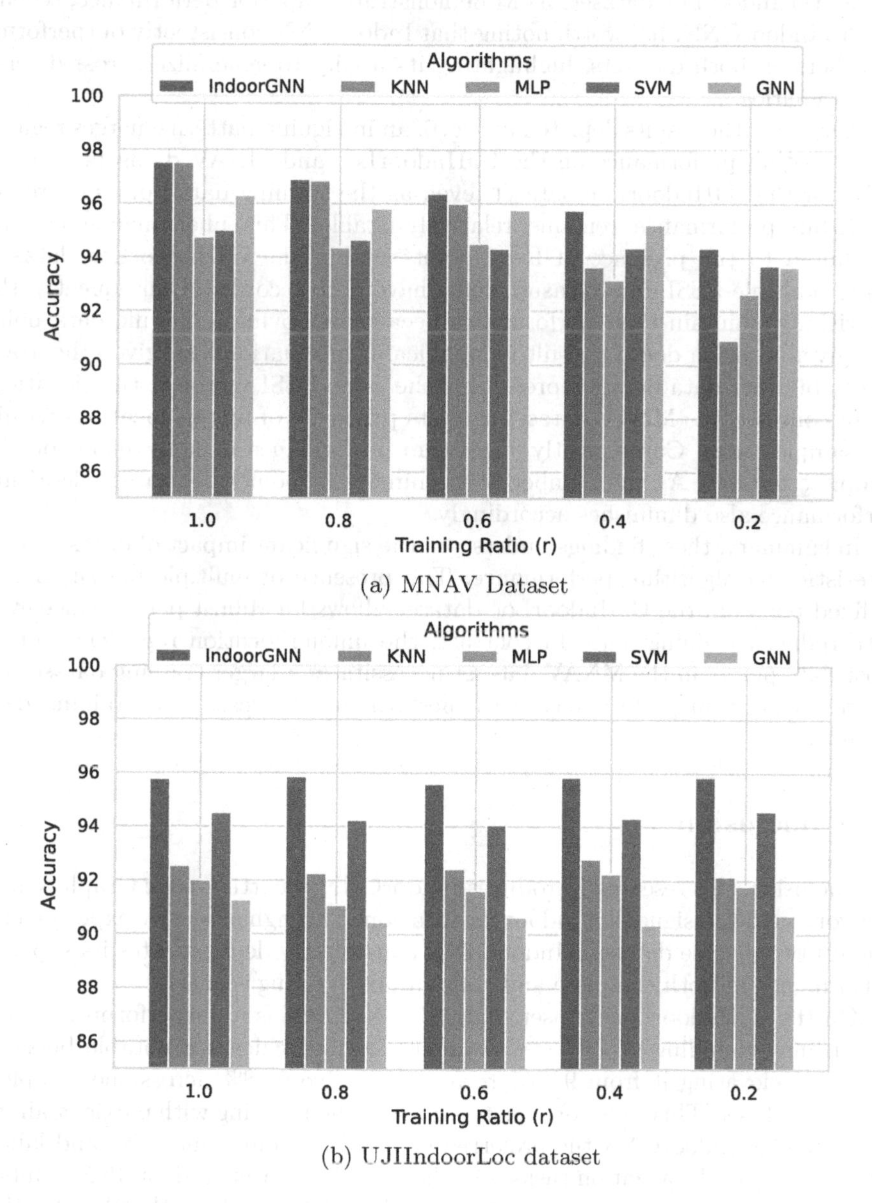

(a) MNAV Dataset

(b) UJIIndoorLoc dataset

Fig. 6. Comparison of Classification Accuracy for different Train Ratio on the MNAV and UJIIndoorLoc datasets.

different training dataset sizes. Evidently, IndoorGNN either matches or surpasses the performance of other algorithms when trained on a fraction of the dataset.

Figure 6 further emphasizes these observations. For the MNAV dataset, besides IndoorGNN, kNN and GNN also exhibit strong performance. Conversely, on the UJIIndoorLoc dataset, SVM demonstrates superior performance, second only to IndoorGNN. It's worth noting that IndoorGNN consistently outperforms all others on both datasets, highlighting its ability to generalize across diverse data scenarios.

Analyzing the results depicted in Fig. 6, an intriguing pattern emerges regarding algorithm performance on the UJIIndoorLoc and MNAV datasets. Specifically, for the UJIIndoorLoc dataset, even as the training data points decrease, algorithm performance remains relatively stable. This phenomenon can be attributed to the presence of fixed points within the UJIIndoorLoc dataset, where multiple RSSI data measurements have been recorded. Consequently, the algorithms maintain their performance because removing a specific data point at a given location doesn't result in significant information loss, given the availability of other data points representing the same RSSI vector at that location.

In contrast, the MNAV dataset consists primarily of unique locations within the sampled area. Consequently, algorithm performance is more reliant on the complete dataset. As the number of training data points decreases, algorithm performance also diminishes accordingly.

In summary, these findings underscore the significant impact of dataset characteristics on algorithm performance. The presence of multiple measurements at fixed points in the UJIIndoorLoc dataset allows for robust performance even with reduced training data. In contrast, the unique location representation of most data points in the MNAV dataset necessitates a larger training dataset for optimal algorithm performance, with performance decreasing as training data diminishes.

6 Conclusion

In conclusion, this research introduces IndoorGNN, a cutting-edge Graph Neural Network model designed for indoor localization. Through extensive experimentation on two diverse datasets, IndoorGNN consistently demonstrates its superior performance in both complete and constrained training scenarios.

On the UJIIndoorLoc dataset, IndoorGNN shines as it outperforms all other algorithms, including GNN-based counterparts. It achieves a notable boost in accuracy, elevating it from 94.5% to an impressive 95.8% across the complete training dataset. This superiority extends even when dealing with partial training data, making IndoorGNN the preferred choice for resource-intensive and labor-intensive indoor localization tasks. On the MNAV dataset, IndoorGNN exhibits a marked improvement over the GNN-based model, matching the classification accuracy of the top-performing kNN algorithm at a remarkable 97.5% across the entire dataset. Furthermore, it surpasses all alternative algorithms when applied to a limited dataset. These findings firmly establish IndoorGNN's effectiveness across a spectrum of scenarios.

In summary, IndoorGNN emerges as a compelling solution for indoor localization, consistently delivering superior performance even with smaller training datasets. Its capacity to outshine competing algorithms in challenging scenarios underscores its potential for real-world applications that demand both efficiency and precision.

Looking ahead, our future research will explore several exciting avenues:

- Investigating methods for precise latitude and longitude location determination.
- Tackling scenarios with limited labeled data but an abundance of unlabeled data.
- Developing strategies to effectively handle missing RSSI values.
- Investigating localization methods in private settings using Graph Neural Networks [14].

These endeavors will further enhance the capabilities of IndoorGNN and extend its applicability in the field of indoor localization.

References

1. Adege, A.B., Lin, H.P., Tarekegn, G.B., Jeng, S.S.: Applying deep neural network (DNN) for robust indoor localization in multi-building environment. Appl. Sci. **8**(7), 1062 (2018)
2. Al Nuaimi, K., Kamel, H.: A survey of indoor positioning systems and algorithms. In: 2011 International Conference on Innovations in Information Technology, pp. 185–190. IEEE (2011)
3. Bekkelien, A., Deriaz, M., Marchand-Maillet, S.: Bluetooth indoor positioning. Master's thesis, University of Geneva (2012)
4. Bracco, A., Grunwald, F., Navcevich, A., Capdehourat, G., Larroca, F.: Museum accessibility through Wi-Fi indoor positioning (2020)
5. Chriki, A., Touati, H., Snoussi, H.: SVM-based indoor localization in wireless sensor networks. In: 2017 13th International Wireless Communications and Mobile Computing Conference (IWCMC), pp. 1144–1149. IEEE (2017)
6. Dabove, P., Di Pietra, V., Piras, M., Jabbar, A.A., Kazim, S.A.: Indoor positioning using ultra-wide band (UWB) technologies: positioning accuracies and sensors' performances. In: 2018 IEEE/ION Position, Location and Navigation Symposium (PLANS), pp. 175–184. IEEE (2018)
7. Dedes, G., Dempster, A.G.: Indoor GPS positioning-challenges and opportunities. In: VTC-2005-Fall. 2005 IEEE 62nd Vehicular Technology Conference, 2005. vol. 1, pp. 412–415. Citeseer (2005)
8. Eksen, K., Serif, T., Ghinea, G., Grønli, T.M.: InLoc: location-aware emergency evacuation assistant. In: 2016 IEEE International Conference on Computer and Information Technology (CIT), pp. 50–56. IEEE (2016)
9. El-Sheimy, N., Li, Y.: Indoor navigation: state of the art and future trends. Satell. Navig. **2**(1), 1–23 (2021)

10. Fernandez-Llatas, C., Lizondo, A., Monton, E., Benedi, J.M., Traver, V.: Process mining methodology for health process tracking using real-time indoor location systems. Sensors **15**(12), 29821–29840 (2015)
11. He, S., Chan, S.H.G.: Wi-fi fingerprint-based indoor positioning: recent advances and comparisons. IEEE Commun. Surv. Tutor. **18**(1), 466–490 (2015)
12. Huang, C.N., Chan, C.T.: ZigBee-based indoor location system by k-nearest neighbor algorithm with weighted RSSI. Procedia Comput. Sci. **5**, 58–65 (2011)
13. Joshi, R.B., Mishra, S.: Learning graph representations. In: Biswas, A., Patgiri, R., Biswas, B. (eds.) Principles of Social Networking. SIST, vol. 246, pp. 209–228. Springer, Singapore (2022). https://doi.org/10.1007/978-981-16-3398-0_10
14. Joshi, R.B., Mishra, S.: Locally and structurally private graph neural networks. Digital Threats (2023). https://doi.org/10.1145/3624485
15. Kalbandhe, A.A., Patil, S.C.: Indoor positioning system using Bluetooth low energy. In: 2016 International Conference on Computing, Analytics and Security Trends (CAST), pp. 451–455. IEEE (2016)
16. Khatab, Z.E., Hajihoseini, A., Ghorashi, S.A.: A fingerprint method for indoor localization using autoencoder based deep extreme learning machine. IEEE Sens. Lett. **2**(1), 1–4 (2017)
17. Lee, C.K.M., Ip, C., Park, T., Chung, S.: A Bluetooth location-based indoor positioning system for asset tracking in warehouse. In: 2019 IEEE International Conference on Industrial Engineering and Engineering Management (IEEM), pp. 1408–1412. IEEE (2019)
18. Lezama, F., González, G.G., Larroca, F., Capdehourat, G.: Indoor localization using graph neural networks. In: 2021 IEEE URUCON, pp. 51–54. IEEE (2021)
19. Lezama, F., Larroca, F., Capdehourat, G.: On the application of graph neural networks for indoor positioning systems. In: Tiku, S., Pasricha, S. (eds.) Machine Learning for Indoor Localization and Navigation, pp. 239–256. Springer, Cham (2023). https://doi.org/10.1007/978-3-031-26712-3_10
20. Liu, C., Wang, C., Luo, J.: Large-scale deep learning framework on FPGA for fingerprint-based indoor localization. IEEE Access **8**, 65609–65617 (2020)
21. Liu, Y., et al.: Inferring gender and age of customers in shopping malls via indoor positioning data. Environ. Plann. B Urban Anal. City Sci. **47**(9), 1672–1689 (2020)
22. Luo, X., Meratnia, N.: A geometric deep learning framework for accurate indoor localization. In: 2022 IEEE 12th International Conference on Indoor Positioning and Indoor Navigation (IPIN), pp. 1–8. IEEE (2022)
23. Qian, W., Lauri, F., Gechter, F.: Supervised and semi-supervised deep probabilistic models for indoor positioning problems. Neurocomputing **435**, 228–238 (2021)
24. Rezgui, Y., Pei, L., Chen, X., Wen, F., Han, C.: An efficient normalized rank based SVM for room level indoor WiFi localization with diverse devices. Mob. Inf. Syst. **2017**, 6268767 (2017)
25. Roy, P., Chowdhury, C.: A survey of machine learning techniques for indoor localization and navigation systems. J. Intell. Robot. Syst. **101**(3), 63 (2021)
26. Sakpere, W., Adeyeye-Oshin, M., Mlitwa, N.B.: A state-of-the-art survey of indoor positioning and navigation systems and technologies. S. Afr. Comput. J. **29**(3), 145–197 (2017)
27. Sánchez-Rodríguez, D., Hernández-Morera, P., Quinteiro, J.M., Alonso-González, I.: A low complexity system based on multiple weighted decision trees for indoor localization. Sensors **15**(6), 14809–14829 (2015)
28. Segura, M., Mut, V., Sisterna, C.: Ultra wideband indoor navigation system. IET Radar, Sonar Navig. **6**(5), 402–411 (2012)

29. Shum, L.C., et al.: Indoor location data for tracking human behaviours: a scoping review. Sensors **22**(3), 1220 (2022)
30. Singh, N., Choe, S., Punmiya, R.: Machine learning based indoor localization using Wi-Fi RSSI fingerprints: an overview. IEEE Access **9**, 127150–127174 (2021)
31. Subhan, F., Hasbullah, H., Rozyyev, A., Bakhsh, S.T.: Indoor positioning in Bluetooth networks using fingerprinting and lateration approach. In: 2011 International Conference on Information Science and Applications, pp. 1–9. IEEE (2011)
32. Ting, S., Kwok, S., Tsang, A.H., Ho, G.T.: The study on using passive RFID tags for indoor positioning. Int. J. Eng. Bus. Manage. **3**(1), 9–15 (2011)
33. Torres-Sospedra, J., et al.: UJIindoorLoc: a new multi-building and multi-floor database for WLAN fingerprint-based indoor localization problems. In: 2014 International Conference on Indoor Positioning and Indoor Navigation (IPIN), pp. 261–270. IEEE (2014)
34. Torteeka, P., Chundi, X.: Indoor positioning based on Wi-Fi fingerprint technique using fuzzy k-nearest neighbor. In: Proceedings of 2014 11th International Bhurban Conference on Applied Sciences & Technology (IBCAST) Islamabad, Pakistan, 14th-18th January, 2014, pp. 461–465. IEEE (2014)
35. Wang, Y., Sun, Y., Liu, Z., Sarma, S.E., Bronstein, M.M., Solomon, J.M.: Dynamic graph CNN for learning on point clouds. ACM Trans. Graph. **38**(5), 1–2 (2019). https://doi.org/10.1145/3326362
36. Xu, H., Wu, M., Li, P., Zhu, F., Wang, R.: An RFID indoor positioning algorithm based on support vector regression. Sensors **18**(5), 1504 (2018)
37. Yang, C., Shao, H.R.: Wifi-based indoor positioning. IEEE Commun. Mag. **53**(3), 150–157 (2015)
38. Zhang, M., Fan, Z., Shibasaki, R., Song, X.: Domain adversarial graph convolutional network based on rssi and crowdsensing for indoor localization. arXiv preprint arXiv:2204.05184 (2022)

Ensemble-Based Road Surface Crack Detection: A Comprehensive Approach

Rajendra Kumar Roul[1(✉)], Navpreet[1], and Jajati Keshari Sahoo[2]

[1] Thapar Institute of Engineering and Technology, Patiala, Punjab, India
raj.roul@thapar.edu

[2] Department of Mathematics, BITS Pilani, K.K.Birla Goa Campus, Sancoale, India
jksahoo@goa.bits-pilani.ac.in

Abstract. The existence of road surface cracks erodes the structural robustness of the infrastructure and casts shadows of risks for countless motorists and walkers. The timely and efficient detection of road cracks is of utmost importance for maintenance and mitigating further deterioration. Currently, existing techniques to identify cracks entail physical examinations rather than deploying automated image-based techniques, which leads to costly and labor-intensive operations. Incorporating automated crack detection systems is necessary to optimize processes, reduce costs, and enable proactive maintenance efforts to enhance road safety and durability. This paper presents a comprehensive study on road crack detection, aiming to develop an accurate and efficient system to identify cracks on road surfaces. The proposed approach employs a two-phase Convolutional Neural Network (CNN) in conjunction with the Extreme Learning Machine (ELM) by harnessing advanced deep learning techniques. The model showcases outstanding performance in classifying road cracks, as evidenced by thorough experimentation and evaluation on a well-known CCIC dataset. The proposed approach contributes to the advancement of preventive maintenance strategies and the augmentation of road safety measures. The findings highlight the potential of the combined Conv-ELM approach to automate road crack detection, paving the way for improved infrastructure management and streamlined maintenance practices. This research marks a significant advancement in fostering dependable and resilient transportation infrastructures.

Keywords: CNN · CCIC · Deep Learning · ELM · Machine Learning · Road Condition

1 Introduction

Road networks are the lifelines of transportation, fostering economic growth by connecting urban centers, rural areas, and remote regions. However, the continuous use of roads and exposure to environmental factors such as temperature fluctuations and moisture pose significant challenges to their integrity. Inadequate maintenance practices further exacerbate these issues, necessitating

V. Goyal et al. (Eds.): BDA 2023, LNCS 14418, pp. 166–184, 2023.
https://doi.org/10.1007/978-3-031-49601-1_12

effective crack detection and maintenance strategies. Early detection of cracks is vital for road safety and cost-effective maintenance, with technologies like drones and machine learning playing a crucial role. Sustainable road infrastructure, incorporating eco-friendly materials and practices, is also essential for long-term viability while preserving the environment. To ensure the continuous flow of goods, services, and people, strategic planning, budget allocation, and innovative maintenance methods are essential for the upkeep of road networks [6,36]. Reserving road infrastructure integrity is paramount to ensure safe and efficient mobility [18,39]. However, this infrastructure faces significant challenges, including road cracks, which compromise structural integrity and contribute to increased maintenance costs and reduced lifespan [15]. Detection of crack can be performed either manually or automatically, depending on the available technology and resources [33,34]. Manual crack detection processes are susceptible to human error, fatigue, and limited capacity to handle the increasing volume of data [20]. These limitations hinder the effectiveness and efficiency of crack detection, emphasizing the need for automated solutions. Contrastingly, machine learning techniques, notably those employing neural networks, have found widespread application in object detection [2]. The widespread adoption of neural network-based models for object detection is propelled by their capacity to effectively learn intricate patterns and features from data, resulting in significantly enhanced accuracy and performance in the crucial realm of computer vision tasks [24,27,35]. Among all neural networks, Convolutional Neural Networks (CNN) stand out as one of the most celebrated and powerful architectures, renowned for their exceptional accuracy in extracting significant features from data, particularly in image-related tasks [4,13,16,30,32]. The fully connected layer and backpropagation (BP) can significantly impact the training time of CNNs, rendering them relatively time-consuming.

On the other hand, Extreme Learning Machines (ELM) provide an efficient alternative as classifiers, showcasing exceptional generalization performance without relying on backpropagation and demanding minimal human intervention [12,26]. Several studies have furnished substantial evidence highlighting ELM's swifter training times when juxtaposed with traditional machine learning (ML) methods, including Naive-Bayes, Decision Trees, and Support Vector Machines (SVM), underscoring its efficiency in processing data and facilitating faster model development [9,11,14]. ELM and its various iterations find widespread applications across diverse domains, encompassing pattern recognition, facial recognition, text classification, and image processing [1,17,28]. The combination of ELM with other machine learning techniques has consistently produced remarkable performance outcomes [7,23,29].

Indeed, by harnessing the strengths and successes of CNN and ELM, a novel and efficient approach known as Conv-ELM has been introduced for road crack detection, consisting of two distinct phases: feature extraction and classification. This hybrid model aims to enhance road crack detection by integrating CNN's powerful feature extraction capabilities, capturing intricate crack-related patterns and textures, and ELM's efficient classification performance, distinguishing between non-crack and crack regions. The combination of CNN and

ELM in this integrated approach offers high accuracy, robustness, and efficiency, rendering it well-suited for real-world applications and proactive infrastructure maintenance.

The main contributions of the paper are:

i. The paper introduces a novel hybrid Conv-ELM approach for road crack detection, effectively leveraging the strengths of both CNN and ELM. By merging CNN's adept feature extraction capabilities with ELM's swift learning and generalization properties, this approach presents a robust and efficient solution for accurate road crack detection tasks.
ii. The proposed hybrid Conv-ELM approach undergoes a comprehensive evaluation on the well-established Concrete crack images for classification (CCIC) dataset, affirming its remarkable efficacy in road crack detection. It outperforms traditional methods when CNN is combined with ELM. Additionally, the approach's generalization capability is rigorously tested through exposure to images from diverse scenarios and datasets, illuminating its adaptability and robustness across a spectrum of road crack detection tasks.
iii. The paper conducts a comprehensive analysis of the Conv-ELM approach in comparison to other hybrid models that integrate CNN with various machine learning classifiers, such as SVM, Xgboost, and others. Through rigorous comparisons, the Conv-ELM approach demonstrates its superior performance in road crack detection tasks. It also undergoes thorough benchmarking against several state-of-the-art methods for road crack detection, consistently exhibiting higher levels of accuracy and efficiency, reinforcing its effectiveness as a cutting-edge solution in this domain.

The paper's structure is organized as follows: Sect. 2 outlines the proposed Conv-ELM method. Section 3 presents the experimental results, including performance evaluations. Lastly, Sect. 4 offers the conclusion summarizing the findings and potential future enhancements to the approach.

2 Approach and Procedures

2.1 Dataset Origin and Details

The approach employed in this study utilizes the widely recognized Concrete Crack Images for Classification (CCIC) dataset[1] is a publicly available dataset that contains images of concrete surfaces with cracks and non-cracked areas, serving as a valuable resource for model development. Employing the CCIC dataset in the experiments ensures a comprehensive and diverse representation of road crack images for the training and testing of the model within the context of road crack detection. This dataset has been collected from multiple METU Campus Buildings, encompassing a range of real-world scenarios and environmental conditions. Sample images from the CCIC dataset are visually presented

[1] https://www.kaggle.com/datasets/arnavr10880/concrete-crack-images-for-classification.

in the Figure 1. The CCIC dataset is comprised of concrete images that exhibit cracks and is categorized into negative, representing images without cracks, and positive, representing images with cracks. Each class contains 20,000 images, resulting in a total of 40,000 images within the dataset. These images have dimensions of 227 × 227 pixels and consist of RGB channels. The dataset's construction involved utilizing 458 high-resolution images with dimensions of 4032 × 3024 pixels as the foundational source material. In all likelihood, this dataset creation process entailed resizing and processing the high-resolution images to conform to the specified dimensions, thereby producing a more manageable dataset for subsequent tasks. Zhang et al. [37] created individual images using the dataset creation method. One of the notable strengths of this dataset lies in its comprehensive coverage of diverse surface finishes and varying illumination conditions within the high-resolution images. However, it is important to mention that no data augmentation techniques, such as random rotation or flipping, were applied to this dataset during its creation.

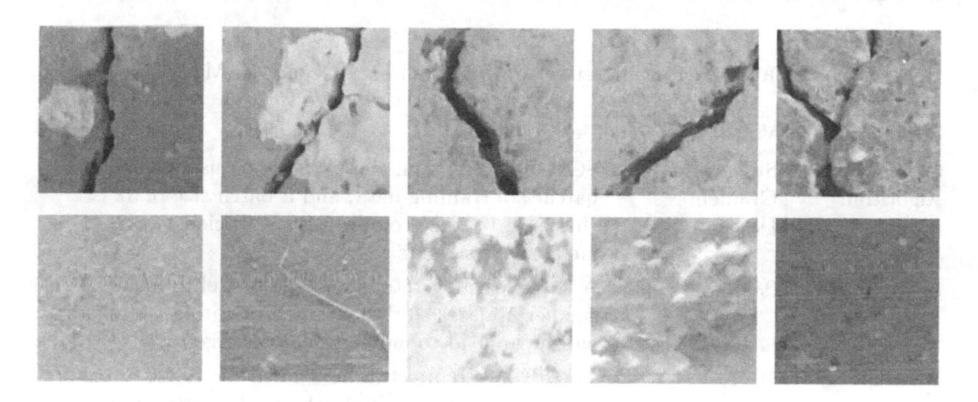

Fig. 1. Sample Images from CCIC Dataset: top row indicates Crack and bottom row indicates No-Crack

2.2 Data Preparation Procedures

A series of preprocessing techniques were applied to ensure data consistency and uniformity prior to the model training and testing. It included image resizing, conversion from RGB to Grayscale, normalization, and augmentation. These steps were crucial in eliminating noise and artifacts and preparing the data for analysis. The dataset's images underwent standardization, and all resized to a consistent resolution of 227 × 227 pixels. Conv-ELM architecture was tailored for grayscale images, and a necessary preprocessing step involves converting the RGB images from both datasets into grayscale format prior to their input into the CNN feature extractor. This conversion helps reduce computational complexity and enables the model to focus on relevant image features effectively. Normalization was carried out on the grayscale images to standardize the pixel values. It involved scaling the pixel values to fit within the range of [0, 1], achieved by dividing each pixel value by 255. Data augmentation techniques were employed to augment the training dataset's diversity and mitigate overfitting. It is crucial to emphasize that data augmentation was intentionally omitted for the CCIC dataset, as indicated in its description. These preprocessing steps made the datasets suitable for training and testing the Conv-ELM architecture. The images were standardized in terms of resolution and intensity values, converted to grayscale, and augmented to create a diverse and robust dataset for training the model effectively.

2.3 Training the Conv-ELM Model

Implementing the Conv-ELM architecture for road crack classification is conducted in Python 3.9.11, utilizing Visual Studio Code as the Integrated Development Environment. This combination offers an efficient and effective platform for developing and refining the Conv-ELM architecture for road crack classification. Table 1 illustrates the key components involves during the training process.

Table 1. Parameter set during Training of Conv-ELM

Parameter	Value	Description
Optimization Algorithm	Stochastic Gradient Descent (SGD)	SGD updates the model parameters using small batches of training data, and a batch size of 32 is chosen to balance computational efficiency and memory constraints.
Learning Rate	0.001	A lower learning rate allows the model to make smaller updates to the parameters during training, which can lead to more stable convergence
Epochs	10–40	Conv-ELM architecture is trained for 10, 20, 30, and 40 epochs, respectively. Each epoch represents a complete iteration over the entire training dataset.
Validation Split	3:1	The training set is used to train the Conv-ELM architecture to learn the patterns and features relevant to road crack classification. The cross-validation set serves as an independent validation set during the training phase.

2.4 Conv-ELM: A Dual-Phase Ensemble Approach

(i) A sequential CNN model is meticulously built during the first phase of COnv-ELM. This model is central in extracting essential features from the input dataset. Convolutional Neural Networks (CNNs) are widely recognized for their remarkable efficacy in tasks related to images, particularly in image classification. Feature extraction encompasses applying convolutional operations on input images by utilizing adaptable filters referred to as kernels. These filters are purposefully designed to identify many patterns and features within localized regions of the images. Mathematically, the convolution operation can be represented using Eq. 1.

$$Z[a,b] = (X * K)[a,b] = \sum_{m=1}^{P} \sum_{n=1}^{Q} X[a-m, b-n] \cdot K[m,n] \tag{1}$$

where: $Z[a,b]$ represents the value at position (a,b) in the output feature map, X denotes the input image, K is the learnable kernel (filter), P and Q are the dimensions of the kernel. Subsequent to the convolution operation, an activation function is employed to introduce non-linearity, thereby improving the network's capacity to capture intricate patterns within the data. In this work, Rectified Linear Unit (ReLU)[2] activation function is used, which is defined as:

$$f(x) = \max(0, x)$$

ReLU stands as a widely utilized activation function within deep neural networks. This function transforms negative input values to zero while preserving positive values unchanged. This intrinsic characteristic of ReLU introduces essential non-linearity to the neural network's computations. ReLU's non-saturating nature allows gradients to flow more readily, mitigating the vanishing gradient problem and facilitating faster and more stable training of deep neural networks. The output of the CNN consists of feature maps that represent the presence of specific patterns or features in different regions of the input images. These feature maps preserve spatial information and are crucial for further processing.

(ii). The ELM algorithm is harnessed for the classification process in the subsequent phase [23]. ELM is renowned for its rapid learning capabilities and outstanding generalization performance, rendering it highly compatible with diverse classification tasks [25]. Unlike conventional iterative training methods, ELM adopts a single-layer architecture and computes the output weights directly, bypassing the need for iterative processing of the training data [22]. Given the feature maps obtained from the CNN, each feature map is transformed into a one-dimensional vector, which is then used as input for the ELM model. The conversion of a 2-D feature map to a 1-D vector involves concatenating the rows or columns of the feature map to form a long vector.

For a binary classification problem (crack or no-crack), the output can be represented as $L = \{l_1, l_2, ..., l_K\}$, where K is the number of training samples. The ELM algorithm aims to find the output weights $W = \{w_1, w_2, ..., w_N\}$ that minimize the training error. Mathematically, the ELM output can be computed using Eq. 2.

$$Y = X \cdot W \tag{2}$$

[2] https://keras.io/api/layers/activations/#relu-function.

where: x is the input feature matrix containing the 1-D vectors of feature maps for all training samples, W is the output weight matrix. To find the output weights W, ELM uses the Moore-Penrose generalized inverse[3], also known as the pseudoinverse, of the input feature matrix H as shown in Eq. 3.

$$W = X^{+} \cdot T \tag{3}$$

where: X^{+} is the pseudoinverse of X, and T is the target matrix containing the true labels of the training samples. Once the output weights are obtained, the ELM model is used to classify the road images as either cracked or not.

The hybrid Conv-ELM architecture mitigates the inherent higher training time limitations of CNNs by capitalizing on the advantageous traits of ELM. In this paradigm, CNNs excel in feature extraction from input data, while ELM takes charge of the ultimate classification endeavor. This amalgamation in the hybrid Conv-ELM approach is instrumental in diminishing the risk of overfitting. It achieves this by solely updating the output weights within the ELM model while the hidden layer biases and input weights are randomly generated. Through this approach, the ELM component seamlessly functions as a swift and efficient classifier, harnessing its strong generalization capabilities and reducing the need for extensive human intervention [10]. Concurrently, the training process capitalizes on CNN's outstanding feature extraction prowess. The CNN adeptly extracts high-level features from the input data and is fed into the ELM for the classification task. This fusion enables the model to benefit from the strengths of both the CNN's capacity to acquire intricate representations and the ELM's rapid learning speed. Consequently, the hybrid Conv-ELM approach attains an expedited overall learning pace without compromising its robust generalization performance. This duality renders it a highly effective solution for the road crack detection task.

2.5 Conceptual Design of the Conv-ELM Framework

The proposed Conv-ELM architecture consists of a well-structured architecture comprising several essential layers. This section provides a detailed overview of the Conv-ELM architecture, depicted in Fig. 2.

- **Input Layer:** This layer is the first layer of Conv-ELM, which receives grayscale images of size 32 × 32 × 1. The first component represents the image's width followed by the second component represents its height, and finally the number of channels (1 for grayscale images) is represented by the third component.
- **Convolutional Layer:** The model architecture incorporates a series of convolutional layers, each utilizing trainable filters to compute dot products over local regions of input images. The initial trio of convolutional layers employs

[3] https://mathworld.wolfram.com/Moore-PenroseMatrixInverse.html.

32 filters with kernel sizes of (3, 3) for first and second layers and (5, 5) for the third layer. Subsequently, another set of three convolutional layers consists of 64 filters each, with kernel sizes of (3, 3) for the fourth and fifth layers and (5, 5) for the final convolutional layer. These sequential convolutional layers are designed to extract a hierarchical range of features from the input data, encompassing both basic and intricate patterns. This approach enables the model to acquire intricate representations, effectively enhancing its capacity for road crack detection.

- **ReLU Layer:** After each convolutional layer, the ReLU activation function is applied element-wise to the output feature maps. The ReLU activation function is represented as f(x) = max(0, x), setting all negative values to zero and leaving positive values unchanged. This non-linear activation introduces the ability to capture complex patterns effectively and helps prevent the vanishing gradient problem during training.
- **Batch Normalization Layer:** The model incorporates Batch Normalization layers after each convolutional layer to enhance convergence and accelerate the training process. Batch normalization standardizes the features extracted by the convolutional layers within batches, enhancing training stability and efficiency. The normalization process can be mathematically represented using Eq. 4.

$$x' = \frac{x - \bar{x}}{\sqrt{\mathrm{v}(x) + \epsilon_0}} \tag{4}$$

where x' denotes the new value of the component, $\bar{x}$ and $\mathrm{v}(x)$ denote the mean and variance within a batch, respectively. The small constant ϵ_0 is added to the denominator for numerical stability.

- **Flatten Layer:** After the batch normalization step, the output is passed to the flatten layer, which transforms the multi-dimensional feature maps obtained from the previous layers into a one-dimensional vector. This flattened vector serves as the input for the subsequent layers.
- **Dropout Layer:** To prevent overfitting, a dropout layer is introduced. Within the dropout layer, specific input contributions are stochastically deactivated during training through a random process. This regularization technique encourages the model to learn robust features by preventing it from relying too heavily on specific input elements.
- **Fully Connected Layer:** The model includes a fully connected layer, also known as a dense layer, comprising 256 neurons. This layer serves as an intermediary for further feature transformation.
- **Softmax Layer:** The final layer of the Conv-ELM architecture is the softmax layer, which is responsible for obtaining the probability distribution of the output classes. The Softmax activation function[4] computes the probabilities for each class based on the inputs and is defined using Eq. 5.

$$\mathrm{softmax}(z_i) = \frac{e^{z_i}}{\sum_{j=1}^{K} e^{z_j}} \tag{5}$$

[4] https://keras.io/api/layers/activations/#softmax-function.

where z_i represents the input value corresponding to class i, and K is the total number of classes (2 in the proposed approach).
In the Conv-ELM architecture, a notable design choice is using two consecutive 3×3 convolutional layers instead of a single 5×5 convolutional layer. This decision introduces a more nonlinear structure, allowing the model to capture complex patterns effectively. Furthermore, the pooling layer is substituted with a learnable convolutional layer with strides of 2, offering enhanced adaptability compared to conventional pooling layers. Integrating batch normalization and data augmentation techniques further enhances the model's accuracy, contributing to improved performance in road crack detection tasks. These modifications and enhancements in the CNN component are crucial in optimizing feature extraction and overall model performance.

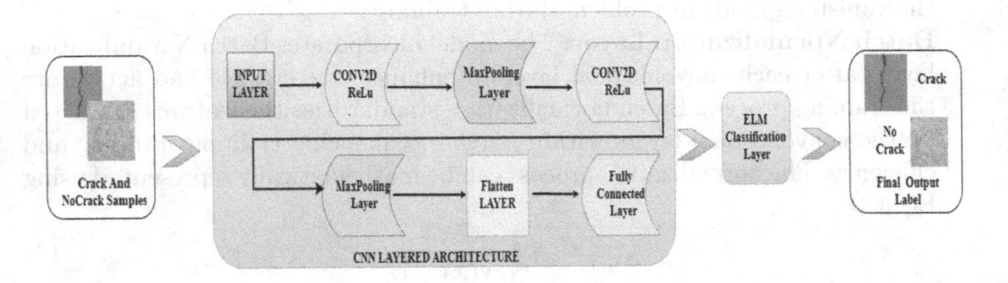

Fig. 2. Conv-ELM Architecture

3 Empirical Investigation

3.1 Experimental Setup

To execute the experiments concerning the classification of road cracks using the proposed Conv-ELM architecture, we utilized an Intel Core i11 processor, a 24 GB GPU, and 32 GB of RAM. The implementation of the Conv-ELM model was carried out using the TensorFlow library in conjunction with the high-level Keras API. In addition, standard data manipulation and analysis libraries were employed to preprocess the concrete crack images and prepare the dataset for both training and testing. The entire training and evaluation processes were conducted within Python 3.9, using the VSCode integrated development environment. TensorFlow and Keras provided a robust and efficient platform for the construction, training, and assessment of the Conv-ELM architecture, ensuring accurate classification of concrete crack images into positive (with cracks) and negative (without cracks) categories.

3.2 Analysis of Results

Evaluation via Confusion Matrix: The performance of the Conv-ELM architecture was evaluated using the CCIC dataset. The model underwent training for a range of epochs, from 10 to 40, with variations in the number of neurons in the hidden layer throughout the training process. Figure 3 presents the confusion matrix for Conv-ELM at different epochs. This matrix provides valuable insights into the model's performance, showcasing its ability to predict each class at various stages of training. At ten epochs, the confusion matrix reveals the model's initial performance, highlighting areas for potential improvement in specific classes. As training progresses, the model's proficiency increases, and by the 20th epoch, it reaches its peak accuracy, as indicated by the enhanced diagonal alignment within the confusion matrix. Beyond the 20-epoch mark, the model's performance tends to stabilize, suggesting that additional training may not yield substantial gains in accuracy. The confusion matrices reinforce this observation at 30 and 40 epochs, which exhibit similar performance patterns compared to the 20-epoch matrix.

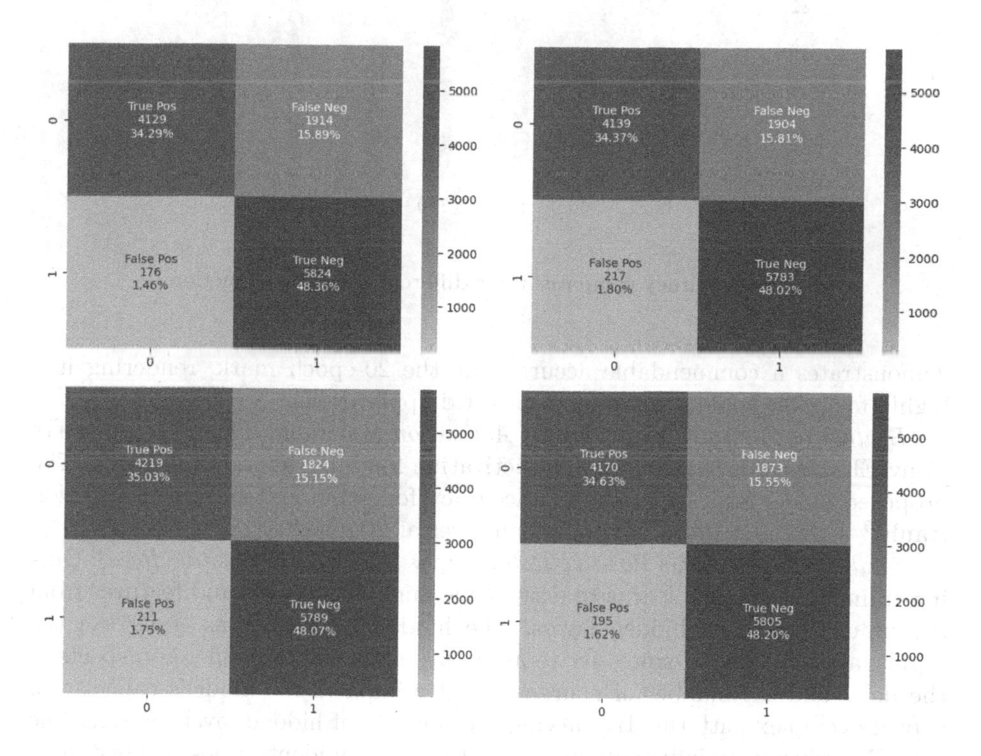

Fig. 3. Analysis of Confusion Matrix for Conv-ELM at different Epochs

Comparing Performance Across Epochs: Table 2 depicts evaluating various performance metric parameters for different epochs, considering the variation in hidden layer neurons. These metrics offer significant insights into the model's performance, aiding in comprehending its capabilities and limitations in categorizing concrete crack images as negative (no-crack) or positive (crack). Overall, the Conv-ELM architecture, particularly with 250 neurons in the hidden layer,

Table 2. Performance Metric comparison for Epochs

Epochs	Hidden Layer Neurons	Precision (%)	Sensitivity (%)	Specificity (%)	F1-Score (%)
20	250	75.26	96.38	68.49	84.50
30	250	75.23	96.48	69.81	85.05
40	250	75.60	96.75	69.00	84.88

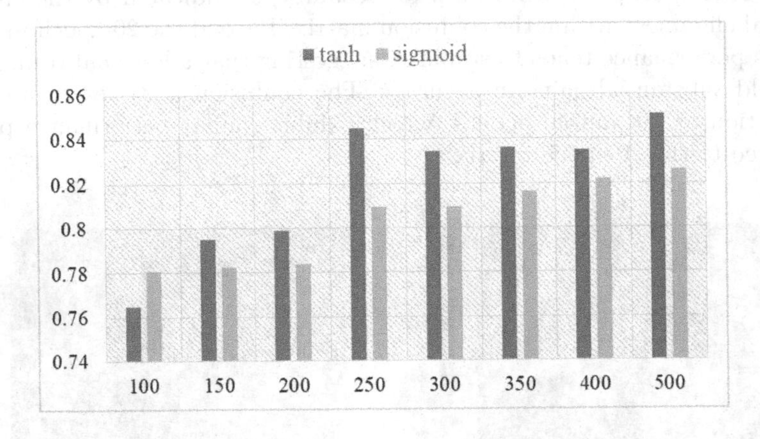

Fig. 4. Accuracy comparison for different activation function

demonstrates a commendable accuracy at the 20-epoch mark, rendering it a highly favorable configuration for real-world applications.

Evaluating Accuracy with Various Activation Functions: The performance of Conv-ELM is compared for different activation function, as shown in Fig. 4. The proposed model achieves maximum accuracy for non-linear activation function (tanh)[5] with 250 neurons in the hidden layer at 20 epochs.

Analyzing TPR (True Positive Rate) versus FPR (False Positive Rate): During training, the model learns to extract meaningful patterns and features from the input data in its hidden layers. The hidden layer neurons transform the input data into a more expressive representation. It captures different aspects of the data, and the number of neurons directly impacts the model's capacity to capture complex patterns. By varying the number of hidden layer neurons, the model's architecture influences its ability to correctly identify positive instances (true positive) from the dataset. The model achieves the highest TPR at 250 neurons for 20 epochs, as depicted in Fig. 5.

3.3 Comparative Assessment with Other ML Classifiers

The performance of Conv-ELM is benchmarked against various machine learning techniques, including XGBoost [3], Support Vector Machines (SVM) [31], Extra

[5] https://keras.io/api/layers/activations/#tanh-function.

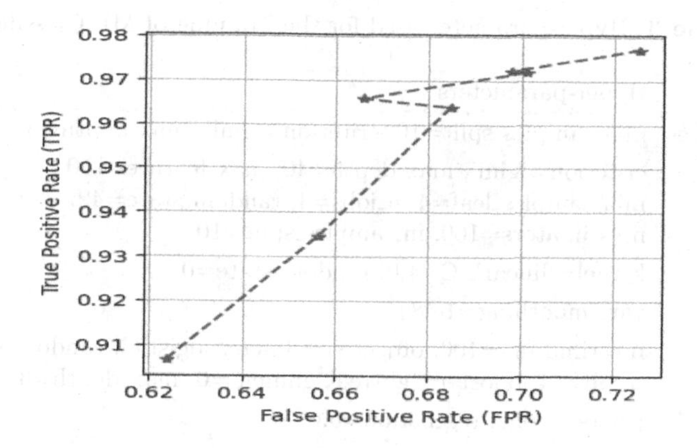

Fig. 5. TPR vs FPR for different hidden layer neurons

Trees (ET) [5], Gaussian Naive Bayes (GNB) [19], Decision Tree [21], and others, when combined with CNN. All models have been trained on the CCIC dataset for 20 epochs with a learning rate of 0.001 and a batch size of 32. The hyperparameters set for training the ML classifier are listed in Table 3. The empirical results revealed that Conv-ELM outperforms other ML classifiers in terms of accuracy, as shown in Table 4 (bold indicates maximum). The CNN-QDA hybrid classifier has the highest specificity among various classifiers because it efficiently organizes features into distinct classes and models the variance-covariance matrices for each class. By integrating CNN for feature extraction and QDA for classification, this approach captures intricate patterns through CNN's hierarchical learning, while QDA effectively accounts for class-specific variations in the data. This synergy enhances classification accuracy by leveraging the separate arrangement of features and the covariance information within distinct classes. Performance metric parameters such as Precision, Recall, and F1-Score are also compared with various ML classifiers, as shown in Figs. 6, 7, and 8, respectively. Results show that Conv-ELM has the highest Precision, but Recall and F1-Score for the Conv-ELM are comparable with CNN combined with ML classifiers. CNN-SVM has the highest F-measure because the combination of CNN's feature extraction and SVM's non-linear decision boundaries allows the model to capture complex relationships within the data.

Table 3. Hyper-parameter used for the Training of ML Classifiers

Classifier	Hyper-parameters
Decision Trees	min_samples_split=10, criterion='gini', max_depth=5
Extra Trees	criterion='gini', max_depth=40, max_features=50, min_samples_leaf=4, n_jobs=4, random_state=42, n_estimators=100, in_samples_split=10
SVM	kernel='linear', C=1.0, random_state=0
GNB	var_smoothing=1e-8
XgBoost	n_estimators=100, objective='binary:logistic', random_state=42, n_jobs=4, booster='gbtree', gamma=0, max_depth=3
QDA	priors=None, reg_param=0.0

Table 4. Comparing Performance: CNN Combined with ML Classifiers

Model	Specificity(%)	Accuracy(%)
CNN-DT	69.20	83.33
CNN-ET	62.33	79.68
CNN-SVM	89.87	84.37
CNN-GNB	85.26	84.30
CNN-XGBoost	51.24	73.21
CNN-QDA	**98.92**	55.16
Conv-ELM (Proposed)	68.49	**84.98**

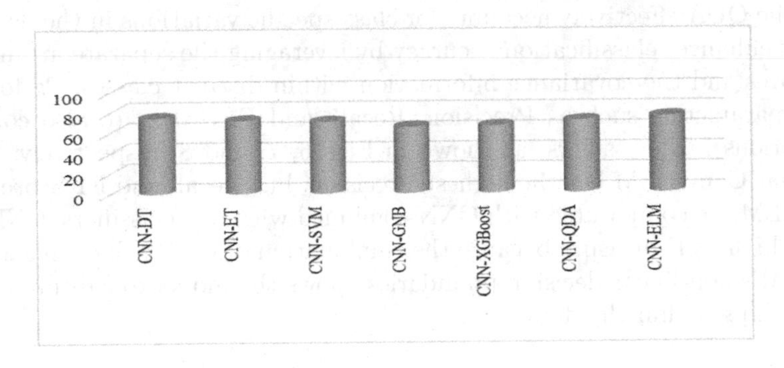

Fig. 6. Comparing Precision: CNN Combined with ML Classifiers

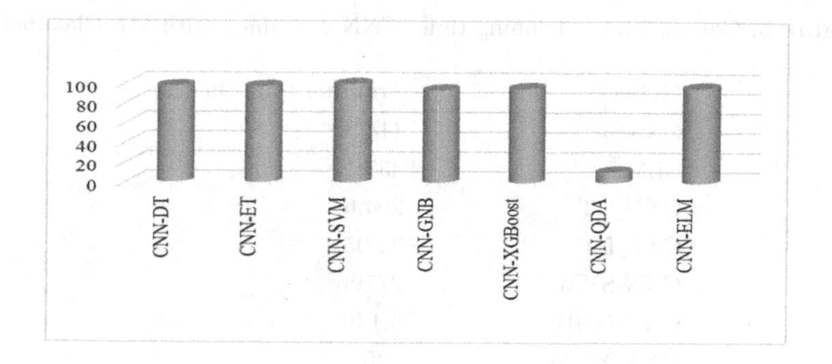

Fig. 7. Comparing Recall: CNN Combined with ML Classifiers

3.4 Evaluation of Training Time

During the training of the CNN, the process of refining the parameters within the convolutional layers through fine-tuning can lead to an extended computational time, as stated in the Eq. 6.

$$\sum_{a=1}^{m} p_{a-1} \cdot q_a^2 \cdot p_a \cdot r_a^2 \tag{6}$$

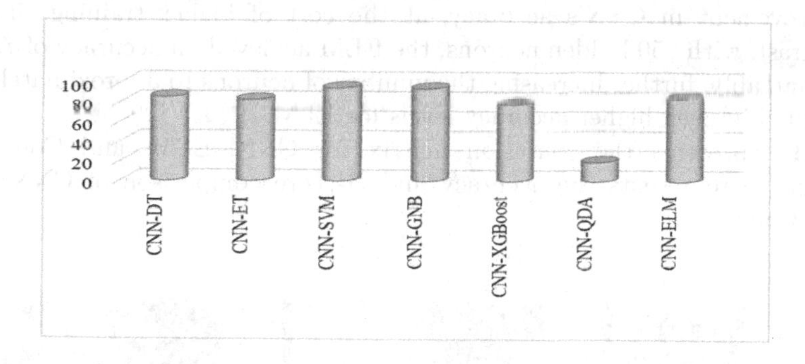

Fig. 8. Comparing F1-Score: CNN Combined with ML Classifiers

where a is the convolutional layer index, m is the number of convolutional layers, p_a is the number of filters at the ath layer, p_{a-1} is the number of input channels of the ath layer, q_a is size of filter, and r_a is output feature size. According to He et al., [8], the fully connected layer of the CNN consumes approximately 10% of the total computation time, thereby contributing to longer training periods. It has been seen that ELM does not involve backpropagation, resulting in less computation time than other ML classifiers [10,38]. Hence, the proposed approach replaces the fully connected layer with ELM, and due to this, the proposed model exhibits the minimum training time compared to all other models. Table 5 showcases the comparison of training time between Conv-ELM and other ML classifiers (with bold figures indicating the minimum values).

Table 5. Comparison of training time: CNN combined with ML Classifiers

Model	Training time (in sec)
CNN	11443
ELM	357.1
CNN-DT	204.6
CNN-ET	32.95
CNN-SVM	27796
CNN-GNB	5.149
CNN-XgBoost	1207
CNN-QDA	2537
Conv-ELM (Proposed)	**3.587**

3.5 Analysis and Discourse

The experimental evaluation was conducted on CCIC dataset and the resulting accuracy values were compared.

- After 20 epochs of training, the CNN attained an accuracy of 51.90%. Notably, increasing the number of training epochs to 50 or more, led to a further improvement in CNN's accuracy, at the cost of higher training times. In contrast, with 250 hidden neurons, the ELM achieved an accuracy of 79.56%. Remarkably, further increasing the number of neurons to approximately 2000 resulted in even higher accuracy levels for ELM.
- Fig. 9 represents the confusion matrix for CNN, ELM, and Conv-ELM. Figure 10 represents the accuracy and f1-score comparison of CNN, ELM, and Conv-ELM.

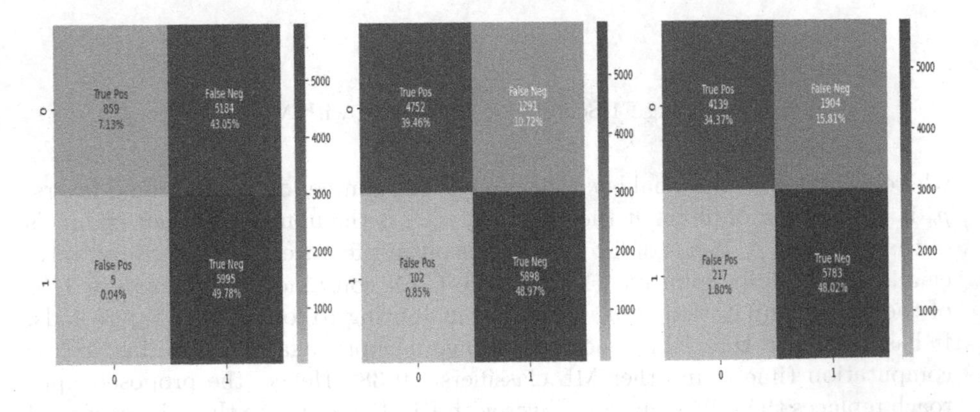

Fig. 9. Confusion Matrix for CNN(left), ELM (middle) and Conv-ELM (right)

. The proposed model harmoniously combines the strengths of CNN and ELM, achieving a remarkable peak accuracy of 84.98% in only 20 training epochs, with the utilization of 250 hidden layer neurons. It's noteworthy that the model accomplished this with notably reduced training time, as elaborated in Sect. 3.4. These results strongly imply that the integrated Conv-ELM architecture is significantly more efficient for road crack detection when compared to using CNN or ELM classifiers in isolation.

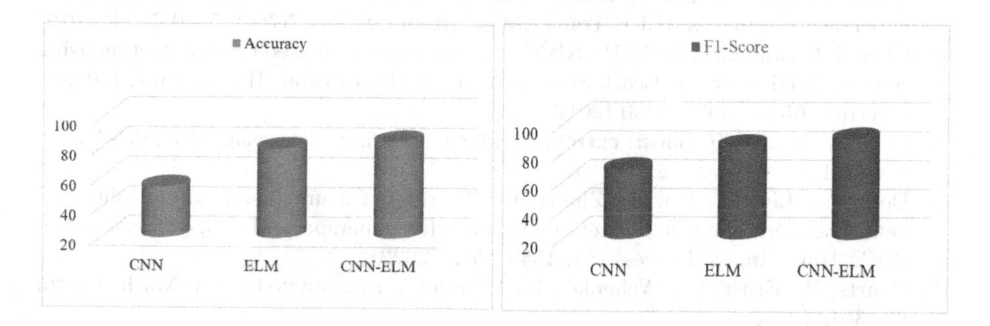

Fig. 10. Performance Comparison of CNN, ELM, and Conv-ELM

4 Conclusion

This research work has thoroughly investigated the vital role of road crack detection in bolstering the safety and efficiency of road infrastructure management. Among the arsenal of automated methods, deep learning techniques, with a particular emphasis on Convolutional Neural Networks (CNN), have emerged as formidable tools for crack detection. Their capacity to autonomously learn and extract pertinent features from road images is a key advantage. Moreover, the synergy of CNN with the efficient Extreme Learning Machine (ELM) has been explored, underscoring its potential to elevate the accuracy and efficiency of road crack detection. This fusion of CNN's feature extraction prowess and ELM's rapid training process for classification presents a promising avenue for proactive infrastructure maintenance and targeted crack repair. Throughout this comprehensive review, numerous research papers have showcased remarkable progress and accomplishments in road crack detection via deep learning methods, including CNN and transfer learning. These investigations have consistently underscored the effectiveness of contemporary neural network architectures in precisely identifying road cracks across diverse road surfaces and conditions. In summary, the integration of CNN and ELM holds great promise for the development of a precise, efficient, and automated road crack detection system. Such a model has the potential to streamline infrastructure maintenance, curtail maintenance expenses, and elevate road safety and beyond. Future extensions of this work could involve fine-tuning the number of hidden neurons required within the ELM for the classification process and the addition of more hidden layers to enhance classification robustness. Additionally, leveraging pre-trained CNN models for feature extraction offers the prospect of further reducing training time.

References

1. Cao, F., Yang, Z., Ren, J., Chen, W., Han, G., Shen, Y.: Local block multilayer sparse extreme learning machine for effective feature extraction and classification of hyperspectral images. IEEE Trans. Geosci. Remote Sens. **57**(8), 5580–5594 (2019)
2. Chen, F.C., Jahanshahi, M.R.: NB-CNN: deep learning-based crack detection using convolutional neural network and Naïve Bayes data fusion. IEEE Trans. Industr. Electron. **65**(5), 4392–4400 (2017)
3. Chen, T., et al.: XGBoost: extreme gradient boosting. R Packag. Vers. 0.4-2 **1**(4), 1–4 (2015)
4. Dong, Y., Liu, Q., Du, B., Zhang, L.: Weighted feature fusion of convolutional neural network and graph attention network for hyperspectral image classification. IEEE Trans. Image Process. **31**, 1559–1572 (2022)
5. Geurts, P., Ernst, D., Wehenkel, L.: Extremely randomized trees. Mach. Learn. **63**, 3–42 (2006)
6. Guo, F., Qian, Y., Liu, J., Yu, H.: Pavement crack detection based on transformer network. Autom. Constr. **145**, 104646 (2023)
7. Gurpinar, F., Kaya, H., Dibeklioglu, H., Salah, A.: Kernel ELM and CNN based facial age estimation. In: Proceedings of the IEEE Conference on Computer Vision and Pattern Recognition Workshops, pp. 80–86 (2016)
8. He, K., Sun, J.: Convolutional neural networks at constrained time cost. In: Proceedings of the IEEE Conference on Computer Vision and Pattern Recognition (CVPR), June 2015
9. Huang, G.B.: An insight into extreme learning machines: random neurons, random features and kernels. Cogn. Comput. **6**(3), 376–390 (2014)
10. Huang, G.B., Ding, X., Zhou, H.: Optimization method based extreme learning machine for classification. Neurocomputing **74**(1–3), 155–163 (2010)
11. Huang, G.B., Zhou, H., Ding, X., Zhang, R.: Extreme learning machine for regression and multiclass classification. IEEE Trans. Syst. Man Cybern. Part B (Cybern.) **42**(2), 513–529 (2011)
12. Huang, G.B., Zhu, Q.Y., Siew, C.K.: Extreme learning machine: theory and applications. Neurocomputing **70**(1–3), 489–501 (2006)
13. Kaur, R., Roul, R.K., Batra, S.: A hybrid deep learning CNN-ELM approach for parking space detection in smart cities. Neural Comput. Appl. **35**, 13665–13683 (2023)
14. Kaur, R., Roul, R.K., Batra, S.: Multilayer extreme learning machine: a systematic review. Multimed. Tools App. **82**, 1–39 (2023). https://doi.org/10.1007/s11042-023-14634-4
15. Kheradmandi, N., Mehranfar, V.: A critical review and comparative study on image segmentation-based techniques for pavement crack detection. Constr. Build. Mater. **321**, 126162 (2022)
16. Kujur, A., Raza, Z., Khan, A.A., Wechtaisong, C.: Data complexity based evaluation of the model dependence of brain MRI images for classification of brain tumor and alzheimer's disease. IEEE Access **10**, 112117–112133 (2022). https://doi.org/10.1109/ACCESS.2022.3216393

17. Li, H., Zhao, H., Li, H.: Neural-response-based extreme learning machine for image classification. IEEE Trans. Neural Netw. Learn. Syst. **30**(2), 539–552 (2018)
18. Mei, Q., Gül, M.: A cost effective solution for pavement crack inspection using cameras and deep neural networks. Constr. Build. Mater. **256**, 119397 (2020)
19. Murphy, K.P., et al.: Naive Bayes classifiers. Univ. B. C. **18**(60), 1–8 (2006)
20. Oliveira, H., Correia, P.L.: Automatic road crack detection and characterization. IEEE Trans. Intell. Transp. Syst. **14**(1), 155–168 (2012)
21. Phalak, P., Bhandari, K., Sharma, R.: Analysis of decision tree-a survey. Int. J. Eng. Res. **3**(3), 1–6 (2014)
22. Roul, R.K.: Detecting spam web pages using multilayer extreme learning machine. Int. J. Big Data Intell. **5**(1–2), 49–61 (2018)
23. Roul, R.K.: Suitability and importance of deep learning feature space in the domain of text categorisation. Int. J. Comput. Intell. Stud. **8**(1–2), 73–102 (2019)
24. Roul, R.K.: Impact of multilayer elm feature mapping technique on supervised and semi-supervised learning algorithms. Soft. Comput. **26**(1), 423–437 (2022)
25. Roul, R.K., Agarwal, A.: Feature space of deep learning and its importance: comparison of clustering techniques on the extended space of ML-ELM. In: Proceedings of the 9th Annual Meeting of the Forum for Information Retrieval Evaluation (2017)
26. Roul, R.K., Asthana, S.R., Kumar, G.: Study on suitability and importance of multilayer extreme learning machine for classification of text data. Soft. Comput. **21**(15), 4239–4256 (2017)
27. Roul, R.K., Bhalla, A., Srivastava, A.: Commonality-rarity score computation: a novel feature selection technique using extended feature space of ELM for text classification. In: Proceedings of the 8th Annual Meeting of the Forum for Information Retrieval Evaluation (2016)
28. Roul, R.K., Satyanath, G.: A novel feature selection based text classification using multi-layer ELM. In: Roy, P.P., Agarwal, A., Li, T., Krishna Reddy, P., Uday Kiran, R. (eds.) Big Data Analytics. BDA 2022. LNCS, vol. 13773, pp. pp. 33–52. Springer, Cham (2022). https://doi.org/10.1007/978-3-031-24094-2_3
29. Rujirakul, K., So-In, C.: Histogram equalized deep PCA with ELM classification for expressive face recognition. In: 2018 International Workshop on Advanced Image Technology (IWAIT), pp. 1–4. IEEE (2018)
30. Satyanath, G., Sahoo, J.K., Roul, R.K.: Smart parking space detection under hazy conditions using convolutional neural networks: a novel approach. Multimed. Tools Appl. **82**(10), 15415–15438 (2023)
31. Suthaharan, S.: Support vector machine. In: Machine Learning Models and Algorithms for Big Data Classification. ISIS, vol. 36, pp. 207–235. Springer, Boston, MA (2016). https://doi.org/10.1007/978-1-4899-7641-3_9
32. Vishnoi, V.K., Kumar, K., Kumar, B., Mohan, S., Khan, A.A.: Detection of apple plant diseases using leaf images through convolutional neural network. IEEE Access **11**, 6594–6609 (2022)
33. Xiao, S., Shang, K., Lin, K., Wu, Q., Gu, H., Zhang, Z.: Pavement crack detection with hybrid-window attentive vision transformers. Int. J. Appl. Earth Obs. Geoinf. **116**, 103172 (2023)
34. Yang, F., Zhang, L., Yu, S., Prokhorov, D., Mei, X., Ling, H.: Feature pyramid and hierarchical boosting network for pavement crack detection. IEEE Trans. Intell. Transp. Syst. **21**(4), 1525–1535 (2019)
35. Zakeri, H., Nejad, F.M., Fahimifar, A.: Image based techniques for crack detection, classification and quantification in asphalt pavement: a review. Arch. Comput. Methods Eng. **24**, 935–977 (2017)

36. Zhang, K., Zhang, Y., Cheng, H.D.: CrackGAN: pavement crack detection using partially accurate ground truths based on generative adversarial learning. IEEE Trans. Intell. Transp. Syst. **22**(2), 1306–1319 (2020)
37. Zhang, L., Yang, F., Daniel Zhang, Y., Zhu, Y.J.: Road crack detection using deep convolutional neural network. In: 2016 IEEE International Conference on Image Processing (ICIP), pp. 3708–3712 (2016). https://doi.org/10.1109/ICIP.2016.7533052
38. Zhong, H., Miao, C., Shen, Z., Feng, Y.: Comparing the learning effectiveness of BP, ELM, I-ELM, and SVM for corporate credit ratings. Neurocomputing **128**, 285–295 (2014)
39. Zou, Q., Zhang, Z., Li, Q., Qi, X., Wang, Q., Wang, S.: Deepcrack: learning hierarchical convolutional features for crack detection. IEEE Trans. Image Process. **28**(3), 1498–1512 (2018)

Potpourri

Fast Similarity Search in Large-Scale Iris Databases Using High-Dimensional Hashing

Abhishek Pratap Singh, Debanjan Sadhya(✉), and Santosh Singh Rathore

ABV-Indian Institute of Information Technology and Management Gwalior, Gwalior, India
{imt_2018007,debanjan,santoshs}@iiitm.ac.in

Abstract. Iris recognition is a widely used biometric identification technology that relies on the accurate and efficient matching of iris images. However, fast matching in large databases poses a significant challenge due to the increasing search time for a given query. To address this problem, this paper proposes an end-to-end hashing framework for iris recognition tasks based on the *DenseFly* algorithm. The presented approach utilizes a deep convolutional neural network to extract features from iris images and then applies hashing to map the features into compact binary codes. This process enables efficient retrieval of the query iris templates by reducing the whole search space. To evaluate and compare the proposed method with the existing IHashNet approach, we conduct experiments on three publicly available iris datasets namely CASIA-Irisv4-Thousand, UBIRIS.v2 and CASIA-Irisv1-Lamp. Our simulation results demonstrate that the proposed method outperforms IHashNet in terms of retrieval accuracy and equal error rate (EER). Furthermore, our method achieves significantly lower query time over all the datasets, thereby vindicating its usage over large iris datasets.

Keywords: Iris · Hashing · Searching · CNN · DenseFly

1 Introduction

Biometric recognition systems have become increasingly popular for their ability to provide secure and efficient personal identification. Among various biometric modalities, iris recognition has garnered considerable attention due to its high accuracy, non-invasive nature, and unique patterns that remain stable over time [4]. Image acquisition in iris recognition systems entails capturing high-quality images of the iris, while preprocessing involves steps such as segmentation and normalization to generate a consistent representation of the iris pattern. Thereafter, feature extraction methods are employed to convert the processed iris image into a compact binary code, which serves as a unique identifier for the individual [4]. These iris codes can then be compared for matching purposes, enabling the system to verify or identify individuals based on their iris patterns.

V. Goyal et al. (Eds.): BDA 2023, LNCS 14418, pp. 187–200, 2023.
https://doi.org/10.1007/978-3-031-49601-1_13

Iris recognition systems can be classified into two categories based on the type of matching performed: one-to-one (1:1) and one-to-many (1:N). In 1:1 matching, also known as verification, the system compares the input iris code against a single reference iris code associated with a claimed identity. In contrast, 1:N matching, also known as identification, involves comparing the input iris code against a database of multiple reference iris codes to determine the identity of the individual. As the size of the iris database increases, the search time for a given query becomes longer, posing challenges for real-time biometric applications such as access control systems, border security, and e-governance [6]. To address this issue, researchers have explored various methods for accelerating biometric matching, such as approximate nearest neighbor (ANN) search techniques. Locality-sensitive hashing (LSH) is one such technique that has been extensively studied for its potential to speed up nearest-neighbor search in high-dimensional spaces [2]. LSH works by partitioning data points into multiple hash buckets, where similar data points are likely to fall into the same bucket. This approach significantly reduces the search space and, consequently, the time required for nearest-neighbor search. However, the performance of LSH can be limited when dealing with high-dimensional data, such as iris codes, due to the curse of dimensionality [2].

DenseFly hashing, introduced by Sharma *et al.* [7], is a hierarchical approach to ANN search that has demonstrated superior performance over traditional LSH methods in terms of query time and accuracy for high-dimensional data. The hashing process involves a normalization step to remove the mean and adjust the variance of the input data. It is subsequently followed by a random projection step to map the data onto a low-dimensional space. The resulting binary codes are then hashed into multiple levels of buckets to construct a hierarchical index structure. During the query phase, the hierarchical index structure is traversed to narrow down the search space and locate the approximate nearest neighbors. DenseFly hashing has been successfully employed in various applications, including image retrieval and face recognition [7,9]. However, the use of DenseFly hashing for large-scale iris recognition has not been thoroughly investigated (to the best of the knowledge of the authors).

1.1 Contributions

In this study, we explore the potential of using DenseFly hashing for large-scale iris recognition and retrieval systems. Our proposed method combines the advantages of deep convolutional neural networks (CNNs) for feature extraction and DenseFly hashing for efficient retrieval. Our model initially utilizes a CNN to extract high-dimensional features from iris images, which are then transformed into compact binary codes using DenseFly hashing. The hierarchical index structure created by DenseFly enables efficient nearest-neighbor search by narrowing down the search space and locating the approximate nearest neighbors. The main contributions of this study are highlighted as follows:

- We propose an end-to-end method for iris recognition based on the DenseFly algorithm. Our approach significantly reduces the search space and improves retrieval performance for high-dimensional iris data.
- We conduct extensive experiments on three benchmark iris datasets to evaluate and compare the proposed method with existing methods. Our method outperforms IHashNet [8] in terms of query time and matched the performance in terms of equal error rate, hit rate, and penetration rate.

2 Related Work

This section discusses recent advancements in large-scale iris recognition. Recently, deep learning-based methods have been proposed for iris recognition, which leverages the power of convolutional neural networks (CNNs) for feature extraction and various hashing techniques for efficient retrieval. Wen *et al.* [11] proposed a deep learning-based method that combines CNNs for feature extraction and product quantization for efficient ANN search. Their method demonstrated improved performance compared to traditional methods in terms of retrieval accuracy and query time. Zhang *et al.*, [12] proposed an efficient iris recognition algorithm that uses a compact representation of iris images to reduce computational complexity. The algorithm was tested on a large dataset of over 1.8 million iris images and showed a significant reduction in computational complexity without sacrificing recognition accuracy.

In order to reduce the computational complexity of high-dimensional data, several variants of locality-sensitive hashing (LSH) had been proposed, such as Multi-probe LSH [3] and Spectral hashing [10]. Spectral hashing aims to minimize the computational complexity of high-dimensional data by utilizing a hashing function that map the data onto a lower-dimensional space. This enables fast and efficient indexing and retrieval of data while preserving its inherent structure. However, this method faced challenges, such as sensitivity to the choice of the number of hash bits used, which could impact the retrieval accuracy of the system. Additionally, spectral hashing did not account for data correspondence estimation. These methods have demonstrated improved performance on high-dimensional datasets but still suffer from limitations such as high query times or memory requirements in image retrieval.

Ahmed *et al.* [1] proposed a hash-based space partitioning approach for iris biometric data indexing. The proposed approach utilizes a combination of multiple techniques, including locality-sensitive hashing (LSH), space partitioning, and vector quantization. By leveraging these techniques, the authors aim to achieve a more efficient indexing process and improve retrieval accuracy for high-dimensional iris data. Their method involves partitioning the feature space into smaller regions and employing LSH to create compact binary codes, which are used as indexing keys. Singh *et al.*proposed a technique called IHashNet [8], an end-to-end system for nearest neighbor search in large-scale iris recognition. IHashNet employed a deep neural network-based approach to generate hash codes for iris codes, which were then indexed using optimized multi-index hashing. The hash codes were generated by thresholding the output of the network, resulting in a binary code that was more efficient to store and index. The multi-index hashing scheme used in IHashNet partitioned the hash codes into smaller subspaces indexed separately, reducing the search space and improving the efficiency of the similarity search. IHashNet has shown promising results in accuracy and speed when compared to other state-of-the-art methods.

Despite the advances in large-scale iris recognition, there still exist challenges regarding high-dimensional data, computational complexity, and retrieval accuracy. Many existing methods are sensitive to the choice of parameters or are limited by high query times and memory requirements, which hinder their practical deployment in large-scale applications. Furthermore, some of these techniques do not take into account data correspondence estimation, which may lead to suboptimal performance in real-world scenarios. Our framework addresses these gaps by combining the strengths of deep convolutional neural networks for feature extraction and DenseFly hashing for efficient retrieval. The hierarchical indexing structure employed in this approach significantly reduces the search space, thus improving retrieval performance for high-dimensional iris data. All these design principles make our approach suited for large-scale iris recognition systems.

3 DenseFly Details

DenseFly [7] introduced a unique approach to generate hash codes that preserve similarity information in the original feature space. Unlike previous methods that rely on pre-trained deep neural networks to embed feature vectors into a lower-dimensional space, DenseFly directly projects real-valued feature vectors onto a dense subspace, effectively addressing the curse of dimensionality. This results in more efficient similarity searches in high-dimensional spaces. DenseFly's hashing technique requires only one hash table to index a database, which significantly reduces the computational complexity of the system.

3.1 Calculating Hash Values

The first stage in the DenseFly algorithm is computing the hash values. Real-valued feature vectors are projected onto a dense mk-dimensional space by multiplying them with a random binary matrix of dimension mk. Hash values are then calculated by thresholding the projected vectors. For each element in the projected vector, if the value is greater than or equal to 0, the corresponding index in the hash code is set to 1; otherwise, it is set to 0. This process effectively transforms the real-valued feature vectors into binary hash vectors. DenseFly generates both high-dimensional $\mathbf{h}_1$ and low-dimensional $\mathbf{h}_2$ hashes for more efficient nearest neighbor searches. The low-dimensional pseudo-hash function $\mathbf{h}_2$ helps reduce the search space and the time required to find the nearest neighbor. The high-dimensional hash, on the other hand, is employed to rank the candidates for a query.

3.2 Nearest Neighbor Search

The second stage of DenseFly hashing is binning. It is performed using a query's $\mathbf{h}_2$ hash to obtain the nearest neighbors. These nearest neighbors are then ranked according to their Hamming distance from the query in the high-dimensional hash space $\mathbf{h}_1$, in decreasing order.

4 Methodology

This work proposes an end-to-end feature extraction, indexing, and matching system to maximize accuracy and efficiency in subject matching using iris images. The system is designed to extract feature vectors from iris images, index the database according to the proposed system, and match a query to its nearest neighbors. This work aims to minimize the query time and improve the efficiency and accuracy of subject matching in large-scale iris image databases. The entire framework is diagrammatically illustrated in Fig. 1.

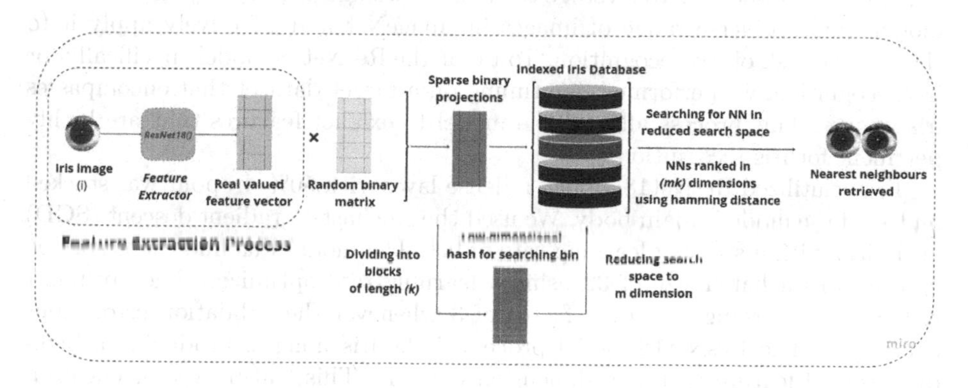

Fig. 1. Diagrammatic representation of the proposed end-to-end iris recognition system with emphasis on efficient searching.

4.1 Pre-processing

We initially conducted essential pre-processing steps to enhance the quality and discriminability of the iris images. We applied adaptive histogram equalization for enhancing the overall image quality and accentuating the intricate details within the iris region. This pre-processing step is crucial in addressing illumination variations and improving the discriminative features of the iris. Furthermore, our pre-processing pipeline encompassed additional measures to ensure the quality and consistency of the input data. Firstly, we employed min-max scaling to normalize the pixel values of the iris images, guaranteeing a uniform range conducive to efficient processing. Additionally, we utilized contour detection methods from the OpenCV library to accurately delineate the iris region in the input images. Lastly, data augmentation techniques, such as random rotations and translations were incorporated to further enhance the robustness and generalizability of the model. Following the pre-processing stage, the iris images were resized to a standardized size of 224×224 pixels. This resizing operation ensured that the images possessed a consistent dimensionality, a prerequisite for feeding them into ResNet18.

4.2 Feature Extraction

The feature extraction stage assumes a critical role in iris recognition systems as it directly impacts the quality and discriminative power of the extracted features. In this study, we adopted the ResNet architecture. ResNet has established its efficacy as a feature extraction model in numerous studies [1,3]. Specifically, we employed the ResNet18 model, which has demonstrated exceptional performance in tasks involving image analysis and understanding. To leverage the capabilities of the ResNet18 model, we employed transfer learning by utilizing a pre-trained model initially trained on the extensive ImageNet database. This transfer learning approach allowed us to leverage the rich knowledge acquired by the ResNet18 model from a diverse range of images in ImageNet and effectively apply it to the specific task of iris recognition. To tailor the ResNet18 model specifically for iris recognition, we performed fine-tuning on a target dataset that encompasses iris images. This process allowed the model to extract features that are highly pertinent for iris recognition.

In our utilized ResNet18 model, a dense layer with 50% dropout was stacked on top of the model's main body. We used the stochastic gradient descent (SGD) optimizer with a starting learning rate of 0.1. The model was fine-tuned for 300 epochs with a batch size of 32 using a learning rate optimizer. The optimizer reduced the learning rate by a factor of 3 whenever the validation error stops decreasing. The ResNet18 model processed the iris images, producing a high-dimensional feature vector of dimension (512, 1). This feature vector encapsulated the salient characteristics of the iris, crucial for accurate recognition.

To assess the performance of our feature extractor, we employed Stratified K-fold cross-validation. Our dataset was divided into five folds, ensuring that each fold maintained a balanced representation of iris images from different subjects for each iris. We partitioned the data into two sets: 80% for training and 20% for validation. The resulting high-dimensional feature vectors effectively captured the distinctive characteristics of the iris, thereby augmenting the accuracy of iris recognition systems. The baseline metrics obtained from the ResNet model are presented in Table 1.

Table 1. Baseline accuracy obtained on the pre-trained ResNet18 based classifier.

Model	Accuracy	F1-score	Precision	Recall
ResNet18	98.49%	0.9648	98.54%	97.89%

4.3 Indexing and Matching

Algorithm 1 discusses in detail the process for efficient and scalable image retrieval based on DenseFly hashing. The algorithm takes as input a dataset of images D, the number of bits in the hash code m, the winner-take-all (WTA) factor k, the sampling rate α, and the threshold for similarity comparison τ. The output of the algorithm is the top n nearest neighbors for a given query image.

We begin by computing the feature vector for each image in the dataset using a fine-tuned ResNet18 model. Next, we generate mk sparse binary random projections, which serve as the basis for computing high-dimensional hashes ($\mathbf{h}_1$) and low-dimensional pseudo-hashes ($\mathbf{h}_2$) for each feature vector, using the DenseFly algorithm. The images are subsequently inserted into a hash table H with keys corresponding to their pseudo-hashes. When presented with a query image q, the algorithm computes its hash codes $\mathbf{h}_1(q)$ and $\mathbf{h}_2(q)$ using DenseFly. It then iteratively retrieves image IDs from the hash table H for each hash code in the query image q and calculates the Hamming distance between the high-dimensional hash of the query image and the high-dimensional hash of the candidate images. If the Hamming distance is less than or equal to the similarity threshold τ, the candidate image is added to the set of candidate matches S.

A numerical representation of the calculation of the hash codes is provided in Fig. 2. Here, a randomly generated sample of k bits is first projected into the sparse mk dimension using a random binary matrix with a 10% sampling rate. The resultant matrix of this projection is $d \times mk$ dimension matrix; in this case, it is a 1D matrix of length mk. Sparse binary projections are then calculated by setting the index of all the negative elements to 0. This hash code is then subdivided into m blocks of length k. If the sum or mean of the block ≥ 0, the corresponding index of the low-dimensional hash is set to 1; otherwise 0.

4.4 Complexity Analysis

For analyzing the time complexity of our holistic model, we examine each individual phase of Algorithm 1 as follows. First, for a dataset of $|D|$ images, the complexity of computing feature vectors using ResNet18 is $\mathcal{O}(|D| \cdot C_{ResNet18})$, where $C_{ResNet18}$ denotes the time complexity of processing an image using the ResNet model. Next, generating the sparse binary random projections for a feature vector takes a time complexity of $\mathcal{O}(mk \cdot d)$. Subsequently, the complexity of computing the hash codes for each image is $\mathcal{O}(|D| \cdot C_{DenseFly})$, where $C_{DenseFly}$ represents the time complexity of the proposed system algorithm for a single image. Assuming that the hash table insert operation has an average-case complexity of $\mathcal{O}(1)$, inserting the database has a complexity of $\mathcal{O}(|D|)$.

Computing hash codes for the query image yields a complexity of $\mathcal{O}(C_{DenseFly})$. In the worst case, all images in the dataset could be candidates, retrieving image IDs and computing Hamming distance yields a complexity of $\mathcal{O}(|D| \cdot mk)$. Hence, the complexity of sorting candidate images is $\mathcal{O}(|D| \cdot \log |D|)$. Thus, the overall time complexity of the algorithm becomes $\mathcal{O}(|D| \cdot (C_{ResNet18} + C_{DenseFly} + mk + \log |D|))$, where m is the number of bits

Algorithm 1. Approach to index and find the first n nearest neighbors of a query iris image using DenseFly.

Input : Dataset of images D
number of bits in hash code m
WTA factor k
sampling rate α
threshold for similarity comparison τ
Output: Top n nearest neighbors for a given query image

$S \leftarrow \emptyset$;

$H \leftarrow \emptyset$;

Compute feature vector using a pre-trained ResNet18;
foreach $i \in D$ **do**
  $\mathbf{v}_i = \mathbf{ResNet18(i)}$;
end
Generate dense binary random projections;
$S = S_i \mid S_i = \text{rand}(\lfloor \alpha d \rfloor, d)$, where $|S| = mk$;
Compute high-dimensional hash $\mathbf{h}_1$ and low-dimensional pseudo-hash $\mathbf{h}_2$, for each feature vector;
foreach $i \in D$ **do**
  $\mathbf{h}_1(i) = \text{DenseFly}(\mathbf{v}_i, m, k, \alpha)$;
  $\mathbf{h}_2(i) = \text{PseudoHash}(\mathbf{h}_1(i))$
  Insert i into hash table H with key $\mathbf{h}_2(i)$
end
Compute hash codes $\mathbf{h}_1(q)$ and $\mathbf{h}_2(q)$ for query image q using the DenseFly algorithm;

Retrieve image IDs from hash table H for each hash code in query image q and compute Hamming distance;

foreach $i \in H[\mathbf{h2}(q)]$ **do**
  $d_H(\mathbf{h1}(q), \mathbf{h1}(i)) = \sum j = 1^{mk} |\mathbf{h1}j(q) - \mathbf{h1}j(i)|$**if** $d_H(\mathbf{h}_1(q), \mathbf{h}_1(i)) \leq \tau$ **then**
    Add i to set of candidate matches S
  end
end
Sort images in S by ascending order of $d_H(\mathbf{h}_1(q), \mathbf{h}_1(i))$;

Return top n images in S;

in the hash code, k is the WTA factor, and d is the dimensionality of the feature vector. The time complexity is linear with respect to the size of the dataset, making it suitable for large-scale image retrieval tasks. However, it should be noted that in practice, the number of candidate images ($|S|$) is typically much smaller than the dataset size ($|D|$), resulting in faster retrieval times.

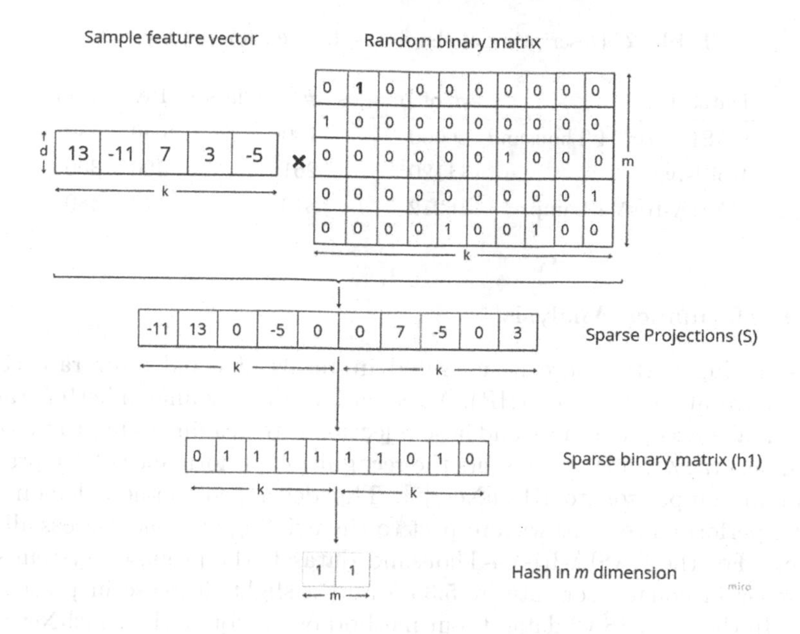

Fig. 2. A numerical illustration of a sample query in DenseFly.

5 Experimental Results and Analysis

The proposed model is implemented in the PyTorch library[1]. The model and ImageNet weights used are open-source and are available in the PyTorch library. Evaluation and training of the proposed system and the previous approach are performed on a Linux system having Intel's i9 processor (13th generation) with 32 GB memory with an NVIDIA A4000 GPU.

5.1 Datasets

In this study, we utilized three iris databases to evaluate the performance of our proposed method. Table 2 presents an overview of these datasets, which include the CASIA-IrisV4-Thousand, CASIA-IrisV4-Lamp[2] and UBIRIS.v2 [5] databases. All three databases are large-scale, contain high-quality images, and have been carefully annotated. The CASIA-IrisV4-Thousand and CASIA-IrisV4-Lamp datasets are both captured using near-infrared (NIR) imaging, while the UBIRIS.v2 dataset is captured using a visible-light imaging system. These databases have been used in a number of published works and have been widely used in large-scale iris recognition applications due to their large subject population and their comprehensive collection of iris images captured under varying illumination conditions.

[1] https://github.com/pytorch/vision.

[2] CASIA Iris Image Database, http://biometrics.idealtest.org/.

Table 2. Description of the datasets used for simulations.

Dataset	# of Images	# of Classes	Resolution
CASIA-IrisV4-Thousand	20000	1000	640 × 480
Ubiris.v2	11120	261	400 × 300
CASIA-IrisV4-Lamp	16212	819	640 × 480

5.2 Performance Analysis

The matching performance is measured in terms of equal error rate (EER), precision, recall, and hit rate (HR). A decrease in EER signifies a better trade-off between false acceptance rate and false rejection rate, leading to improved overall system accuracy. Table 3 represents the recognition performance of the proposed system in comparison to IHashNet [8]. The developed method demonstrates varying performance gains with respect to the existing approach across all three datasets. For the CASIA-IrisV4-Thousand dataset, the proposed system shows a decrease in equal error rate by 5.36% and a slight decrease in precision by 0.26%. In the UBIRIS.v2 dataset, our method outperforms the IHashNet with a 15.16% improvement in EER and a minor decrease in precision by 0.49%. Lastly, for the CASIA-IrisV4-Lamp dataset, the proposed method exhibits a significant improvement with a 44.33% gain in EER while showing a slight decrease in precision by 0.33%. Overall, the method displays competitive performance across all datasets. The associated FAR vs FRR curves and score distribution graphs are illustrated in Fig. 3 and Fig. 4 respectively.

Hit Rate is calculated as the percentage of correct query retrievals out of the total number of iris images in the dataset. For the UBIRIS.v2 and CASIA-Irisv4-Thousand datasets, the maximum HR was achieved as 99.28% and 98.25% respectively. While there is a slight decrease in HR compared to the previous approach, the proposed method maintains competitive matching performance. Thus, the improvements in all these metrics demonstrate the superiority of the proposed method over IHashNet for iris recognition tasks. The corresponding Receiver operating characteristic (ROC) curves are presented in Fig. 5.

5.3 Timing Analysis

In this section, we conduct a thorough analysis of the query resolution time and penetration rate of the proposed system. The DenseFly-based approach significantly reduces the query time across all datasets, leading to more efficient search performance. For the large-scale CASIA-IrisV4-Thousand, the query time was reduced from 195.54 s to 13.97 s. Similarly, for the UBIRIS.v2 dataset, the query time was reduced from 107.65 s to 10.45 s. Finally, for the CASIA-IrisV4-Lamp dataset, the query time went from 181.50 s to 10.85 s. Hence, we observed a

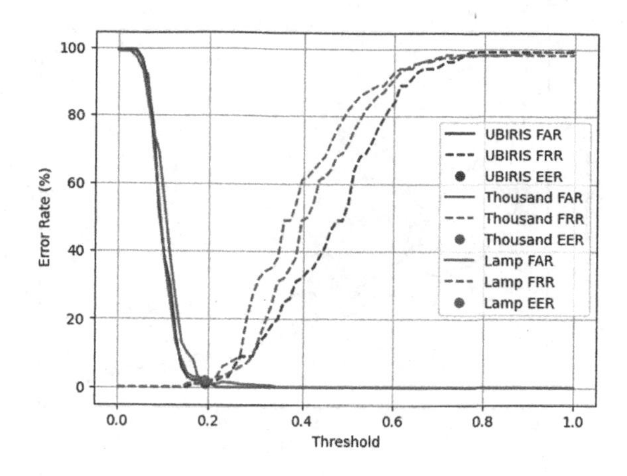

Fig. 3. Relation between the False Non-Match Rate (FNMR) and False Match Rate (FMR) for the three datasets.

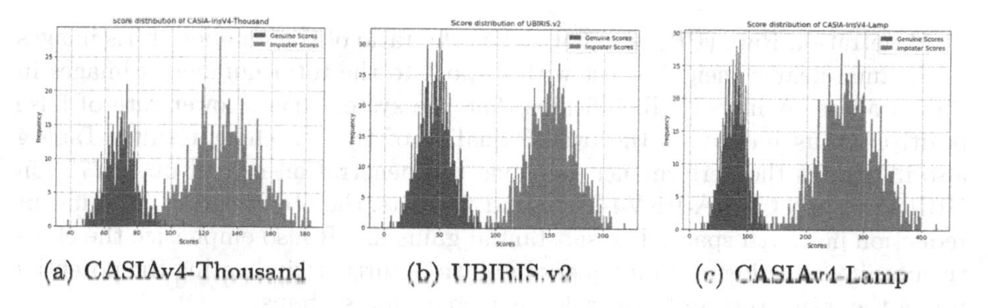

(a) CASIAv4-Thousand (b) UBIRIS.v2 (c) CASIAv4-Lamp

Fig. 4. Distribution of the Genuine and Imposter scores over the three iris datasets.

Table 3. Comparative analysis of the proposed model with IHashNet in terms of EER, Precision and HR.

Dataset	Method	EER	Precision	HR
CASIAv4-Thousand	IHashNet	2.57	98.19	99.64
	Proposed	1.91	98.64	98.25
UBIRIS.v2	IHashNet	2.17	98.74	99.16
	Proposed	1.25	99.29	98.28
CASIAv4-Lamp	IHashNet	1.25	99.55	98.12
	Proposed	2.0	99.87	98.15

substantial improvement in query time across all three datasets, with reductions of up to 94%. Importantly, this significant reduction in query time contributes to the practicality and usability of the framework, particularly for large-scale applications.

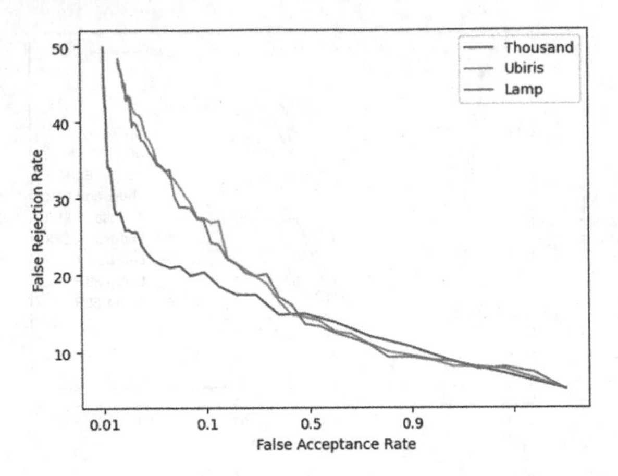

Fig. 5. The ROC curve obtained for the proposed DenseFly-based model.

Penetration Rate (PR) is calculated as the ratio of the number of iris images in the final nearest neighbor list with respect to the total number of images in the database. A higher PR indicates that the system has a lower rate of false positives, thus improving the overall quality of the matching results. Table 4 also illustrates the performance gains via the penetration rate of up to 77% on UBIRIS.v2 and CASIA-IrisV4-Thousand datasets, thereby ensuring a significant reduction in search space. The substantial gains in PR also emphasize the effectiveness in delivering accurate query resolution, further vindicating the potential use of the framework in large-scale iris recognition systems.

Table 4. Comparison of query time and penetration rate (PR) across the three iris datasets. The reported query time is computed as the average query time over 1000 queries.

Dataset	Method	Time (s)	PR
CASIA.v4-Thousand	IHashNet	195.54	1.892
	Proposed	13.97	1.26
UBIRIS.v2	IHashNet	107.65	2.74
	Proposed	10.45	0.63
CASIA.v4-Lamp	IHashNet	181.50	2.76
	Proposed	10.85	0.63

All the simulation results suggest that DenseFly is more effective in achieving better retrieval accuracy and lower query time for iris recognition tasks. Improvements in the PR suggest that the system effectively narrows down the search space for iris images. Hence, the proposed DenseFly-based hashing technique is effective in achieving better retrieval accuracy and lower query time compared to the existing IHashNet mechanism for the iris recognition and retrieval task over large-scale datasets.

6 Conclusion

In this work, we propose an end-to-end framework for iris recognition and searching using the DenseFly algorithm. The hierarchical indexing structure of DenseFly significantly reduces the search space during query time and consequently improves the retrieval performance. We evaluated our method on three benchmark datasets, CASIA-IrisV4-Thousand, UBIRIS.v2, and CASIA-IrisV4-Lamp. The results demonstrated that the proposed method using DenseFly outperforms the existing IHashNet in terms of query time and PR, and matched the performance in terms of EER and HR. The combination of reduced query time, competitive HR, and improved PR makes our developed framework an effective method for enhancing the iris recognition model's speed and accuracy. The performance gains directly impact the nearest neighbor retrieval of iris images, making it an ideal choice for large-scale iris applications.

Acknowledgements. This work is supported by the Start-up Research Grant (SRG), Science and Engineering Research Board (SERB), Government of India [Grant Number(s): SRG/2021/000051 and SRG/2021/000173].

References

1. Ahmed, T., Sarma, M.: Hash-based space partitioning approach to iris biometric data indexing. Expert Syst. Appl. **134**, 1–13 (2019). https://doi.org/10.1016/j.eswa.2019.05.026, https://www.sciencedirect.com/science/article/pii/S0957417419303550
2. Indyk, P., Motwani, R.: Approximate nearest neighbors: towards removing the curse of dimensionality. In: Proceedings of the Thirtieth Annual ACM Symposium on Theory of Computing, pp. 604–613. STOC 1998, Association for Computing Machinery, New York, NY, USA (1998). https://doi.org/10.1145/276698.276876
3. Lv, Q., Josephson, W., Wang, Z., Charikar, M., Li, K.: Multi-probe LSH: efficient indexing for high-dimensional similarity search. In: 33rd International Conference on Very Large Data Bases, VLDB 2007 - Conference Proceedings, pp. 950–961. Association for Computing Machinery, Inc. (2007)
4. Nazmdeh, V., Mortazavi, S., Tajeddin, D., Nazmdeh, H., Asem, M.M.: Iris recognition; from classic to modern approaches. In: 2019 IEEE 9th Annual Computing and Communication Workshop and Conference (CCWC), pp. 0981–0988 (2019). https://doi.org/10.1109/CCWC.2019.8666516
5. Proenca, H., Filipe, S., Santos, R., Oliveira, J., Alexandre, L.A.: The UBIRIS.V2: a database of visible wavelength iris images captured on-the-move and at-a-distance. IEEE Trans. Pattern Anal. Mach. Intell. **32**(8), 1529–1535 (2010). https://doi.org/10.1109/TPAMI.2009.66
6. Proenca, H.: Iris recognition: on the segmentation of degraded images acquired in the visible wavelength. IEEE Trans. Pattern Anal. Mach. Intell. **32**(8), 1502–1516 (2010). https://doi.org/10.1109/TPAMI.2009.140

7. Sharma, J., Navlakha, S.: Improving similarity search with high-dimensional locality-sensitive hashing (2018)
8. Singh, A., Gaurav, P., Vashist, C., Nigam, A., Yadav, R.P.: Ihashnet: Iris hashing network based on efficient multi-index hashing. In: 2020 IEEE International Joint Conference on Biometrics (IJCB), pp. 1–9 (2020). https://doi.org/10.1109/IJCB48548.2020.9304925
9. Wang, H., et al.: CosFace: large margin cosine loss for deep face recognition. In: 2018 IEEE/CVF Conference on Computer Vision and Pattern Recognition, pp. 5265–5274 (2018). https://doi.org/10.1109/CVPR.2018.00552
10. Weiss, Y., et al. (eds.) Advances in Neural Information Processing Systems. vol. 21. Curran Associates, Inc. (2008). https://proceedings.neurips.cc/paper_files/paper/2008/file/d58072be2820e8682c0a27c0518e805e-Paper.pdf
11. Wen, Y., Zhang, K., Li, Z., Qiao, Yu.: A discriminative feature learning approach for deep face recognition. In: Leibe, B., Matas, J., Sebe, N., Welling, M. (eds.) ECCV 2016. LNCS, vol. 9911, pp. 499–515. Springer, Cham (2016). https://doi.org/10.1007/978-3-319-46478-7_31
12. Zhang, W., Lu, X., Gu, Y., Liu, Y., Meng, X., Li, J.: A robust iris segmentation scheme based on improved u-net. IEEE Access **7**, 85082–85089 (2019). https://doi.org/10.1109/ACCESS.2019.2924464

Explaining Finetuned Transformers on Hate Speech Predictions Using Layerwise Relevance Propagation

Ritwik Mishra[1(✉)], Ajeet Yadav[1], Rajiv Ratn Shah[1], and Ponnurangam Kumaraguru[2]

[1] Indraprastha Institute of Information Technology, Delhi, India
{ritwikm,ajeet19010,rajivratn}@iiitd.ac.in
[2] International Institute of Information Technology, Hyderabad, India
pk.guru@iiit.ac.in

Abstract. Explainability of model predictions has become imperative for architectures that involve fine-tuning of a pretrained transformer *encoder* for a downstream task such as hate speech detection. In this work, we compare the explainability capabilities of three post-hoc methods on the HateXplain benchmark with different *encoders*. Our research is the first work to evaluate the effectiveness of Layerwise Relevance Propagation (LRP) as a post-hoc method for fine-tuned transformer architectures used in hate speech detection. The analysis revealed that LRP tends to perform less effectively than the other two methods across various explainability metrics. A random rationale generator was found to be providing a better interpretation than the LRP method. Upon further investigation, it was discovered that the LRP method assigns higher relevance scores to the initial tokens of the input text because fine-tuned *encoders* tend to concentrate the text information in the embeddings corresponding to early tokens of the text. Therefore, our findings demonstrate that LRP relevance values at the input of fine-tuning layers are not a good representative of the rationales behind the predicted score.

Keywords: LRP · LIME · SHAP · Hate Speech · Explainability

1 Introduction

Neural networks have gained extensive use in diverse applications such as natural language processing, speech recognition, and image recognition. Despite their widespread applicability, a significant critique of Deep Neural Networks is their opaque nature, which renders it challenging to comprehend how they arrive at their predictions [11]. Furthermore, there have been reports suggesting that these

Disclaimer: This study includes quotes of text considered profane or offensive to illustrate the model's workings but does not reflect the authors' views. The authors condemn online harassment and offensive language.

V. Goyal et al. (Eds.): BDA 2023, LNCS 14418, pp. 201–214, 2023.
https://doi.org/10.1007/978-3-031-49601-1_14

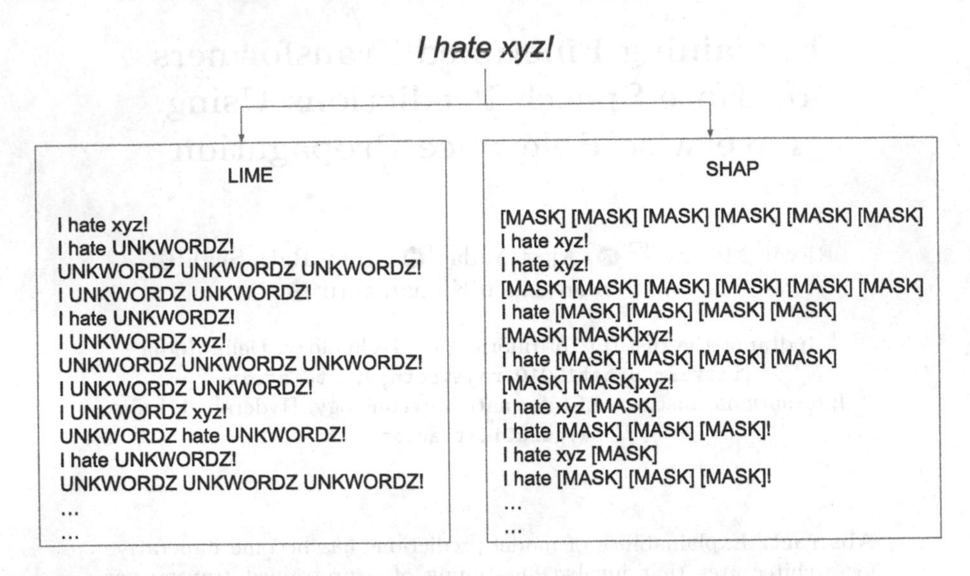

Fig. 1. Manipulation of text in LIME and SHAP techniques. Observe that LIME employs word-level masking within the provided string using the `UNKWORDZ` token, while SHAP employs subword-level masking using the `[MASK]` token. In the SHAP method, every altered text is fed to a fine-tuned model, and the relevance of subwords is calculated by the resulting change in model predictions. However, for the LIME method, a linear model is trained based on the modified text and its corresponding model output. This linear model is subsequently used to assign relevance values to each word.

models may exhibit biases towards individuals of particular races [17], genders [13,26], or ethnicity [1].

We used HateXplain benchmark [22], to evaluate the effectiveness of various post-hoc explainability methods. The benchmark includes not only gold class labels but also gold rationales. These gold labels are binary annotations provided by human annotators, classifying a given text as either containing hate speech (1) or not containing hate speech (0). The objective of hate speech detection revolves around predicting a score closer to 1 if the text includes hate speech and closer to 0 otherwise. Gold rationales are exclusively available for texts that annotators classify as containing hate speech. Originally introduced by [32], a gold rationale consists of a binary label vector with a length matching the number of words in the respective text. Therefore, our research inquiry for this study is as follows: *How much explainability LRP provides on the HateXplain benchmark compared to other post-hoc methods?*

In this study, we present a comparison of three Explainable AI (XAI) methods: Local Interpretable Model-agnostic Explanations (LIME) proposed by [25], SHapley Additive exPlanations (SHAP) by [19], and Layer-wise Relevance Propagation (LRP) presented by [6]. These methods were selected for their ability to generate post-hoc explanations (rationales) for models that are trained without access to explicit gold *rationales*.

The LIME method constructs a linear model by altering the provided text and generating additional samples in its vicinity. This learned model is subsequently employed to predict relevance values for each input feature. Conversely, the SHAP method utilizes partial dependency plots to compute Shapley values (as a proxy for relevance values) for each input feature. Drawing from game theory, a Shapley value for an input feature is established based on the alteration in model output when the input feature is present or absent in the manipulated text. Both approaches modify the given text to calculate relevance values for the input features. LIME treats words as input features, whereas SHAP considers subwords when the input is text and the model is a transformer-based fine-tuned model. Figure 1 illustrates the distinct ways text manipulation is carried out in LIME and SHAP.

To the best of our knowledge, no previous studies have utilized LRP or SHAP methods to assess the explainability of transformer-based hate speech prediction models on the HateXplain benchmark. Our study is the first to compare the three aforementioned XAI methods on HateXplain benchmark, and we implemented LRP method as a constituent of this work.

The present study concentrates on prediction models for detecting hate speech that employs pretrained transformer-based encoders (or simply *encoders* henceforth) to generate text embeddings. Such models have become popular because they have exhibited state-of-the-art performances in hate speech detection [2,20]. Furthermore, many works have emphasized the importance of interpretability in transformer-based architectures [8,26,28]. To the best of our knowledge, LRP method has not been applied to hate speech prediction models that utilizes pretrained *encoders*.

The LRP method operates by backpropagating relevance values. Specifically, the output layer of the hate speech model is used to determine the relevance values, which are then propagated backward through the network. The relevance value for a node j in layer L is determined by considering three factors: (a) the relevance values of all nodes in the succeeding layer $(L+1)$, (b) the learned parameters between L and $L+1$, and (c) the activation values in layer L. We refer readers to [24] for a detailed explanation of LRP.

Numerous previous studies have emphasized the societal implications of Explainable AI (XAI). The significant applications of XAI have been adequately outlined in the exhaustive survey conducted by [11]. Furthermore, a comparative investigation of two XAI techniques provides future researchers with an empirical rationale for selecting a particular method based on their task. Our research endeavors to investigate which method generates superior rationales and to present the insights derived from these rationales.

We release[1] our implementation of the LRP method for transformer-based text classification models. It provides word-level rationales for a predicted class in a multi-class classification setting. To the best of our knowledge, it is the first implementation of LRP method for fine-tuned transformer architectures.

[1] github.com/ritwikmishra/hateXplain-metrics-calculation.

2 Related Works

The literature has widely employed the LIME method to explain hate speech predictions [3,23]. However, using the LRP method to explain hate speech predictions has not been common. Karim et al. [16] have employed the LRP method to explain hate speech predictions for the *Bengali* language, but their model was based on the Long Short-Term Memory (LSTM) architecture, a variant of Recurrent Neural Networks (RNN). Previously, Arras et al. [4] used the LRP method to explain sentiment predictions by a RNN model. Similarly, the LRP method's explanatory potential has been demonstrated in intent detection with Bidirectional LSTMs [15]. The LRP method has been applied to transformer-based Neural Machine Translation (NMT) models to analyze the contributions of source and target tokens in the translation process [30]. However, to the best of our knowledge, there is no prior work that implements LRP to explain class predictions from a fine-tuned transformer-based model.

Our decision to employ the LRP method in this study is rooted in its ability to produce post-hoc explanations for generated predictions. Furthermore, it has been observed that LRP yields meaningful explanations in various tasks, including question classification and semantic role labeling [9]. The relevance values derived from LRP have also found utility in refining pretrained word embeddings [29]. Moreover, the LRP method has been applied in Layerwise Relevance Visualization (LRV) for sentence classifiers based on Graph Convolution Networks (GCN) [27].

Numerous studies have analyzed the performance of two or more XAI methods. In comparison to the LIME method, SHapley Additive exPlanations (SHAP) [19] has shown to provide better explanations for disease classification [12]. However, for various models from the finance domain, LIME rationales were found to be more stable than SHAP [21]. Balkir et al. [7] proposed scores based on *necessity* and *sufficiency* to explain the predictions of a hate speech detection model. After comparing the performance of LIME and SHAP with their proposed scores, LIME failed to generate relevant rationales for false-positive predictions. In a sentiment analysis task, Jørgensen et al. [14] compared the rationales generated by SHAP and LIME. It was observed that SHAP is more successful in selecting relevant spans from the text, whereas LIME rationales align better with the ranking of words in human rationales.

3 Experimental Setup

In this study, we used the *transformers* library [31] to load various *encoders*. We used the Ferret [5] library to predict rationales by LIME and SHAP methods because it is built on top of the official implementation of these methods and the tool generates subword-level rationales. We calculated word-level rationales from subword-level rationales using the `.word_ids()` function from Huggingface

```
>>> from transformers import AutoTokenizer
>>> tokenizer = AutoTokenizer.from_pretrained("bert-base-cased")
>>> print(tokenizer.tokenize('I hate xyz!'))
['I', 'hate', 'x', '##y', '##z', '!']
>>> print(tokenizer('I hate xyz!',add_special_tokens=False).word_ids())
[0, 1, 2, 2, 2, 3]
```

Fig. 2. Python code illustrating the output of the `.word_ids()` function from Huggingface (transformers) tokenizers. The provided example text is *I hate xyz!*. It can be observed that the word *xyz* is tokenized into three subwords: *x*, *##y*, and *##z*. Hence, the output of the `.word_ids()` function includes three instances of the index 2.

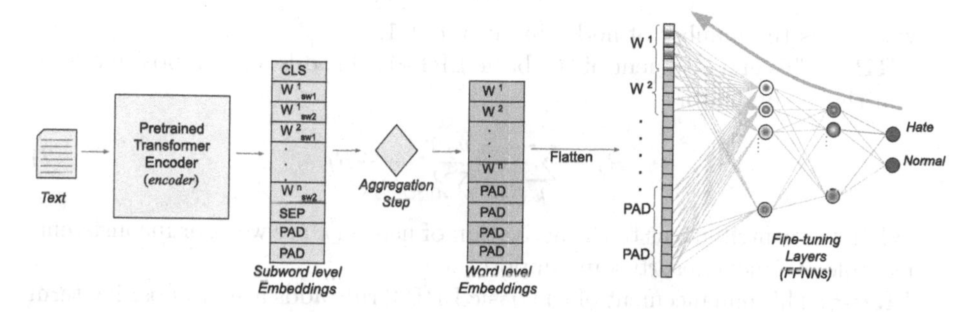

Fig. 3. Overall architecture of Hatespeech prediction model and relevance calculation using LRP method. The confidence score for the *Hate* label is highlighted in red. The predicted values are adopted as relevance scores at the output layer and subsequently propagated backward up to the initial fine-tuning layers (denoted by the red arrow), following the formulations introduced in the work by [24].

tokenizers. The output of the `.word_ids()` function is depicted in Fig. 2. We implemented different variants of LRP for the fine-tuning layers of the hate speech classifier due to the unavailability of any such tool. To test the validity of our LRP implementation, we ensured that the sum of relevance values remains consistent across each layer during the relevance backpropagation.

As shown in Fig. 3, we generated word-level embeddings by computing an unweighted average of the corresponding subword embeddings. This approach was adopted due to the presence of groundtruth rationales at the word level rather than the subword level. As a result, embeddings of special tokens (e.g., CLS and SEP) that indicate sentence boundaries were omitted, as there are no groundtruth rationales available for such tokens.

The flattened word-level embeddings are then passed through the fine-tuning layers, which consist of multiple Feed-forward Neural Networks (FFNNs). We employed three linear layers with a dropout rate of 0.1 and relu activation applied in between. The relevance values of the flattened word embeddings are summed up to obtain the relevance value of the corresponding word. Due to the difficulty in backpropagating relevances through the transformer architecture containing multi-head attention, we calculate relevance values only till the input layer of the fine-tuning module in this work. We implemented the following LRP variants by using formulations of Montavon et al. [24].

- **LRP-0**: It redistributes relevance in proportion to how much each input contributed to the activation of the neurons. Relevance of node j in layer L is calculated as:
$$R_j = \sum_k \frac{a_j w_{j,k}}{\sum_{j=0}^{j} a_j w_{j,k}} R_k \tag{1}$$
where k is the number of nodes in layer $L+1$.
- **LRP-ϵ**: This enhancement of the basic LRP-0 rule adds a small positive term ϵ in the denominator:
$$R_j = \sum_k \frac{a_j w_{jk}}{\epsilon + \sum_0^j a_j w_{jk}} R_k \tag{2}$$
When the contributions to the activation of neuron k are weak or inconsistent, the role of ϵ is to absorb some importance.
- **LRP-γ**: This enhancement of the basic LRP-0 rule adds a small positive term γ in the denominator:
$$R_j = \sum_k \frac{a_j.(w_{jk} + \gamma w_{jk}^{+})}{\sum_0^j a_j.(w_{jk} + \gamma w_{jk}^{+})} \cdot R_k \tag{3}$$
The parameter γ controls by how much positive contributions are favored. As γ increases, negative contributions start to disappear.

In their work, Montavon et al. [24] suggested utilizing gradients to compute relevance values. We discovered that gradient-based relevance values would not yield the desired results if bias is enabled in the fine-tuning layers. We refer readers to Appendix A for further details. Our LRP implementation supports the absence or presence of bias in the fine-tuning layers.

In all three explainability methods, a token is classified as relevant (1) or not-relevant (0) based on a threshold on its calculated relevance value. Similar to [22], we set the threshold value to 0.5 for all the methods considered in this study.

For the experimental investigation, we utilized two different *encoders*, namely *bert-base-cased* (BERT) by [10] and *roberta-base* (RoBERTA) by [18]. These encoders were chosen due to their prevalence in the literature for hate speech detection in English [2,20]. To explore the impact of the encoder on rationale prediction, we trained our hate speech prediction model while keeping the underlying encoder frozen or allowing it to be fine-tuned. All models were trained for

Table 1. A comparison of different post-hoc rationale generation methods (LIME, SHAP, and LRP) on hate speech prediction architectures with different *encoders*. The architectures where the underlying encoder was fine-tuned during the training phase are represented by the subscript $f-t$, whereas architectures where the encoder was frozen during training, are represented by the subscript fr. We report numbers on the official test set of the HateXplain benchmark. The numbers inside round brackets represent the mean and std over 3-fold cross-validation. An upward arrow signifies that higher values are preferable, while a downward arrow indicates that lower values are preferable. It is observed that fine-tuned RoBERTA achieves the best Performance metrics whereas fine-tuned BERT achieves the best Explainability metrics.

Encoder	Method	Performance		Explainability				
				Plausibility			Faithfulness	
		Accuracy↑	Macro-F1↑	IOU F1↑	Token F1↑	AUPRC↑	Compr.↑	Suff.↓
$BERT_{f-t}$	LRP	74 (74±0)	72 (71.7±0.6)	0.10 (0.11±0.0)	0.17 (0.178±0.0)	0.45 (0.469±0.01)	0.11 (0.12±0.0)	0.22 (0.217±0.01)
	LIME			**0.26 (0.25±0.01)**	**0.30 (0.30±0.0)**	**0.64 (0.64±0.0)**	**0.30 (0.29±0.01)**	0.09 (0.11±0.01)
	SHAP			**0.26 (0.26±0.01)**	**0.30 (0.30±0.01)**	**0.64 (0.64±0.0)**	0.29 (0.29±0.01)	0.09 (0.11±0.01)
$BERT_{fr}$	LRP	67 (67.6±1.1)	63 (63.7±1.2)	0.13 (0.13±0.0)	0.23 (0.23±0.0)	0.49 (0.50±0.01)	0.07 (0.08±0.01)	0.09 (0.10±0.01)
	LIME			0.14 (0.15±0.01)	0.24 (0.25±0.01)	0.54 (0.54±0.0)	0.09 (0.10±0.01)	0.08 (0.09±0.01)
	SHAP			0.17 (0.18±0.0)	0.26 (0.27±0.01)	0.56 (0.57±0.01)	0.10 (0.10±0.0)	0.08 (0.09±0.01)
$RoBERTA_{f-t}$	LRP	**75 (75.7±0.6)**	**73 (73.7±0.6)**	0.11 (0.12±0.01)	0.17 (0.18±0.02)	0.46 (0.47±0.01)	0.11 (0.11±0.0)	0.27 (0.26±0.01)
	LIME			0.24 (0.24±0.0)	0.27 (0.27±0.0)	0.61(0.61±0.0)	0.07 (0.07±0.0)	0.07 (0.07±0.0)
	SHAP			0.24 (0.23±0.01)	0.26 (0.26±0.0)	0.61(0.61±0.0)	0.07(0.07±0.0)	0.06(0.06±0.0)
$RoBERTA_{fr}$	LRP	67 (67.3±0.6)	63(63.7±1.2)	0.13 (0.13±0.0)	0.23 (0.23±0.0)	0.49 (0.48±0.01)	0.05 (0.05±0.0)	0.06 (0.06±0.0)
	LIME			0.15 (0.15±0.0)	0.24 (0.24±0.0)	0.55 (0.55±0.0)	0.0 (0.01±0.01)	**0.01 (0.01±0)**
	SHAP			0.16 (0.16±0.0)	0.25 (0.26±0.01)	0.55 (0.56±0.01)	0.0 (0.0±0.0)	**0.01 (0.01±0.0)**
Random		50	50	0.10	0.23	0.47	0.27	0.26

10 epochs with an *encoder* learning rate of 5e-7 and fine-tuning layers learning rate of 1e-6. None of the models exhibited signs of either overfitting or underfitting the data.

The explainability metrics used in this study are classified into two categories: (a) Plausibility, and (b) Faithfulness. The former measures the extend to which the predicted rationales are similar the gold rationales annotated by humans while the later does not rely on gold rationales and expresses the sensitivity of the underlying model with respect to the predicted rationales. Under the plausibility category, different metrics like Intersection over Union (IOU) F1, Token F1, and Area Under Precision Recall Curve (AUPRC) are used. In contrast, faithfulness encompasses two metrics namely comprehensiveness and sufficiency. It's important to note that, unlike other metrics, lower sufficiency values are preferable. Readers are referred to the Hatexplain paper [22] for a detailed explanation of the definitions of the explainability metric used in this study.

		I	d (would)	rather	get	fisted	by	a	nigger	tbh
BERT fine-tuned	LRP	1	0.99	0.95	0.68	0.49	0.37	0.28	0.1	0.13
	LIME	0	0.23	0.10	0.22	0.30	0.21	0.23	1	0.09
	SHAP	0	0.22	0.16	0.13	0.35	0.22	0.19	1	0.16
RoBERTA fine-tuned	LRP	0.55	1	0.98	0.65	0.5	0.28	0.31	0.26	0.18
	LIME	0.07	0.0	0.11	0.16	0.18	0.1	0.02	1	0.14
	SHAP	0	0.06	0.05	0.17	0.1	0.03	0.03	1	0.02

		I	d (would)	rather	get	fisted	by	a	nigger	tbh
BERT frozen	LRP	0.23	0.53	1	0.52	0.88	0.47	0.63	0.84	0
	LIME	0	0.53	1	0.67	0.73	0.40	0.69	1	0.40
	SHAP	0	0.65	1	0.63	0.7	0.7	0.66	0.96	0.65
RoBERTA frozen	LRP	0.46	1	0.8	0.54	0.67	0	0.09	0.09	0.1
	LIME	0	0.21	0.51	0.78	0.94	0.36	0.53	1	0.07
	SHAP	0.48	0.7	0.44	0.77	0.66	0	0.75	1	0.38

Fig. 4. Visualization of the predicted rationales by different methods on an example from the test set of the HateXplain benchmark. It can be seen that LRP rationales on fine-tuned *encoders* tend to give high relevance values to the early tokens, whereas LIME and SHAP focus more on the profane word.

4 Results

The performance of various encoders under different training paradigms is presented in Table 1. Our experimentation showed a lack of substantial performance variation among the different LRP variants. Therefore, we present results based on the LRP-0 variant in this study. The results of different LRP variants are presented in Appendix B.

Our analysis shows that the rationales predicted by LIME exhibit similar performance to SHAP across almost all explainability metrics. Moreover, the explainability power of LRP rationales on fine-tuned *encoders* was less than that of a random rationale generator. Additionally, we observed that while the plausibility scores of LIME and SHAP decrease for architectures with frozen encoders, they increase for LRP.

After a qualitative analysis of the rationales predicted by LRP, LIME, and SHAP for fine-tuned *encoders*, we observed that while LIME and SHAP predicted high relevance values for profane words, the LRP method predicted high relevance values on the early tokens of the sentence. Figure 4 illustrates the rel-

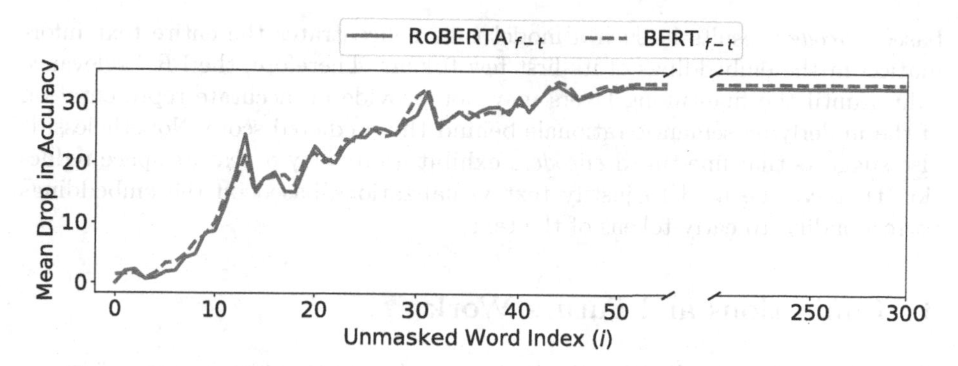

Fig. 5. Mean drop in accuracy over 3-fold cross-validation when word embeddings of ith word are fed to fine-tuning layers while rest of the embeddings are masked to zero. It can be seen that drop in accuracy is negligible (y-axis $\rightarrow$ 0) when the embeddings of the first few tokens (x-axis $\rightarrow$ 0) are fed to the fine-tuning layers.

evance values of different methods on a sentence predicted as hate speech by all the architectures based on fine-tuned *encoders* and the human annotators of HateXplain benchmark.

Since relevance values by LRP represent the contribution of that node while making the prediction, high relevance on early tokens of the sentence indicates that the embeddings of early tokens primarily contribute towards the model prediction. To find out the importance of the embeddings corresponding to a token (i), we set the embeddings corresponding to the rest of the tokens ($\{0, 299\} - \{i\}$) as zero. Figure 5 illustrates that drop in accuracy of architectures with fine-tuned *encoders*. We infer that when encoders are fine-tuned during the training phase, they learn to concentrate all the sentence information in the embeddings of the first few tokens only.

Among the sentences which were predicted as containing hate speech, the LRP method predicts first token as relevant in 75% of sentences[2] in BERT_{f-t} architecture, and 86% of sentences in RoBERTA_{f-t} architecture.

5 Conclusion

Transformers-based architectures have shown state-of-art performances in various tasks ranging from vision to language. However, explainability in such deep neural architecture becomes of utmost importance since feature extraction happens in an end-to-end manner. In this work, we attempt to measure the explainability power of the LRP method using different *encoders* for hate speech detection. We noticed that LIME performs similarly to SHAP across the majority of the explainability metrics in the HateXplain benchmark. However, rationales predicted by LRP led us to conclude that fine-tuning a pretrained transformer

[2] Averaging over the three test sets in 3-fold cross-validation.

based *encoder* results in such a model that concentrates the entire text information in the embeddings of its first few tokens. Therefore, the LRP relevance values until the fine-tuning layers may not provide an accurate representation of the underlying semantic rationale behind the predicted score. Nonetheless, it also suggests that fine-tuned *encoders* exhibit a property of feature space reduction that can be used to justify text visualizations based on the embeddings corresponding to early tokens of the text.

6 Limitations and Future Work

Since the implemented LRP is limited to the fine-tuning layers of transformed-based models, the present study cannot explain the concentration of sentence information in the embeddings of its early tokens. Furthermore, we undertake a performance comparison of various explanation techniques in the context of hate speech detection, utilizing a single benchmark dataset, with the understanding that our findings are constrained to this particular dataset. To make analogous assertions in different domains and with other datasets, additional research is essential. Our goal is to expand upon the formulations presented in the work by Voita et al. [30] and conduct backpropagation of relevance values through the *encoder* block, thereby enhancing our understanding of the fine-tuning procedures applied to these *encoders*.

Acknowledgements. Ritwik Mishra extends his appreciation to the University Grant Commission (UGC) of India, as he receives partial support through the UGC Senior Research Fellowship (SRF) program. Rajiv Ratn Shah acknowledges the partial assistance received from the Infosys Center for AI (CAI) and the Center of Design and New Media (CDNM) at IIIT Delhi.

A Appendix

Montavon et al. [24] mentions the generic rule to calculate the relevance value of node j in layer L as:

$$R_j = \sum_k \frac{a_j.\rho(w_{jk})}{\epsilon + \sum_0^j a_j.\rho(w_{jk})} R_k \tag{4}$$

The relevance are backpropagated using the following four steps:

- Forward pass: $z_k = \epsilon + \sum_0^j a_j.\rho(w_{jk})$
- Division: $s_k = R_k/z_k$
- Backward pass: $c_j = \sum_k \rho(w_{jk}).s_k$
- Relevance: $R_j = a_j c_j$

The paper asserts that c_j can be represented by the gradients attached to a, denoted as c_j =`a.grad`. We intend to explain the reasoning behind it. To simplify the explanation, we assume that $\rho(w_{jk}) = w$.

1. We know that $c_j = w \times s$ where $dim(w) = (j, k)$ and $dim(s) = (k, 1)$.
2. We know that `z = w.forward(a)` which is equivalent to $w^T \times a$ where $dim(a) = (j, 1)$. Therefore $dim(z) = (k, 1)$.
3. We know that $s = R_k/z$ where $dim(R_k) = (k, 1)$. Therefore $dim(s) = (k, 1)$.
4. If we use gradients then `z*s.data` can be written as $t = (w^T \times a) \cdot s$ where $dim(t) = (k, 1)$.
5. When t is summed and the gradients are back-propagated using `.backward()`, the gradients will be attached to w and a both.
6. `a.grad` will be the differentiation of t with respect to a. Therefore, `a.grad`$= \partial t/\partial a = w \times s = c_j$.

The deliberate use of `s.data` by the authors is intended to prevent the flow of gradients into a through a new path, given the dependence of s on a. However, when biases are present in the neural network layers, Eq. 4 will contain biases in both the numerator and the denominator. As differentiation ignores the biases, the gradient attached to a will remain as $w \times s$. Nevertheless, in order to satisfy the LRP constraint that requires an identical sum of relevance values in each layer, the expression of c_j in the backward pass needs to be modified as follows:

$$c_j = \sum_k \left(\rho(w_{jk}) + \frac{b_k}{|j| a_j} \right) \cdot s_k$$

Hence, `a.grad` will not be equivalent to c_j when the fully-connected neural network layers have bias enabled (Table 2).

Table 2. An evaluation of various LRP variants on hate speech prediction architectures with different *encoders*. Results are presented based on the official HateXplain benchmark's test set, and the figures in parentheses indicate the mean and standard deviation derived from 3-fold cross-validation. Architectures with fine-tuned encoders are denoted by the subscript $f-t$, while those with frozen encoders are indicated by the subscript fr. Notably, there is minimal variation in results among different LRP variants.

Encoder	Method	Performance		Explainability				
				Plausibility			Faithfulness	
		Accuracy↑	Macro-F1↑	IOU F1↑	Token F1↑	AUPRC↑	Compr.↑	Suff.↓
$BERT_{f-t}$	LRP-0	74 (74±0)	72 (71.7±0.6)	0.10 (0.11±0.0)	0.17 (0.178±0.0)	0.45 (0.469±0.01)	0.11 (0.12±0.0)	0.22 (0.217±0.01)
	LRP-ϵ			0.10 (0.11±0.0)	0.16 (0.171±0.0)	0.45 (0.464±0.01)	0.11 (0.11±0.0)	0.22 (0.224±0.0)
	LRP-γ			0.10 (0.10±0.0)	0.16 (0.17±0.01)	0.45 (0.466±0.01)	0.11 (0.12±0.0)	0.22 (0.222±0.01)
$BERT_{fr}$	LRP-0	67 (67.6±1.1)	63 (63.7±1.2)	0.13 (0.13±0.0)	0.23 (0.23±0.0)	0.49 (0.50±0.01)	0.07 (0.08±0.01)	0.09 (0.10±0.01)
	LRP-ϵ			0.14 (0.14±0.0)	0.23 (0.23±0.0)	0.48 (0.49±0.01)	0.08 (0.08±0.0)	0.09 (0.09±0.0)
	LRP-γ			0.14 (0.14±0.01)	0.23 (0.23±0.0)	0.49 (0.50±0.01)	0.08 (0.08±0.01)	0.09 (0.09±0.01)
$RoBERTA_{f-t}$	LRP-0	**75 (75.7±0.6)**	**73 (73.7±0.6)**	0.11 (0.12±0.01)	0.17 (0.18±0.02)	0.46 (0.47±0.01)	0.11 (0.11±0.0)	0.27 (0.26±0.01)
	LRP-ϵ			0.11 (0.12±0.01)	0.16 (0.18±0.02)	0.46 (0.47±0.01)	0.11 (0.11±0.0)	0.20 (0.20±0.0)
	LRP-γ			0.11 (0.12±0.01)	0.16 (0.18±0.02)	0.46 (0.47±0.01)	0.12 (0.11±0.01)	0.20 (0.20±0.0)
$RoBERTA_{fr}$	LRP-0	67 (67.3±0.6)	63 (63.7±1.2)	0.13 (0.13±0.0)	0.23 (0.23±0.0)	0.49 (0.48±0.01)	0.05 (0.05±0.0)	0.06 (0.06±0.0)
	LRP-ϵ			0.14 (0.14±0.0)	0.23 (0.23±0.0)	0.49 (0.48±0.01)	0.05 (0.05±0.0)	0.05 (0.06±0.01)
	LRP-γ			0.13 (0.14±0.01)	0.23 (0.23±0.0)	0.49 (0.48±0.01)	0.05 (0.05±0.0)	0.06 (0.06±0.0)

B Appendix

References

1. Ahn, J., Oh, A.: Mitigating language-dependent ethnic bias in BERT. In: Proceedings of the 2021 Conference on Empirical Methods in Natural Language Processing, pp. 533–549 (2021)
2. Aluru, S.S., Mathew, B., Saha, P., Mukherjee, A.: A deep dive into multilingual hate speech classification. In: Machine Learning and Knowledge Discovery in Databases. Applied Data Science and Demo Track: European Conference, ECML PKDD 2020, Ghent, Belgium, September 14–18, 2020, Proceedings, Part V, pp. 423–439. Springer-Verlag, Berlin, Heidelberg (2020). https://doi.org/10.1007/978-3-030-67670-4_26
3. Aluru, S.S., Mathew, B., Saha, P., Mukherjee, A.: A deep dive into multilingual hate speech classification. In: Dong, Y., Ifrim, G., Mladenić, D., Saunders, C., Van Hoecke, S. (eds.) ECML PKDD 2020. LNCS (LNAI), vol. 12461, pp. 423–439. Springer, Cham (2021). https://doi.org/10.1007/978-3-030-67670-4_26
4. Arras, L., Montavon, G., Müller, K.R., Samek, W.: Explaining recurrent neural network predictions in sentiment analysis. In: Proceedings of the 8th Workshop on Computational Approaches to Subjectivity, Sentiment and Social Media Analysis, pp. 159–168. Association for Computational Linguistics, Copenhagen, Denmark (2017). https://doi.org/10.18653/v1/W17-5221, https://aclanthology.org/W17-5221
5. Attanasio, G., Pastor, E., Di Bonaventura, C., Nozza, D.: ferret: a framework for benchmarking explainers on transformers. In: Proceedings of the 17th Conference of the European Chapter of the Association for Computational Linguistics: System Demonstrations. Association for Computational Linguistics (2023)
6. Bach, S., Binder, A., Montavon, G., Klauschen, F., Müller, K.R., Samek, W.: On pixel-wise explanations for non-linear classifier decisions by layer-wise relevance propagation. PLoS ONE **10**(7), e0130140 (2015)
7. Balkir, E., Nejadgholi, I., Fraser, K., Kiritchenko, S.: Necessity and sufficiency for explaining text classifiers: a case study in hate speech detection. In: Proceedings of the 2022 Conference of the North American Chapter of the Association for Computational Linguistics: Human Language Technologies, pp. 2672–2686. Association for Computational Linguistics, Seattle, United States (2022). https://doi.org/10.18653/v1/2022.naacl-main.192, https://aclanthology.org/2022.naacl-main.192
8. Bourgeade, T.: From text to trust: a priori interpretability versus post hoc explainability in natural language processing, Ph. D. thesis, Université Paul Sabatier-Toulouse III (2022)
9. Croce, D., Rossini, D., Basili, R.: Auditing deep learning processes through kernel-based explanatory models. In: Proceedings of the 2019 Conference on Empirical Methods in Natural Language Processing and the 9th International Joint Conference on Natural Language Processing (EMNLP-IJCNLP), pp. 4037–4046 (2019)
10. Devlin, J., Chang, M., Lee, K., Toutanova, K.: BERT: pre-training of deep bidirectional transformers for language understanding. CoRR abs/1810.04805, http://arxiv.org/abs/1810.04805 (2018)
11. Ding, W., Abdel-Basset, M., Hawash, H., Ali, A.M.: Explainability of artificial intelligence methods, applications and challenges: a comprehensive survey. Inf. Sci. **615**, 238–292 (2022)

12. Dolk, A., Davidsen, H., Dalianis, H., Vakili, T.: Evaluation of LIME and SHAP in explaining automatic ICD-10 classifications of Swedish gastrointestinal discharge summaries. In: Scandinavian Conference on Health Informatics, pp. 166–173 (2022)
13. Garimella, A., et al.: He is very intelligent, she is very beautiful? on mitigating social biases in language modelling and generation. In: Findings of the Association for Computational Linguistics: ACL-IJCNLP 2021, pp. 4534–4545 (2021)
14. Jørgensen, R., Caccavale, F., Igel, C., Søgaard, A.: Are multilingual sentiment models equally right for the right reasons? In: Proceedings of the Fifth BlackboxNLP Workshop on Analyzing and Interpreting Neural Networks for NLP, pp. 131–141 (2022)
15. Joshi, R., Chatterjee, A., Ekbal, A.: Towards explainable dialogue system: explaining intent classification using saliency techniques. In: Proceedings of the 18th International Conference on Natural Language Processing (ICON), pp. 120–127 (2021)
16. Karim, M.R., et al.: DeepHateExplainer: explainable hate speech detection in under-resourced Bengali language. In: 2021 IEEE 8th International Conference on Data Science and Advanced Analytics (DSAA), pp. 1–10. IEEE (2021)
17. Kwako, A., Wan, Y., Zhao, J., Chang, K.W., Cai, L., Hansen, M.: Using item response theory to measure gender and racial bias of a BERT-based automated English speech assessment system. In: Proceedings of the 17th Workshop on Innovative Use of NLP for Building Educational Applications (BEA 2022), pp. 1–7 (2022)
18. Liu, Y., et al.: RoBERTa: a robustly optimized BERT pretraining approach. CoRR abs/1907.11692, http://arxiv.org/abs/1907.11692 (2019)
19. Lundberg, S.M., Lee, S.I.: A unified approach to interpreting model predictions. In: Advances in Neural Information Processing Systems, vol. 30 (2017)
20. Maimaitituoheti, A.: ABLIMET@ LT-EDI-ACL2022: a RoBERTa based approach for homophobia/transphobia detection in social media. In: Proceedings of the Second Workshop on Language Technology for Equality, Diversity and Inclusion, pp. 155–160 (2022)
21. Man, X., Chan, E.P.: The best way to select features? comparing MDA, LIME, and SHAP. J. Financ. Data Sci. **3**(1), 127–139 (2021)
22. Mathew, B., Saha, P., Yimam, S.M., Biemann, C., Goyal, P., Mukherjee, A.: HateXplain: a benchmark dataset for explainable hate speech detection. In: Proceedings of the AAAI Conference on Artificial Intelligence, vol. 35, pp. 14867–14875 (2021)
23. Mehta, H., Passi, K.: Social media hate speech detection using explainable artificial intelligence (XAI). Algorithms **15**(8), 291 (2022)
24. Montavon, G., Binder, A., Lapuschkin, S., Samek, W., Müller, K.-R.: Layer-wise relevance propagation: an overview. In: Samek, W., Montavon, G., Vedaldi, A., Hansen, L.K., Müller, K.-R. (eds.) Explainable AI: Interpreting, Explaining and Visualizing Deep Learning. LNCS (LNAI), vol. 11700, pp. 193–209. Springer, Cham (2019). https://doi.org/10.1007/978-3-030-28954-6_10
25. Ribeiro, M.T., Singh, S., Guestrin, C.: "why should i trust you?" explaining the predictions of any classifier. In: Proceedings of the 22nd ACM SIGKDD International Conference on Knowledge Discovery and Data Mining, pp. 1135–1144 (2016)
26. Sarat, P., Kaundinya, P., Mujumdar, R., Dambekodi, S.: Can machines detect if you're a jerk (2020)
27. Schwarzenberg, R., Hübner, M., Harbecke, D., Alt, C., Hennig, L.: Layerwise relevance visualization in convolutional text graph classifiers. In: Proceedings of the Thirteenth Workshop on Graph-Based Methods for Natural Language Processing (TextGraphs-13), pp. 58–62 (2019)

28. Szczepański, M., Pawlicki, M., Kozik, R., Choraś, M.: New explainability method for BERT-based model in fake news detection. Sci. Rep. **11**(1), 23705 (2021)
29. Utsumi, A.: Refining pretrained word embeddings using layer-wise relevance propagation. In: Proceedings of the 2018 Conference on Empirical Methods in Natural Language Processing, pp. 4840–4846 (2018)
30. Voita, E., Sennrich, R., Titov, I.: Analyzing the source and target contributions to predictions in neural machine translation. In: Proceedings of the 59th Annual Meeting of the Association for Computational Linguistics and the 11th International Joint Conference on Natural Language Processing (Volume 1: Long Papers), pp. 1126–1140. Association for Computational Linguistics (2021). https://doi.org/10.18653/v1/2021.acl-long.91, https://aclanthology.org/2021.acl-long.91
31. Wolf, T., et al.: Transformers: state-of-the-art natural language processing. In: Proceedings of the 2020 Conference on Empirical Methods in Natural Language Processing: System Demonstrations, pp. 38–45. Association for Computational Linguistics (2020). https://www.aclweb.org/anthology/2020.emnlp-demos.6
32. Zaidan, O., Eisner, J., Piatko, C.: Using "annotator rationales" to improve machine learning for text categorization. In: Human Language Technologies 2007: The Conference of the North American Chapter of the Association for Computational Linguistics; Proceedings of the Main Conference, pp. 260–267 (2007)

Multilingual Speech Sentiment Recognition Using Spiking Neural Networks

Shreya Parashar[1(✉)] and K G Srinivasa[2]

[1] Department of Computer Science and Engineering, IIIT Naya Raipur, Chhattisgarh 493661, India
meghaparashar20@gmail.com

[2] Department of Data Science and Artificial Intelligence, IIIT Naya Raipur, Chhattisgarh 493661, India
srinivasa@iiitnr.edu.in

Abstract. Speech sentiment and emotion recognition has grown significantly as a research field in recent years as it has potential uses in a variety of domains. Multilingual speech sentiment recognition still remains a challenging task due to the cultural and linguistic differences in the speech. Although several approaches, particularly those based on deep learning techniques (ANNs), have shown encouraging outcomes in speech sentiment recognition, there remain obstacles in accurately representing the temporal and dynamic fluctuations in sentiments conveyed through speech. Spiking Neural Networks (SNNs), which have shown promising result in other machine learning and pattern recognition applications, such as handwriting and facial expression detection, are a potentially successful approach to deal with those challenges. In this research, we present a system for multilingual speech sentiment recognition using spiking neural networks. We have used Mel-frequency-cepstral coefficients (MFCC) features to train the model. Experiments are conducted on the data created by combining four datasets SAVEE (English), EMO-DB (German), EMOVO (Italian), and CaFE (French). We compared the performance of SNNs with ANNs such as MLP and RNN. In our experiments, spiking neural network achieves 76.85% of accuracy and outperforms its counterpart ANN models on this task.

Keywords: Spiking neural networks (SNN) · MFCC · Deep Learning

1 Introduction

Communication is the most efficient way to convey one's thoughts. Speech is not just about conveying words and information, it is also a powerful medium for expressing and conveying various human emotions and sentiments. A speaker's speech is a valuable source of information regarding their emotional state [4]. Speech sentiment recognition refers to the task of categorizing a human speech as positive, negative, or neutral. Speech sentiment recognition is a significant

V. Goyal et al. (Eds.): BDA 2023, LNCS 14418, pp. 215–234, 2023.
https://doi.org/10.1007/978-3-031-49601-1_15

domain of research due to its application in a variety of fields, including healthcare, education, entertainment, and customer services. Sentiments help us communicate and comprehend others' points of view by expressing our feelings and giving feedback [28].

Humans employ diverse methods to express their emotions, utilizing verbal communication and physical activity such as facial expressions [21], gestures, and body language are examples of non-verbal cues used for this purpose. Generally, In conversations, the significance of the words spoken by the person and their emotional state is crucial. The same sentence can have different meanings on how it is delivered, for example, sarcasm. This phenomenon is also evident in emojis in daily text-based conversations. Depending on their emotional state, humans communicate a wide range of emotions through speech, including sadness, happiness, fear, and enthusiasm [15].

The human voice is often incorporated into robots because it provides a natural and intuitive interface for human communication. The development of emotion or sentiment-based human-computer interactions depends on research results in emotion recognition. Recent studies demonstrate that people connect with their computers using interpersonal interaction strategies. Consequently, developing a trustworthy technique for automatically recognizing human emotions and sentiments through speech could benefit in various applications related to speech processing, such as human-machine interfaces and patient monitoring. Analysis of the audio signal is required for speech emotion and sentiment recognition to determine the proper emotion based on the training of its features, such as pitch, formant, and phoneme [16]. Although several approaches, particularly those based on deep learning techniques (ANNs), have shown encouraging outcomes in speech sentiment recognition, there remain obstacles in accurately representing the temporal and dynamic fluctuations in sentiments conveyed through speech.

On the other hand, biological neural networks known as Spiking Neural Networks (SNNs) [19] aim to more accurately resemble the behavior of biological neurons and synapses in the human brain. The design of SNNs is such that these are better at capturing the temporal dynamics of neural computation, which are essential for many brain processes including speech. The fact that SNNs have a temporal dimension and convey information through time as opposed to processing it statically or repeatedly is one of the key differences between SNNs and ANNs. Numerous researchers are actively exploring the potential of SNNs for speech recognition. In line with this trend, our experiment aims to contribute to this field by addressing the existing challenges and limitations. Specifically, the problem lies in developing an effective and accurate model for multilingual speech sentiment recognition using SNNs.

The proposed research aims to investigate the use of SNNs for multilingual speech sentiment recognition. We created a speech sentiment dataset in four languages for training and testing. This dataset is used to evaluate the model's performance in multilingual settings. We extract MFCC features from speech utterances. With these features, we build a model using a spiking neural network for speech sentiment recognition. We also compare the performance of SNN model with traditional artificial neural networks.

The rest of the paper is organized as follows: In Sect. 2, we discuss the research work related to speech sentiment and emotion recognition. Section 3, we explain the dataset and Feature Extraction, research methodology, and the outcomes of this work. Section 4 describes the implementation details and the conducted experiments. It also discusses the results of the experiments. The study is summarised in Sect. 5, along with suggestions for future work that would be helpful for conducting additional research in this area.

2 Related Work

This section discusses the studies that is related to the opportunities and difficulties associated with speech emotion recognition. Speech sentiment recognition (SSR) involves extracting sentiment from speech data, which is challenging due to the diverse nature of human speech, including accents, dialects, and noise.

Various approaches have been employed in SSR. Conventional machine learning methods like support vector machines [12] (SVMs) and decision trees have been used, as well as deep learning-based models such as recurrent neural networks [22] (RNNs) and convolutional neural networks (CNNs) [9]. RNNs excel in capturing long-range dependencies in speech data, while CNNs are adept at identifying spatial patterns.

In their study, Yuan et al. [27] presented a novel approach for speech emotion recognition using MFCC features. Their study focused on using Multilayer Perceptron (MLP) as an input of MFCC features. They successfully demonstrated data augmentation techniques to enhance system performance.

Issa et al. [10] presented a novel architecture in their study to directly extract features from sound files. Their research focused on the extraction of various features such as 'chromagram', 'Tonnetz representation', 'mel-scale spectrogram', and 'spectral contrast'. The researchers employed 1-D CNNs to utilize these features for emotion recognition. In their study, various language datasets were investigated, including the RAVDESS and EMO-DB datasets.

Utilizing a CNN-based classification architecture, Sora et al. [25] proposed a method for accurately differentiating and classifying speech moods and emotions. The model's application extends to understanding public opinion, evaluating citizen contentment through sentiment analysis of social media, and facilitating automated sentiment analysis in the context of smart cities, enabling valuable insights and decision-making.

Buscicchio et al. [4] proposed a method for emotion recognition in audio speech signals. Researchers have investigated the SNNs for Speech emotion recognition. Their approach involves extracting features from spoken word sentences using Brandt's Generalized Likelihood Ratio (GLR) method to detect signal discontinuities. Specifically, they focused on isolating the vowels in the speech signal as vowels carry more emotional information compared to consonants. The speech signal is divided into vowel segments, and then MFCC features are extracted from these segments. Finally, a spiking neural network is employed to categorize the speech into five distinct emotion classes.

In order to automatically recognize speech emotions without the need for feature extraction, Lotfidereshgi et al. propose a novel approach in [18] that combines the source-filter model of human speech production with a biologically-inspired spiking neural network called a Liquid State Machine(LSM). The proposed method involves dividing the speech signal into the source and vocal tract components, which are subsequently processed by two different neuronal reservoirs. Each reservoir's output is then fed into a final classifier.

Although deep learning approaches have been successful in SER, there are still issues with accurately capturing the temporal and dynamic variations of emotions expressed through speech. The paper [20] explores the potential of spiking neural networks (SNNs) for speech emotion recognition (SER). The researcher suggests a novel cross-modal enhancement method to boost recognition ability that is motivated by the brain's processing of audio information. An early cross-modal method uses visual data to improve speech-based emotion recognition.

Human beings can produce diverse vocalizations during communication that involve a multitude of linguistic expressions through the mouth. The comprehension of these lexical items in human communication is easy to understand, but their interpretation in human-machine interaction requires significant speech signal processing [1].

As the world becomes increasingly globalized, people from different cultures and backgrounds are interacting with each other more and more. This can lead to misunderstandings, as people may not be familiar with the cultural norms and expectations of others. Intelligent systems that recognize emotions and sentiments across multiple languages can help bridge this gap and improve communication.

Several challenges are associated with recognizing emotions and sentiments across multiple languages. One challenge is that the same words can have different meanings in different languages. Another challenge is that the way people express their emotions can vary from culture to culture. Furthermore, the similarity in emotional nature poses a challenge regarding the analysis [26]. For instance, 'angry' and 'disgust' reflect a negative sentiment toward a given entity. Therefore, it is necessary to develop an appropriate approach for modeling people's emotional state based on their speech.

Artificial neural networks (ANNs) use real-valued neuron responses, while biological neurons rely on spike trains. Spiking neural networks (SNNs) have the potential for greater representation, especially in time series tasks like speech. However, their progress is hindered by instability of training algorithms and compatible baselines. Through a thorough literature review in [2], focusing on surrogate gradient approaches, a study demonstrates that SNNs can compete with ANNs by combining adaptation, recurrence, and surrogate gradients. These lightweight SNNs remain highly compatible with contemporary frameworks for deep learning. The research concludes that SNNs are suitable for AI research, particularly in speech processing, and may even aid in inferring biological function.

The above literature review provides valuable insights into speech sentiment and emotion recognition using spiking neural networks (SNNs) and other architectures. It emphasizes the potential of SNNs in achieving robust sentiment recognition across different languages. By leveraging the unique capabilities of SNNs, such as spike-based computation and temporal processing, researchers can explore novel approaches for multilingual sentiment analysis.

3 Methodology

In this research, we propose a system for multilingual speech sentiment recognition based on spiking neural networks. We extract MFCC features on the go and train the model in a multilingual setting. Compare the SNN model with traditional artificial neural networks such as MLP and RNN.

3.1 Dataset Description

We have chosen the four most commonly used datasets for speech emotion and sentiment recognition, after acquiring a thorough understanding and accumulating sufficient domain knowledge. These datasets are publicly accessible. The model's performance is compared using these datasets in this research, which aids in developing some intriguing findings.

1. **Surrey Audio-Visual Expressed Emotion (SAVEE)** [11] This database was created by Jackson and Haq in 2014. It contains a total of 480 British English utterances recorded by four male speakers who were researchers and postgraduate students at the University of Surrey. There are seven different emotion categories in the database: anger, disgust, sadness, surprise, happiness, fear, and neutral.

2. **Emotional Database (EMO-DB)** [3] It is an emotional speech database in German, developed by the ICS at Berlin's Technical University. To record the speech utterances, five men and five women speakers participated. It contains 535 utterances distributed in seven emotions: Ärger (anger), Angst (anxiety/fear), Langeweile (boredom), Freude (happiness), Ekel (disgust), Trauer (sadness), and Neutral.

3. **EMOVO Corpus** [6] It is an Italian emotional speech database.. Six professional actors recorded 14 sentences in seven different emotions: disgusto (Disgust), paura (Fear), Rabbia (Anger), gioia (Joy/Happy), Sorpresa (Surprise), triste (Sad), neutro (Neutral).

4. **CaFE** [8] This dataset contains utterances from French emotional speech. Six different sentences were recorded by 12 professional actors (six female and six male). There are six basic emotions: Joie (Happiness), Colère (Anger), Peur (Fear), Surprise (Surprise), Dágoût (Disgust), Tristesse (Sadness), and one Neutre (Neutral) in this dataset.

After finalizing the datasets, The audio files needed to be recognized and interpreted. Each dataset had a unique name scheme.

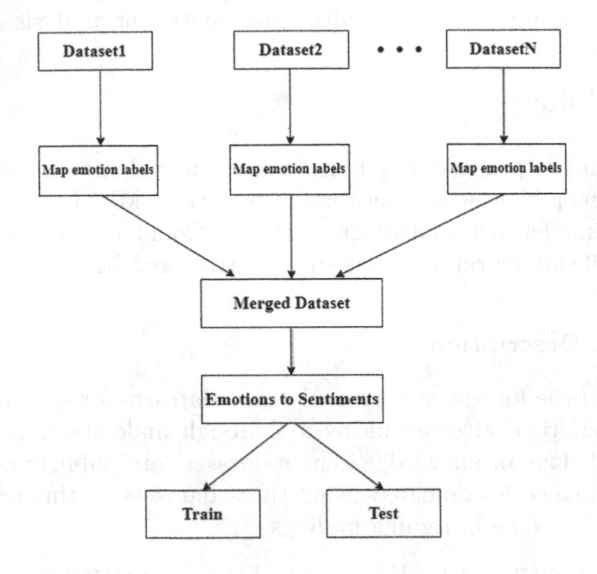

Fig. 1. Data Description

Since these datasets has labels as emotions therefore, first we derived the emotion labels from the file names and then converted the emotions into sentiment labels for classification as shown in Fig. 1. We created the sentiment labels as defined below in Table 1.

Table 1. Mapping of Sentiments

Positive	Negative	Neutral
Happiness	Boredom, Sadness	Neutral
Surprise	Fear, Disgust, Anger	Calm

3.2 Data Preprocessing and Data Transformation

To ensure uniformity in sampling rate across all databases, every utterance was resampled at a frequency rate of 16khz. Audio signals can have different amplitude levels, which can be influenced by a variety of factors, such as recording equipment, distance from the sound source, and environmental noise. Therefore, each audio signal was normalized using "amplitude normalization" or "peak normalization" before extracting the features. The primary objective of amplitude normalization is to ensure consistent and uniform loudness levels across different speech signals or segments. It is particularly beneficial in several applications, such as speech recognition, speaker identification, and speech synthesis as it improve robustness, enhance comparability, mitigates overload and distortion, facilitate concatenative Speech Synthesis, and improves speech intelligibility.

Figure 2. depicts the steps involved in the process of amplitude normalization of an audio signal.

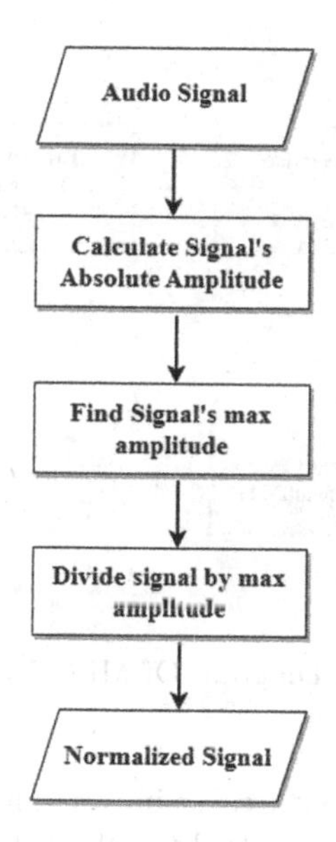

Fig. 2. Flow Diagram of Amplitude Normalization

A comparison of the audio sample waveform has shown in Fig. 3 with normalization and without normalization.

Normalization of the waveform retains its shape, structure, and fundamental characteristics, including frequency content, timing, and overall shape. It also preserves the relative dynamics, variations in loudness and softness, and temporal relationship, pitch, frequency, and speech characteristics between waveform components.

3.3 Feature Extraction

Feature extraction and characteristics generation from the data is a crucial step in Speech sentiment Recognition. Since we are working with multilingual data,

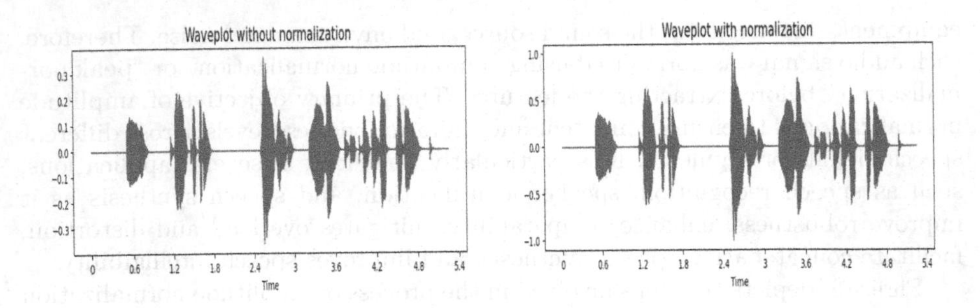

Fig. 3. Transforming Audio: A Comparison of Sample Audio Before and After Normalization

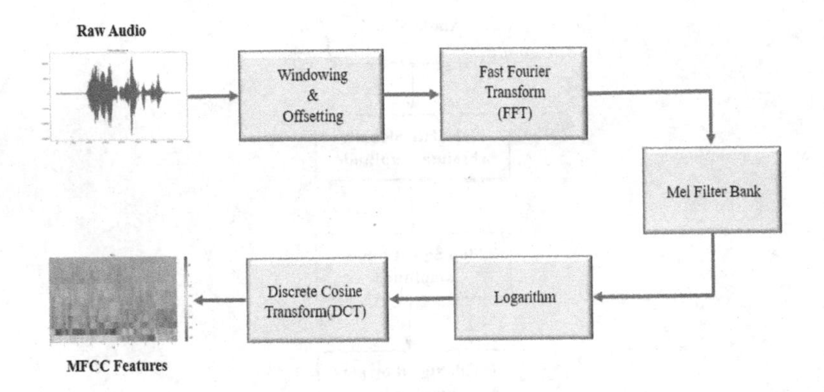

Fig. 4. Block Diagram Of MFCC Computation

having more discriminative features is important. Therefore, for our work we extract the 40 Mel Frequency Cepstral Coefficient (MFCC) as acoustic features. With all 40 coefficients there are more chances of capturing more nuanced variations in the speech signals. The idea behind this is to allow the model to learn more discriminative patterns and make finer-grained distinctions between differ-

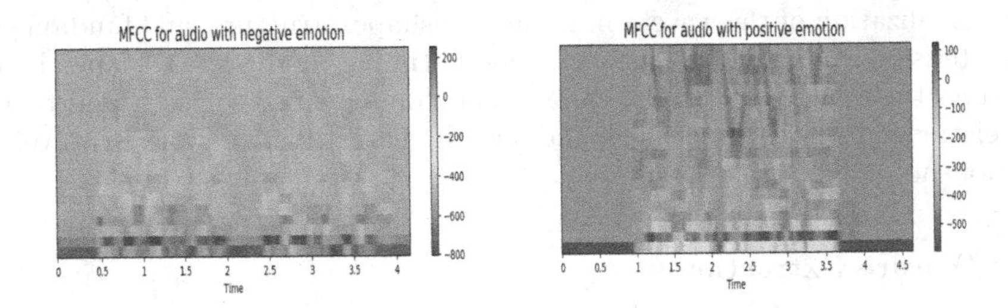

Fig. 5. Transforming Audio: A Comparison of Sample Audio Before and After Normalization

ent sentiments. The richer coefficients can capture more subtle variations in the speech signals, allowing for better discrimination between different senitments. The computation of MFCC is depicted in Fig. 4. Figure 5 shows two dimensional MFCC matrix for negative and positive sentiments extracted from two random audio files.

3.4 Model Architecture

We design a multilingual speech sentiment recognition system illustrated in Fig. 6, which takes audio waveforms as input, does preprocessing, extracts MFCC features, and outputs predicted sentiment (positive, negative or neutral).

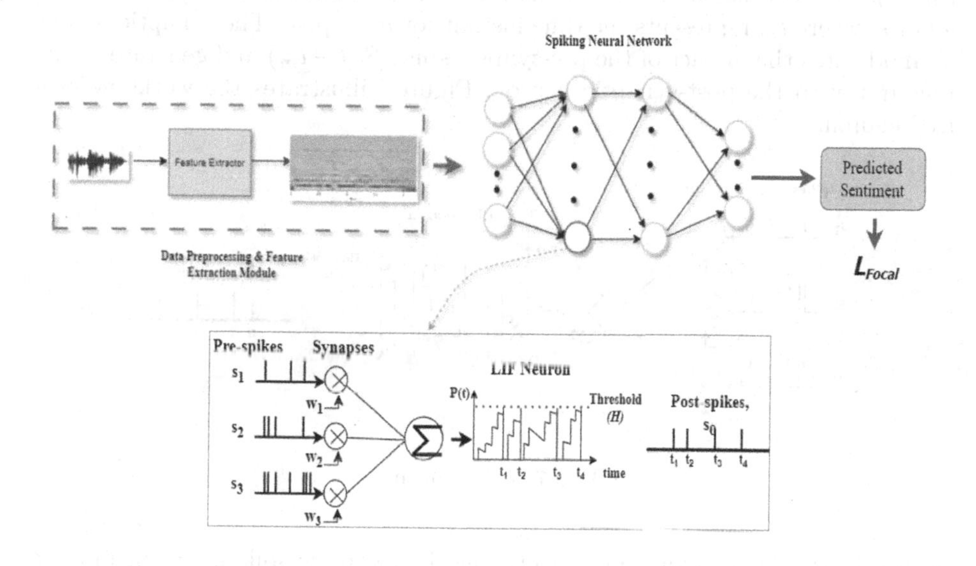

Fig. 6. Proposed System For Speech Sentiment Recognition

A leaky integrate and fire (LIF) neuron model was utilized to generate a baseline model. It is the singal neuron model that is used the most frequently. The membrane potential P, which takes an input signal I and varies in time (t), describes the dynamics of a single neuron. Without any stimuli (when I = 0), P decays to a resting value P_{rest} with the time constant t_p at an exponential rate. When there is an input signal (I != 0), P increases or decreases according to the integration of incoming inputs. The following differential equation is used to represent the single neuron's dynamics in continuous time (Gerstner and Kistler, 2002) [7]:

$$\tau_P \dot{P}(t) = -(P(t) - P_{rest}) + RI(t) \quad (1)$$

R: membrane resistance The signal I is defined as the sum of the pre-synptic spikes

$$I(t) = \sum_{i=1}^{N^l} w_i \sum_m s_i(t - t_m) \quad (2)$$

where N^l is the number of weights and w_i is the weight for the synaptic connection of ith neuron. $S_i(t - t_m)$ represents the spike in the ith pre-neuron at time t_m. We can use a Kronecker delta function to formulate a spike event as follows: where t_m represents the time instant for m^{th} spike. The synaptic weight w_i modulates the impact of the pre-synptic spike $S_i(t - t_m)$ and generate a current inflow to the post-synaptic neuron. Figure 7 illustrates the workflow of a LIF neuron.

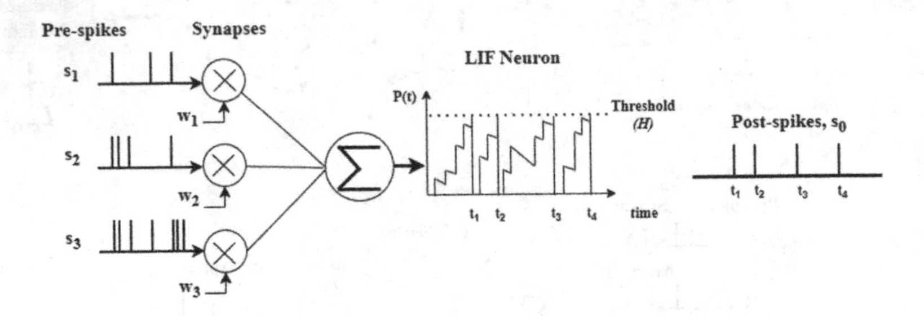

Fig. 7. LIF neuron

A threshold value θ is introduced to the model to create spikes. The membrane potential leaks exponentially with the time constant τ and when the membrane potential hits the threshold, a spike train is released and the membrance is then set to a new value $P_r < \theta$. The $-(P(t) - P_r)$ term is the leak term that drives the potential towards P_r.

$$\begin{aligned} &\text{If } P(t^f) \geq \theta, \text{ then } s(t^f) = 1, \\ &\lim_{\delta \to 0} P(t^f + \delta) = P_r, \text{ where } \delta > 0. \end{aligned} \quad (3)$$

A neuron i in l^{th} layer of the network, receives the pre-synptic inputs from the neurons in the previous layer $(l-1)^{th}$ layer. These pre-synptic inputs come in as the spike trains s_j^{l-1} (0 or 1). So, for the ith neuron in the lth layer the stimulus can be defined as follows:

$$I_i^l[t] = \sum_{j=1}^{N^{l-1}} W_{ji}^l s_j^{l-1}[t] + b_i^l \quad (4)$$

When the bias b^l refers to various resting values of the membrane potential and the matrix W^l to the weights of the synaptic connections. Once a neuron is fired, it enters into a period of refractoriness, preventing it from instantly firing a second spike.

To write a discrete-time approximation of Eq. (1), the first-order Euler exponential integrator method can be used with a time step of size Δt. Assuming $P_{\text{rest}} = P_r$ and substituting $P = \frac{P - P_{\text{rest}}}{\theta - P_{\text{rest}}}$, $I = \frac{RI}{\theta - P_{\text{rest}}}$, the discrete time formulation of Eq. (2) can be rewritten as follows:

$$P[t] = \beta P[t-1] + (1-\beta)I[t] \tag{5}$$

where $\beta = \exp\left(-\frac{\Delta t}{\tau_p}\right)$
In the equation (5), the first term represents the leak and second term the excitation. Equation (5) is vectorized and loop it over layers and time in the forward pass in SNN. Therefore, with the threshold behavior the dynamics of the membrane potential produces binary signals $s^l \in \{0, 1\}$ Instead of analog ones $y^l \in \mathbb{R}$ (as in artificial neural networks).

LIF Neurons as Recurrent Neural Networks Cells
Research have shown that LIF neurons may not be able to replicate many firing patterns including adaptive, delayed, and brusting (Gerstner and Kistler, 2002; Bittar and Philip, 2022) [2,7]. Therefore, we have also used recurrent connections in the baseline (LIF) model.

If we enable the recurrent connections in the network, the neuron i in the lth layer also connected to all other neurons $m = 1, \ldots, N^l, m \neq i$ in the same l^{th} layer and receives spike trains s^l_m from these neuron as well. Therefore, for the ith neuron in the lth layer the stimulus can be written as

$$I_i^l[t] = \sum_{j=1}^{N^{l-1}} W_{ji}^l s_j^{l-1}[t] + \sum_{m=1, m\neq i}^{N^l} V_{mi}^l s_m^l[t-1] + b_i^l \tag{6}$$

where the weights of the recurrent synaptic connections are represented by V^l. Figure 8. demonstrates the neurons' connectivity in RLIF.

For the purpose of comparison of the performance of SNNs to their counterpart standard ANNs, we used the equivalent architectures for MLP and RNN. These networks only differ in the types of neurons.

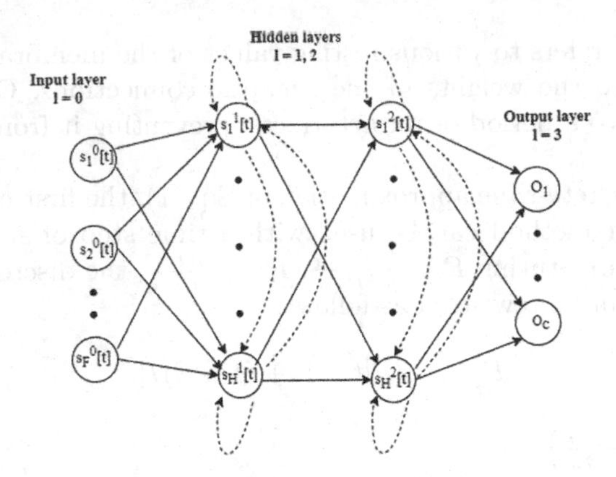

Fig. 8. RLIF neuron

3.5 Loss Function

We have a class imbalance in our data (negative = 56.46%, positive = 24.46%, neutral = 19.08%) in Fig. 9.

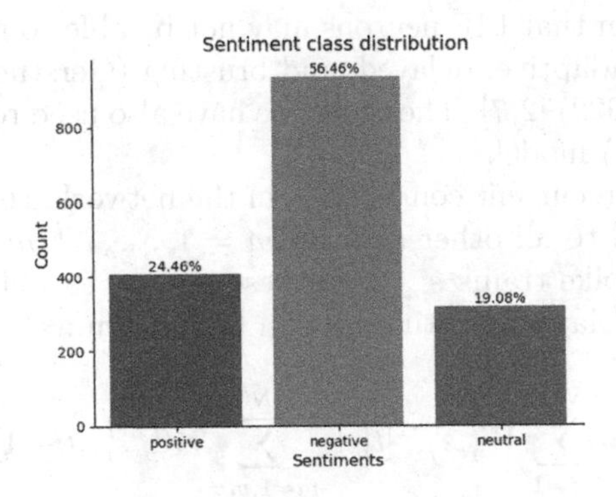

Fig. 9. Data Distribution

Therefore, we need to tackle the class imbalance problem in the data. The data level and algorithm level are the common techniques used to handle imbalance datsets. In the data level techniques, oversampling and undersampling are used to balance the datasets. To address the problem of imbalanced data, Chawla et al. introduced SMOTE, an oversampling technique [5] that creates synthetic samples of the minority class. Due to the synthetic data, oversampling comes with the risk of overfitting, noise, and loss of information [13]. While the algorithm level techniques uses cost-sensitive learning to modify the cost function. One of the most popular and widely applied cost-sensitive learning techniques is focal loss. [23]. Focal loss balances the cross-entropy loss such that the hard negative samples are learned [17]. Since focal loss does not require modifying the dataset, it reduces the risk of overfitting. Also, focal loss improve focus by assigning the different weights to different examples, emphasizing hard-to-classify samples. It can also help generalizing the model which is required in real-world applications. Therefore, we used the weighted focal loss as our loss function to train the model.

The Focal loss technique as proposed in [17], aims to decrease the issue of class imbalance by assigning lower weights to inliers, which are considered easier examples. Let's assume we have a multiclass classification problem with C classes, where the ground truth labels for a given sample are denoted as

$$\boldsymbol{y} = (y_1, y_2, \ldots, y_C) \tag{7}$$

where y_i is 1 if the sample belongs to class i and 0 otherwise. The output of the model is often converted into class probabilities using the softmax function. Let's denote the predicted probabilities for the classes as

$$\mathbf{p} = (p_1, p_2, \ldots, p_C) \tag{8}$$

where p_i is the predicted probability for class i.

The focal loss is described as:

$$\text{FocalLoss}(p, y) = -\sum_{i=1}^{C} (\alpha_i \cdot y_i \cdot (1 - p_i)^{\gamma} \cdot \log(p_i)) \tag{9}$$

where:

- α_i is the weight assigned to class i in order to balance the contribution of different classes.
- γ is the focusing parameter that controls the degree of emphasis on hard, misclassified examples. Higher values of γ will focus more on misclassified examples.
- log is the natural logarithm function.

The focal loss consists of two terms: the modulating factor $(1 - p_i)^{\gamma}$ and the cross-entropy loss $\log(p_i)$. To compute the overall loss for a batch of samples, you can simply take the average over all samples. Let's assume we have a batch of size N, the focal loss for the entire batch can be computed as:

$$Total\ FocalLoss = \frac{1}{N} \sum_{j=1}^{N} \text{FocalLoss}(p^{(j)}, y^{(j)}) \tag{10}$$

where:

$p^{(j)}$ and $y^{(j)}$ represent the predicted probabilities and ground truth labels, respectively, for the jth sample in the batch.

4 Experiments

4.1 Experimental Setup

To examine the model's performance, we trained all the models with the number of hidden layers ranging from 3 to 5 and utilized 64 or 128 hidden units in each layer. We train the model with a 0.001 initial learning rate in each configuration for 100 epochs. All the models are trained using a single NVIDIA RTX A4000 GPU with 16 GB VRAM. Each model training took approximately 2 h to complete. Following are the details about training and evaluation settings:

• **Surrogate Gradient Method:** Stochastic gradient descent (SGD) is the most commonly used method for training artificial neural networks. The predictions of the model are compared to the ground truth on a batch of samples through a loss function. Then back-propagation is used to find the optimal parameters of the network based on the error of the whole batch. The gradient Descent method can not be used with SNNs because the spike function is discrete in nature, and therefore, its derivative is 0 or undefined. Also, the thresholding mechanism creates a lot of discontinuities due to which the search for the global minima becomes very difficult. But to maintain the compatibility with ANN framework, we concentrate on training SNNs using gradient descent. One of the techniques to train SNNs using SGD is surrogate gradients.

Surrogate gradients can be used to solve the behaviour of the non-differentiable threshold (Neftci et al. 2019) [24]. The spike generation's Heaviside step function is smoothed out in this manner during the backward pass while remaining a step function during the forward pass. There are various choices of surrogate gradients including sigmoid, gaussian, multi-gaussian, exponential decay, linear and boxcar. Boxcar is inexpensive in terms of computations shown Bittar Alexandre et al. (2022) [2]. It is defined as:

$$\frac{\partial s[t]}{\partial u[t]} = \begin{cases} 0.5 & \text{if } |u[t] - \theta| \leq 0.5 \\ 0 & \text{otherwise} \end{cases}$$

• **Output layer:** Since the outputs of the SNNs come as a sequence of spikes, therefore, to convert these sequences to a single number we used non-spiking LIF neurons and a cumulative sum of the potential over time to generate outputs [24].

$$o_i = \sum_{t=1}^{T} \text{softmax}(u_i^L[t])$$

• **Optimizer:** In all experiments conducted, we utilized the Adam optimizer, as proposed by Kingma and Ba in [14]. In deep learning, the Adam optimizer is a prominent optimization algorithm known for its effectiveness in training neural networks. It combines the advantages of both adaptive gradient descent and momentum-based optimization methods. The Adam optimizer's initial learning rate for our experiments was 0.001. This selection of learning rate was discovered to find a balance between convergence speed and stability, enabling the model to understand the underlying patterns in the data effectively.

- **Metrics:** We used unweighted accuracy (UA) for evaluation in our experiments. Unweighted accuracy is the average accuracy across classes. We also calculated precision, recall, and F1-score in all conducted experiments.

4.2 Results

LIF Model Results. We conducted experiments using different parameter settings to examine the model's performance. We trained and evaluated the model with the number of hidden layers ranging from 3 to 5 and utilized 64 or 128 hidden units in each layer. Figure 10 depicts the results of the best-performing LIF model with three layers. The model without dropout performs better, achieving an accuracy of 73.5%.

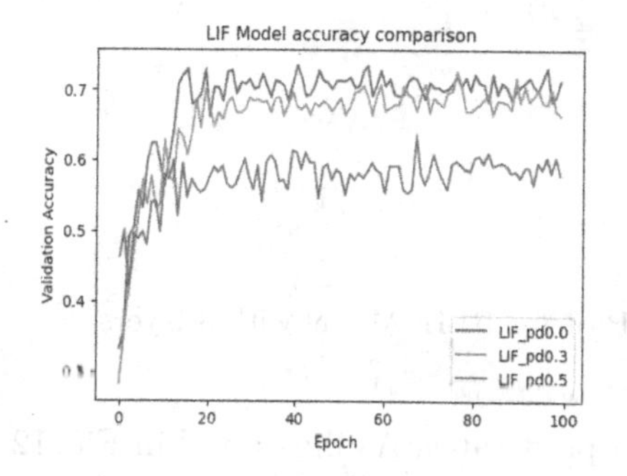

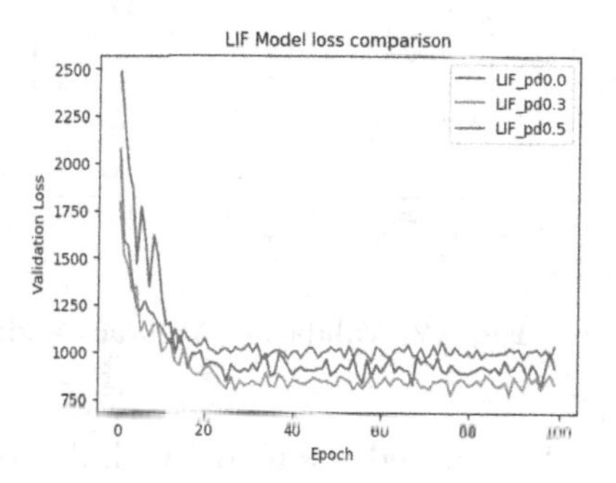

Fig. 10. Validation Accuracy and Loss plot for LIF Model with 3 layers

Increasing dropout regularization penalizes the model. Due to this, the model becomes sparse, and it is unable to capture the underlying pattern of the data properly.

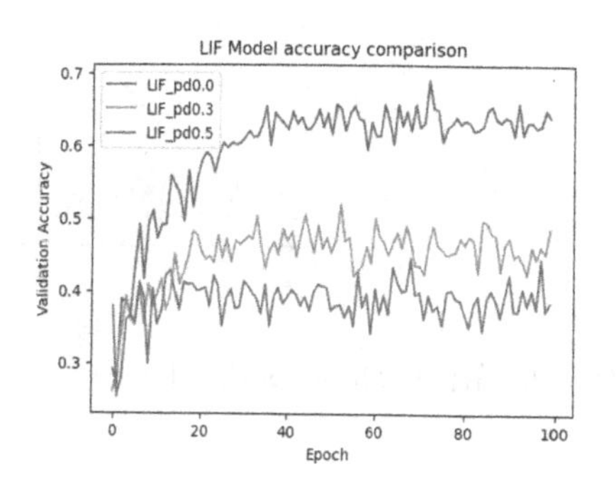

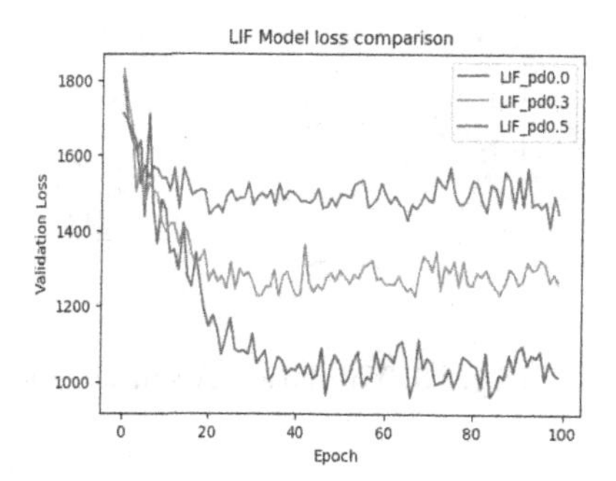

Fig. 11. Validation Accuracy and Loss Plot for LIF Model with 5 layers

In Fig. 11, we illustrate the results of the LIF model with five layers. With five layers, the performance of the model is dropped to 69.15%. This highlights the trade-off between regularization and model performance, suggesting that excessive dropout regularization may hinder the model's ability to accurately learn and predict patterns in the data.

RLIF Model Results. Similar to LIF Model, We changed the number of hidden layers ranging from 3 to 5 and utilized 64 or 128 hidden units in each layer.

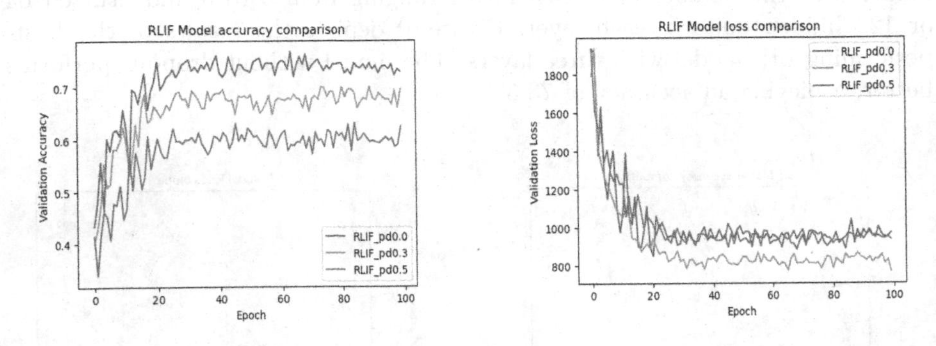

Fig. 12. Validation Accuracy and Loss Plot for RLIF Model with 3 layers

We trained the model with different dropout rates. As illustrated in Fig. 12. RLIF model with 3 layers achieved the best accuracy of 76.85%. Figure 13. represents the results for RLIF model with five layers. This model Achieved 70% accuracy.

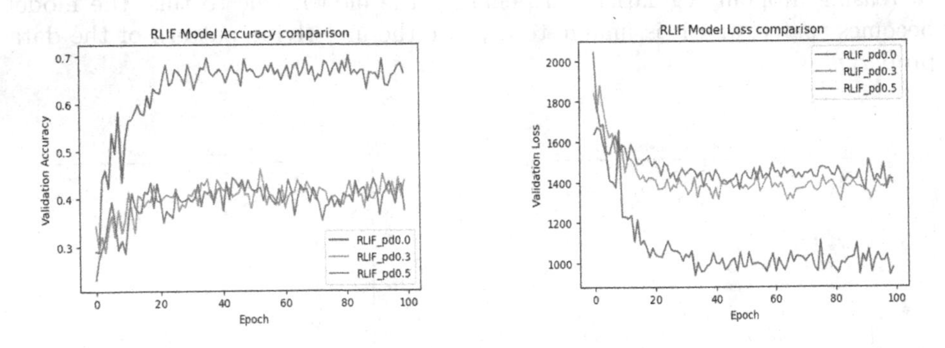

Fig. 13. Validation Accuracy and Loss Plot for RLIF Model with 5 layers

We conducted training experiments utilizing equivalent Multilayer Perceptron (MLP) and Recurrent Neural Network (RNN) architectures in order to compare the performance of Spiking Neural Networks (SNNs) to that of their traditional Artificial Neural Network (ANN) counterparts. The key distinction between these networks lies in the types of neurons employed. Figure 14 provides a visual comparison of the LIF and RLIF models alongside the MLP and RNN architectures, allowing for an insightful analysis of their respective performances and potential advantages.

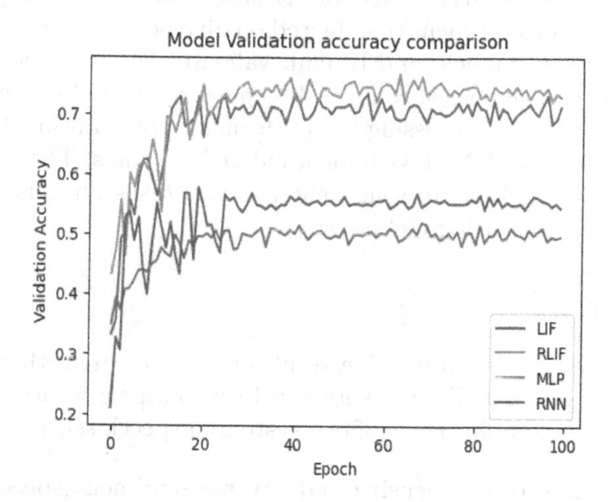

Fig. 14. Comparison of SNN and ANN

As illustrated in Table 2. The results validate the efficiency of the SNN models on this task. Both the SNN models, LIF and RLIF, outperform the ANN models by a significant margin. This may be due to the fact that SNNs are better at capturing the temporal dynamics. Among all the models, RLIF is the best performing.

Table 2. Comparison of our proposed Method with Traditional Artificial Neural Networks.

Model	Unweighted Accuracy (UA)	Precision	Recall	F1-score
MLP	58.07%	53.72%	46.34%	41.55%
RNN	55.4%	57.19%	50.81%	42.62%
LIF*	73.25%	72.24%	75%	72.6%
RLIF*	76.85%	71.69%	73.27%	71.75%

5 Conclusion and Future Work

5.1 Conclusion

In conclusion, our research introduces an innovative end-to-end multilingual speech sentiment recognition system based on Spiking Neural Networks (SNNs). This system demonstrates its capability to automatically extract relevant features from raw audio signals and accurately classify sentiments as positive, negative, or neutral. Our experiments conducted on diverse language datasets, including English, German, French, and Italian, validate the effectiveness of our proposed system across languages. Remarkably, our SNN model achieves an impressive accuracy of 76.85%, surpassing the performance of traditional artificial neural networks (MLP and RNN) with identical architectures. These findings highlight the potential of SNNs in sentiment analysis tasks and encourage further exploration of this promising technology.

5.2 Future Work

Though SNNs have shown promising results in our research, there is still room for further improvements. Future work can focus on incorporating the following steps to improve the performance of our system in speech sentiment recognition:

- Curating a more comprehensive and diverse sentiment speech dataset can contribute to a more robust model. By including a wider range of voices, accents, and cultural contexts, the model can better adapt to various real-world scenarios.
- Expanding the language coverage beyond the currently used datasets (English, German, French, and Italian) by incorporating additional languages like Hindi, Tamil, Telugu, Bengali, and others can lead to a more effective multilingual speech sentiment recognition system.
- Applying different data augmentation techniques can help increase the generalizability of the model. Techniques such as adding background noise, altering pitch and speed, or introducing perturbations can simulate real-world variations and improve the model's ability to handle diverse audio inputs.
- Integrating an attention mechanism within the SNN model can enhance its capability to distinguish and focus on salient features related to sentiments.

By considering these avenues for future research and development, we can strive towards even more accurate and versatile speech sentiment recognition systems based on SNNs.

References

1. Al-onazi, B.B., Nauman, M.A., Jahangir, R., Malik, M.M., Alkhammash, E.H., Elshewey, A.M.: Transformer-based multilingual speech emotion recognition using data augmentation and feature fusion. Appl. Sci. **12**(18), 9188 (2022)

2. Bittar, A., Garner, P.N.: A surrogate gradient spiking baseline for speech command recognition. Front. Neurosci. **16**, 865897 (2022)
3. Burkhardt, F., Paeschke, A., Rolfes, M., Sendlmeier, W.F., Weiss, B., et al.: A database of German emotional speech. In: Interspeech, vol. 5, pp. 1517–1520 (2005)
4. Buscicchio, C.A., Górecki, P., Caponetti, L.: Speech emotion recognition using spiking neural networks. In: Esposito, F., Raś, Z.W., Malerba, D., Semeraro, G. (eds.) ISMIS 2006. LNCS (LNAI), vol. 4203, pp. 38–46. Springer, Heidelberg (2006). https://doi.org/10.1007/11875604_6
5. Chawla, N.V., Bowyer, K.W., Hall, L.O., Kegelmeyer, W.P.: Smote: synthetic minority over-sampling technique. J. Artif. Intell. Res. **16**, 321–357 (2002)
6. Costantini, G., Iaderola, I., Paoloni, A., Todisco, M., et al.: EMOVO corpus: an Italian emotional speech database. In: Proceedings of the ninth international conference on language resources and evaluation (LREC'14), pp. 3501–3504. European Language Resources Association (ELRA) (2014)
7. Gerstner, W., Kistler, W.M.: Spiking Neuron Models: Single Neurons, Populations, Plasticity. Cambridge University Press, Cambridge (2002)
8. Gournay, P., Lahaie, O., Lefebvre, R.: A canadian french emotional speech dataset (2018). https://doi.org/10.5281/zenodo.147876510.5281/zenodo.1478765
9. Huang, Z., Dong, M., Mao, Q., Zhan, Y.: Speech emotion recognition using CNN. In: Proceedings of the 22nd ACM International Conference on Multimedia, pp. 801–804 (2014)
10. Issa, D., Demirci, M.F., Yazici, A.: Speech emotion recognition with deep convolutional neural networks. Biomed. Signal Process. Control **59**, 101894 (2020)
11. Jackson, P., Haq, S.: Surrey Audio-Visual Expressed Emotion (SAVEE) Database. University of Surrey, Guildford, UK (2014)
12. Jain, M., et al.: Speech emotion recognition using support vector machine. arXiv preprint arXiv:2002.07590 (2020)
13. Japkowicz, N., Stephen, S.: The class imbalance problem: a systematic study. Intell. Data Analysis **6**(5), 429–449 (2002)
14. Kingma, D.P., Ba, J.: Adam: a method for stochastic optimization. In: Bengio, Y., LeCun, Y. (eds.) 3rd International Conference on Learning Representations, ICLR 2015, San Diego, CA, USA, May 7–9, 2015, Conference Track Proceedings (2015). http://arxiv.org/abs/1412.6980
15. Koduru, A., Valiveti, H.B., Budati, A.K.: Feature extraction algorithms to improve the speech emotion recognition rate. Int. J. Speech Technol. **23**(1), 45–55 (2020)
16. Likitha, M., Gupta, S.R.R., Hasitha, K., Raju, A.U.: Speech based human emotion recognition using MFCC. In: 2017 International Conference on Wireless Communications, Signal Processing and Networking (WiSPNET). pp. 2257–2260. IEEE (2017)
17. Lin, T., Goyal, P., Girshick, R.B., He, K., Dollár, P.: Focal loss for dense object detection. CoRR abs/1708.02002 (2017). http://arxiv.org/abs/1708.02002
18. Lotfidereshgi, R., Gournay, P.: Biologically inspired speech emotion recognition. In: 2017 IEEE International Conference on Acoustics, Speech and Signal Processing (ICASSP), pp. 5135–5139. IEEE (2017)
19. Maass, W.: Networks of spiking neurons: the third generation of neural network models. Neural Netw. **10**(9), 1659–1671 (1997)
20. Mansouri-Benssassi, E., Ye, J.: Speech emotion recognition with early visual cross-modal enhancement using spiking neural networks. In: 2019 International Joint Conference on Neural Networks (IJCNN), pp. 1–8. IEEE (2019)

21. Mansouri-Benssassi, E., Ye, J.: Generalisation and robustness investigation for facial and speech emotion recognition using bio-inspired spiking neural networks. Soft. Comput. **25**(3), 1717–1730 (2021)
22. Mirsamadi, S., Barsoum, E., Zhang, C.: Automatic speech emotion recognition using recurrent neural networks with local attention. In: 2017 IEEE International Conference on Acoustics, Speech and Signal Processing (ICASSP), pp. 2227–2231. IEEE (2017)
23. Mulyanto, M., Faisal, M., Prakosa, S.W., Leu, J.S.: Effectiveness of focal loss for minority classification in network intrusion detection systems. Symmetry **13**(1), 4 (2020)
24. Neftci, E.O., Mostafa, H., Zenke, F.: Surrogate gradient learning in spiking neural networks. CoRR abs/1901.09948 (2019). http://arxiv.org/abs/1901.09948
25. Sora, C.J., Alkhatib, M.: Speech sentiment analysis for citizen's engagement in smart cities' events. In: 2022 7th International Conference on Smart and Sustainable Technologies (SpliTech), pp. 1–5. IEEE (2022)
26. Yadav, A., Vishwakarma, D.K.: A multilingual framework of CNN and Bi-LSTM for emotion classification. In: 2020 11th International Conference on Computing, Communication and Networking Technologies (ICCCNT), pp. 1–6. IEEE (2020)
27. Yuan, X., Wong, W.P., Lam, C.T.: Speech emotion recognition using multi-layer perceptron classifier. In: 2022 IEEE 10th International Conference on Information, Communication and Networks (ICICN), pp. 644–648. IEEE (2022)
28. Zehra, W., Javed, A.R., Jalil, Z., Khan, H.U., Gadekallu, T.R.: Cross corpus multi-lingual speech emotion recognition using ensemble learning. Complex & Intelligent Systems, pp. 1–10 (2021)

FopLAHD: Federated Optimization Using Locally Approximated Hessian Diagonal

Mrinmay Sen(✉) and C Krishna Mohan

IIT Hyderabad, Hyderabad, India
ai20resch11001@iith.ac.in

Abstract. Federated optimization or Federated learning (FL) is a novel decentralized machine learning algorithm, where a server model or global model is trained collaboratively without collecting local data from the sources or clients. However, the data heterogeneity across the ever-growing data sources affects the FL optimization in terms of slow convergence of the global model, which requires large number of communication rounds between the server and the clients to achieve a targeted performance from the global model. In this paper, we propose FopLAHD, that aims to decrease this communication rounds by accelerating global model convergence. To accelerate convergence, FopLAHD uses locally approximated Hessian diagonal vectors along with local gradients while finding the global update direction. Experimental results on highly heterogeneous data partitions with partial participation of the total clients show that FopLAHD outperforms existing state-of-the-art FL algorithms such that FedProx, SCAFFOLD, DONE and GIANT in terms of communication rounds while reaching to a certain precision of the global model convergence.

Keywords: Federated optimization · Huge communication rounds · Slow convergence · Approximated Hessian diagonal · Data heterogeneity

1 Introduction

The increasing number of edge devices such as smartphones, wearables, sensors, and other Internet-connected devices leads to advancement of huge amount of data, which can be useful for training a machine learning model. Traditional centralized learning requires collection of these distributed data in a common place or server, which is associated with huge communication cost and lack of privacy for the users. To efficiently train a machine learning model with these distributed data, federated learning has emerged recently as a promising solution, which enables distributed clients collaboratively to train a global model under the supervision of a hosting server while protecting private data locally. This collaborative training refers to the problem of optimization of distributed sub-problems, that means optimization of a global loss function, which is the average of all the local loss functions. The most popular and base line of FL algorithm, FedAvg [12], utilizes stochastic gradient descent to optimize local loss

V. Goyal et al. (Eds.): BDA 2023, LNCS 14418, pp. 235–245, 2023.
https://doi.org/10.1007/978-3-031-49601-1_16

function while updating local model parameters. Then the server collects all the locally updated models and aggregates these models to find the updated global model. So, one communication round or one global epoch of FedAvg consists of sharing initial global model to all the participating clients and collecting locally updated models to the server. This communications are the major challenges in federated learning, which needs to be minimized to reduce overall communication cost associated with training the global model. Number of communication rounds in FL depends on how fast the global model gets converged. In FedAvg, the global model gets converged in a few communication rounds, when data are homogeneous across all the clients, which may not be possible in practical applications. When data are heterogeneous, the global model of FedAvg faces problem of objective inconsistency, which leads to convergence of the global model at a stationary point of a mismatched objective function, which may not be same as the optima of the true global objective function [21]. Due to this objective inconsistency in heterogeneous settings, FedAvg requires comparatively large number of communication rounds to find the optima of the global loss function [10,19,24,26].

To overcome this issue, many efforts have been made, which include FedMD [7], FedProx [9], FedNova [21], FOLB [13], FedAdp [23], MOON [8], SCAFFOLD [5], FedInit [18] etc. FedMD utilizes transfer learning and knowledge distillation to train a global model through the collaboration of multiple clients, each of which has its own uniquely designed model. FedProx provides relaxation for local clients to perform varying amounts of work during local training by introducing a proximal term to the local sub-problems. FedNova utilizes normalized averaging of locally updated models to overcome the problem of objective inconsistency in heterogeneous FL settings. FOLB performs random and uniform sampling of the local clients and employs a calibration procedure for aggregating local model updates, assigning weights to estimate their importance to the local models. FedAdp finds the angel between the local gradients and the global gradient and uses a non-linear function to assign the corresponding weight for each client's contribution towards the global model. MOON aims to correct the variability of the local updates by incorporating similarity between model representations to handle the issues of heterogeneous clients with image data. SCAFFOLD utilizes variance reduction to control the weight divergence between the local and global models. In SCAFFOLD, the difference between global and local model updates is added with the local loss function to correct the local updates. In FedInit, the initial local model is initialized by moving away from the current global state in the direction of latest local state. As all these modifications on FedAvg utilize only gradient information while updating local models, the convergence is still slower than the methods, which incorporate second-order Hessian curvature along with gradient while finding local updates [3]. While the Newton method of optimization offers quadratic convergence rates in machine learning, a significant problem arises with computing and storing the full Hessian matrix and its inverse in large-scale settings as the time & space complexities of finding Hessian matrix are both $O(d^2)$ and time & space complexities of finding the inverse Hessian are

$O(d^3)$ and $O(d^2)$respectively, where d is number of model parameters [1,16, 20]. This compels researchers to prioritize Hessian approximation rather than employing the exact Hessian when seeking Newton updates.

Existing second order based FL algorithms include GIANT [22], FedSSO [11], DONE [4], FedNL [15], Basis Matters [14] etc. GIANT employs the conjugate gradient method to approximate local Newton updates through the assistance of the global gradient. It aggregates these local updates using the harmonic mean to determine the global update. Taking inspiration from GIANT, DONE utilizes global gradient to find local Newton direction and aggregates these local Newton directions to find the global Newton direction. To accelerate FL convergence, DONE uses Richardson iteration while finding local updates. Since GIANT and DONE both utilize the global gradient, these methods involve four times of communication between the server and clients, leading to an increase in per-global-epoch time. FedSSO utilizes server side Quasi-Newton method (BFGS) with the aggregated gradients over all the local clients. Utilization of BFGS algorithm leads to the requirement of storing full Hessian matrix, which may not be practical for the resource constraint server. In Basis Matters and FedNL, the approximation of the Hessian for each client's loss function is attained by employing the Hessian from the preceding step. During communication, the Hessians from all clients are compressed and transmitted to the server along with the gradient. As a result, in FedNL and Basis Matters, the storage, computation, and compression of clients' Hessian information place an extra computational load on the clients.

In this paper, we propose FopLAHD, where we use only diagonal elements of the Hessian matrix to avoid forming and storing full Hessian and its inverse while finding global Newton update. As the estimation of Hessian diagonal involves same effort like estimation of full Hessian, we utilize Hessian vector product to approximate the Hessian diagonal with the same concept described in the paper of Bekas et al [2]. In FopLAHD, individual local client calculates the approximated Hessian diagonal and gradient of their respective local loss function and transmits these calculations to the server. The server finds the global Newton direction using the aggregated Hessian diagonal and aggregated gradient derived from all clients. Utilizing these aggregated Hessian diagonal and gradient leads to the convergence of the global model within a relatively smaller number of communication rounds. This has been verified through experimental results on highly heterogeneous data partitions, considering the partial participation of the total clients. As in practical FL applications, all the clients may not be available together during a FL iteration, the data distribution of the set of participated clients in a FL iteration may vary from the set of participated clients of an another FL iteration, which may lead to information loss from the global model trained in previous FL iterations. To handle this issue, FopLAHD utilizes exponential moving averages of the aggregated Hessian diagonal and the gradient (which is same as ADAM [6]) before finding the global Newton update. As the Hessian diagonal can vary significantly over all the model parameters, the performance of the updated model may be deteriorated. To overcome this issue, we reduce the variance of the aggregated Hessian diagonal by detecting the noise and replacing it with the mean value of this Hessian diagonal. To detect the noise, we use a threshold on the mean value of the hessian diagonal. For example, if the mean value of Hessian diagonal is e_m, values which lie below fe_m and above $e_m + (1-f)e_m$ are considered as noise term of the Hessian diagonal, where threshold f $\in (0, 1]$.

2 Problem Formulation

Let, C=$\{C_1, C_2, ..., C_N\}$ is the set of N clients. Each client C_i has its local data D_i and local loss function F_i. F_i is the average loss over all the local data samples D_i. Then the objective of federated optimization is to find the optimal global model parameters w by minimizing the global loss function F as shown in Eq. 1 (the global loss function F is the average of all the local loss functions $\{F_i\}$).

$$\min_{w} F(w) = \frac{1}{N} \sum_{D_i} F_i(w; D_i) \tag{1}$$

where, $w \in R^d$, $F_i(w) = \frac{1}{|D_i|} \sum_{\xi \in D_i} f_{ij}(w; \xi)$ and f_{ij} is the loss associated with j^{th} sample of ith client.

3 Preliminaries

3.1 Newton Method of Optimization

Newton method of optimization [20] is a second-order optimization technique, where the search direction of finding the optima of the loss function is calculated by minimizing the second order Taylor expansion of the loss function. In Newton method of optimization the search direction is calculated by scaling the gradient g with the help of inverse of Hessian H of the loss function, which is shown in Eq. 2 (where, w_t, H_t and g_t are the model parameters, Hessian and gradient respectively at iteration t).

$$w_t = w_{t-1} - H_t^{-1} g_t \tag{2}$$

Use of Hessian curvature of the loss function along with its gradient, while finding the optima, leads to quadratic convergence rate of the model parameters [1,16,17], which is advantageous for training a machine learning model in less time. The property of this faster convergence of Newton method motivates us to use second order Hessian information in federated learning. However, there are certain challenges associated with using the Newton method of optimization. It necessitates the computation and retention of the complete Hessian matrix and its inverse during the search direction determination process. This results in both time and space complexities of O(d^3) and O(d^2), respectively. For large scale machine learning application, it may be difficult to compute and store the full Hessian. To avoid forming and storing full Hessian, we utilize the approximated Hessian diagonal in federated optimization.

Algorithm 1. Hessian diagonal vector approximation using Hessian vector product (Here, $\otimes$ and $\oslash$ represent element wise multiplication and division respectively)

0: **Input:** g : Gradient of the loss function, s : Number of iterations,
p_0=0
q_0=0
1: **for** $k = iter$ **to** s **do**
2: Generate random vector $v_k \in R^d$
3: Find Hessian vector product $Hv_k = \frac{\partial(g \otimes v_k)}{\partial w}$
4: Update $p_k = p_{k-1} + Hv_k \otimes v_k$
5: Update $q_k = q_{k-1} + v_k \otimes v_k$
6: **end for**
7: Approximate the Hessian diagonal $Z = p_k \oslash q_k$

3.2 Hessian Diagonal Vector Approximation

To avoid the issues of Newton method, which are mentioned in Sect. 3.1, we utilizes only the diagonal vector instead of full Hessian while finding global optima. As the estimation of Hessian diagonal involves same effort like estimation of full Hessian, we utilize Hessian vector product to approximate the Hessian diagonal vector with the same concept described in the paper of Bekas et al [2], which is shown in Alg. 1.

Algorithm 2. Removing Noise from Hessian diagonal vector

0: **Input:** Z : Hessian diagonal vector, $f \in (0, 1]$: Smoothing threshold term

1: Find mean value e_m of Z
2: **for** ele $\in$ Z **do**
3: **if** $ele < fe_m$ or $ele > (e_m + (1 - f)e_m)$ **then**
4: $ele = e_m$
5: **end if**
6: **end for**

4 FopLAHD

One FL iteration of FopLAHD is depicted in Alg. 3. At iteration t, FopLAHD initializes local model w_{it} with the global model w_{t-1}. Then, each local client C_i finds Hessian diagonal vector $Z_{it} \in R^{d \times 1}$ by using Alg. 1 and shares this to the server along with the gradient g_{it} computed on the local loss function. Server collects and aggregates all the local Hessian diagonal vectors $\{Z_{it}\}$ and gradients $\{g_{it}\}$ to find global Hessian diagonal vector Z_t and global gradient g_t across all the clients. As the Hessian diagonal can vary significantly over all the model parameters, the performance of the model updated using this Hessian diagonal may be deteriorated. To overcome this issue, we reduce the variance of the aggregated Hessian diagonal by detecting the noise based on a threshold f$\in (0, 1]$ and replacing it with the mean value of this Hessian diagonal, which has been shown in Alg. 2. As in practical FL applications, all the clients may not be available together during a FL iteration, the data distribution of the set of participated clients in a FL iteration may vary from the set of participated clients of an another FL iteration, which may lead to information loss from the global model trained in previous FL iterations. To handle this issue, FopLAHD utilizes first moment M_1 and second moment M_2 of global gradient and global Hessian diagonal respectively (same as ADAM [6]), while finding the global Newton update. First moment M_1 is the exponential moving average of global gradient with decay rate β_1 and second moment M_2 is the exponential moving average of squared Hessian diagonal vector with decay rate β_2. M_1 and M_2 are biased towards zero as these are initialized as (vectors of) 0's. To correct this, we use bias-corrected estimates $\widehat{M_1}$ and $\widehat{M_2}$ instead of M_1 and M_2 while finding updated global model as shown in Alg. 3.

Algorithm 3. FopLAHD

0: **Input:** T : Number of FL iterations or global epochs, w_0: Initial global model, η: Learning rate, $\rho > 0$: Hessian diagonal regularization term, $f \in (0, 1]$: Threshold term for diagonal smoothing, $\{\beta_1, \beta_2\} \in [0, 1)$: Exponential decay rates for the moment estimates, $\{M_1^0 \leftarrow 0, M_2^0 \leftarrow 0\}$: Initial moment vectors which are initialized with zeros

1: **for** $t = 1$ **to** T **do**
2: Sample subset C^t from total clients C
3: **for** client $i \in C^t$ **do**
4: $w_{it} \leftarrow w_{t-1}$
5: Find stochastic gradient $g_{it} = \frac{\partial F_i(w_{it}, D_i)}{\partial w_{it}} \in R^d$
6: Compute Hessian diagonal $Z_{it} \in R^d$ using Alg. 1.
7: Share g_{it} and Z_{it} to the server
8: **end for**
9: Server collects all g_{it} and aggregates these to find the average gradient or global gradient $g_t = \frac{1}{|C^t|} \sum_{i=1}^{|C^t|} g_{it}$
10: Similarly server finds average Hessian diagonal vector $Z_t = \frac{1}{|C^t|} \sum_{i=1}^{|C^t|} Z_{it}$
11: Remove noise from Z_t by using Alg. 2
12: Find $M_1^t = \beta_1 M_1^{t-1} + (1 - \beta_1) g_t \in R^d$
13: Find $M_2^t = \beta_2 M_2^{t-1} + (1 - \beta_2) Z_t Z_t \in R^d$
14: Find $\widehat{M_1}^t = \frac{M_1^t}{1-\beta_1^t}$ and $\widehat{M_2}^t = \frac{M_2^t}{1-\beta_2^t}$, here β_j^t is β_j to the power t
15: Update global model $w_t = w_{t-1} - \eta(\sqrt{\widehat{M_2}^t} + vect(\rho))^{-1} \otimes \widehat{M_1}^t$
16: **end for**

4.1 Finding Updated Global Model

To find the updated global model, FopLAHD uses the following variant of Newton method of optimization as shown in Eq. 3 (Where I is an identity matrix and $Diag(Z_t)$ is the diagonal matrix whose diagonal elements are obtained from the vector Z_t). In Eq. 3, FopLAHD replaces $Diag(Z_t)$ with the square-root of the bias corrected estimate $\widehat{M_2}^t$ and replaces g_t with the bias corrected estimate $\widehat{M_1}^t$ as shown in Eq. 4, here $vect(\rho) \in R^d$ is a vector whose elements are ρ's and $\otimes$ represents element wise multiplication between two vectors. FopLAHD uses $\sqrt{\widehat{M_2}^t}$ instead of $\widehat{M_2}^t$ as, $\widehat{M_2}^t$ is the second moment of the Hessian diagonal. As, $[Diag(Z_t) + \rho I]$ is a diagonal matrix, the inverse of $(\sqrt{\widehat{M_2}^t} + vect(\rho)) \in R^d$ can be obtained by performing element wise reciprocal of this vector.

$$w_t = w_{t-1} - \eta[Diag(Z_t) + \rho I]^{-1} g_t \tag{3}$$

$$w_t = w_{t-1} - \eta(\sqrt{\widehat{M_2}^t} + vect(\rho))^{-1} \otimes \widehat{M_1}^t \tag{4}$$

4.2 Complexity Analysis

The time complexity of finding gradient is O(d) and the time complexity of approximating Hessian diagonal is O(sd), where, s is number of iterations required for this approximation. So the overall time complexity for each cient is O(sd). As each client shares Hessian diagonal along with gradient, the overall space complexity for each client is O(2d). In FopLAHD, server needs to recall the previous global Hessian and gradient to find their momentums, the overall space complexity for the sever is O(4d). The time complexity for finding momentum estimates is O(d) and for finding the model updates is O(d) time complexity. So, the overall time complexity for the server is O(d).

5 Experimental Setup

We validate our proposed algorithm by doing an extensive experiments on heterogeneously partitioned MNIST and FashionMNIST datasets with partial clients participation. We use multinomial logistic regression (MLR) model with crossentropy loss function for federated image classification tasks. To compare the performance of FopLAHD, we use state-of-the-art FL algorithms such that FedProx, SCAFFOLD, DONE and GIANT. For all the algorithms, we use same initialization and same FL settings. Taking inspiration from the paper of Yurochkin et al [25], we utilizes Dirichlet distribution to create heterogeneous data partitions. Same as the paper of Yurochkin et al., we simulate $P_i \sim Dir_K(0.2)$ and then create a partition by assigning a $P_{(i,j)}$ proportion of the samples of ith class to the jth client (C_j). With this, each client may not get samples from all the classes as we use very small value of Dirichlet distribution's concentration parameter (0.2), which indicates a highly heterogeneous data partitions across all the clients. In our experiments, we use total N=200 number of clients with 40% partial participation per global epoch to make a more realistic FL settings where, in each FL iteration, there may be a chance of facing new participants. For all the methods including existing and proposed, we did extensive experiments with several sets of hyper-parameters and choosed the best performing model for each method by considering minimum train & test losses and maximum test accuracy. We use learning rate $\in \{1, 0.5, 0.1, 0.01, 0.001, 0.0001\}$, FedProx proximal term & Hessian diagonal regularization term $\in \{1, 0.5, 0.1, 0.01\}$, number of Rechardson iterations of DONE and CG iteration of GIANT $\in \{5, 10\}$, $\alpha_{DONE} \in \{0.1, 0.01\}$, mini-batch size=512 and number of iterations for FopLAHD's Hessian diagonal approximation $s = 5$. From our experiments on FopLAHD, we observed that like ADAM, FopLAHD can perform well with $\beta_1 = 0.9$, $\beta_2 = 0.99$.

We did further experiments with CNN model of two convolutional layers and two fully connected layers (total 431080 trainable parameters). For CNN based federated image classification, we use the same heterogeneously partitioned FashionMNIST data as mentioned earlier. While doing experiments with CNN based image classification, we can not able to find the suitable hyper-parameters for DONE and GIANT which may be because of their assumption of strongly convex loss function. So, we compare the performance of FopLAHD with FedProx and SCAFFOLD.

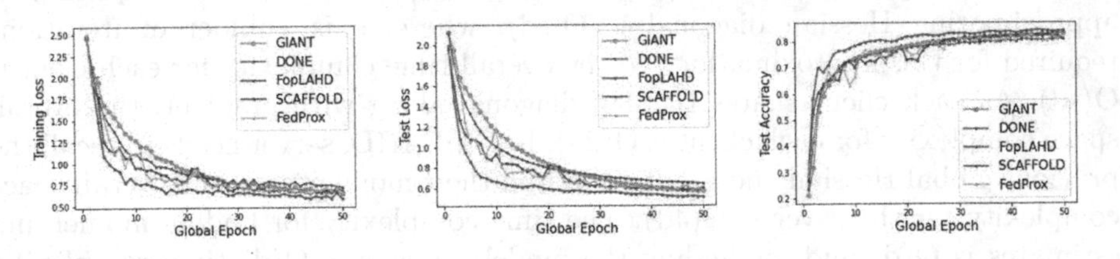

Fig. 1. Comparisons of training loss, test loss and test accuracy on MNIST image classification using MLR

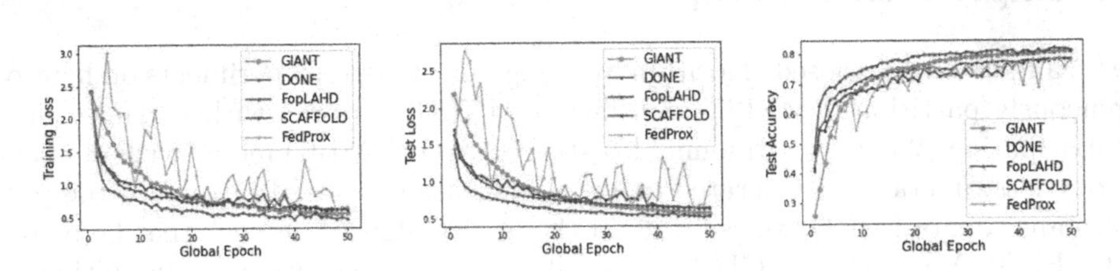

Fig. 2. Comparisons of training loss, test loss and test accuracy on FashionMNIST image classification using MLR

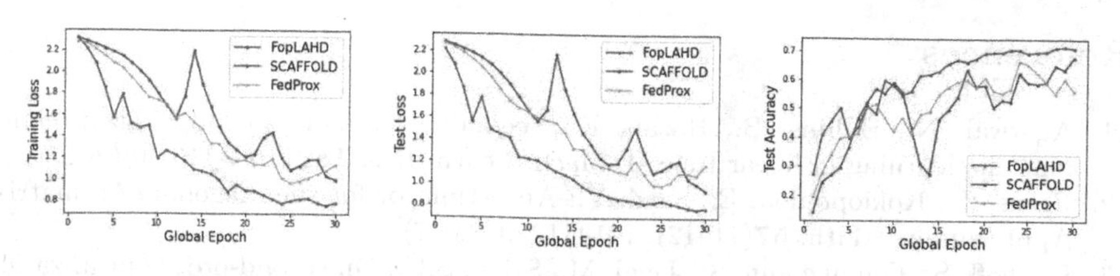

Fig. 3. Comparisons of training loss, test loss and test accuracy on FashionMNIST image classification using CNN

5.1 Results

Our experimental results are depicted in Figs. 1, 2 and 3. From the figures, it may be noticed that FopLAHD can achieve a faster reduction in the global epoch-wise training loss and test loss compared to FedProx, SCAFFOLD, DONE, and GIANT. From the figures, it may also be observed that FopLAHD can achieve better test accuracy than FedProx, SCAFFOLD, DONE, and GIANT. As we use same initialization and same settings for all the methods while doing comparisons, the faster reduction of train & test losses and faster increase in test accuracy indicate a better convergence of the global model in FopLAHD, which results in reduction of communication rounds while achieving a certain precision of the convergence from the global model.

6 Conclusions

In this paper, we proposed FopLAHD, which aims to improve FL convergence or reduce FL iterations in heterogeneous data distribution with partial clients participation. To accelerate FL convergence, FopLAHD utilizes locally approximated Hessian diagonal vectors along with local gradients computed on local loss functions and uses these to obtain global Hessian diagonal and global gradient across all the available clients. With these global Hessian diagonal and global gradient, FopLAHD finds the global Newton update. FopLAHD uses exponential moving averages of the global Hessian diagonal and global gradient to avoid the issues with partial participation in FL. As the Hessian diagonal may contain some noise, we use a threshold based noise detection and replacement technique to remove these noise. Results of our extensive experiments on image classification tasks of various datasets (with both the MLR and CNN models) show that FopLAHD can reduce number of FL rounds better than existing state-of-the-art FL algorithms such that FedProx, SCAFFOLD, DONE, and GIANT, while achieving a certain precision of convergence from the global model.

References

1. Agarwal, N., Bullins, B., Hazan, E.: Second-order stochastic optimization for machine learning in linear time. J. Mach. Learn. Res. **18**, 116:1-116:40 (2017)
2. Bekas, C., Kokiopoulou, E., Saad, Y.: An estimator for the diagonal of a matrix. Appl. Numer. Math. **57**(11–12), 1214–1229 (2007)
3. Bischoff, S., Günnemann, S., Jaggi, M., Stich, S.U.: On second-order optimization methods for federated learning. CoRR abs/2109.02388 (2021)
4. Dinh, C.T., et al.: DONE: distributed approximate newton-type method for federated edge learning. IEEE Trans. Parallel Distributed Syst. **33**(11), 2648–2660 (2022)
5. Karimireddy, S.P., Kale, S., Mohri, M., Reddi, S.J., Stich, S.U., Suresh, A.T.: SCAFFOLD: stochastic controlled averaging for federated learning. In: Proceedings of the 37th International Conference on Machine Learning, ICML 2020, 13–18 July 2020, Virtual Event. Proceedings of Machine Learning Research, vol. 119, pp. 5132–5143. PMLR (2020)
6. Kingma, D.P., Ba, J.: Adam: a method for stochastic optimization. In: Bengio, Y., LeCun, Y. (eds.) 3rd International Conference on Learning Representations, ICLR 2015, San Diego, CA, USA, May 7–9, 2015, Conference Track Proceedings (2015)
7. Li, D., Wang, J.: Fedmd: Heterogenous federated learning via model distillation. CoRR abs/1910.03581 (2019)
8. Li, Q., He, B., Song, D.: Model-contrastive federated learning. In: IEEE Conference on Computer Vision and Pattern Recognition, CVPR 2021, virtual, June 19–25, 2021, pp. 10713–10722 (2021)
9. Li, T., Sahu, A.K., Zaheer, M., Sanjabi, M., Talwalkar, A., Smith, V.: Federated optimization in heterogeneous networks. In: Proceedings of Machine Learning and Systems 2020, MLSys 2020, Austin, TX, USA, March 2–4, 2020. mlsys.org (2020)
10. Li, X., Huang, K., Yang, W., Wang, S., Zhang, Z.: On the convergence of FedAvg on Non-IID data. In: 8th International Conference on Learning Representations, ICLR 2020, Addis Ababa, Ethiopia, April 26–30, 2020 (2020)
11. Ma, X., et al.: FedSSO: a federated server-side second-order optimization algorithm. CoRR abs/2206.09576 (2022)
12. McMahan, B., Moore, E., Ramage, D., Hampson, S., y Arcas, B.A.: Communication-efficient learning of deep networks from decentralized data. In: Proceedings of the 20th International Conference on Artificial Intelligence and Statistics, AISTATS 2017, 20–22 April 2017, Fort Lauderdale, FL, USA, vol. 54, pp. 1273–1282. PMLR (2017)
13. Nguyen, H.T., Sehwag, V., Hosseinalipour, S., Brinton, C.G., Chiang, M., Poor, H.V.: Fast-convergent federated learning. CoRR abs/2007.13137 (2020)
14. Qian, X., Islamov, R., Safaryan, M., Richtárik, P.: Basis matters: better communication-efficient second order methods for federated learning. In: Camps-Valls, G., Ruiz, F.J.R., Valera, I. (eds.) International Conference on Artificial Intelligence and Statistics, AISTATS 2022, 28–30 March 2022, Virtual Event. Proceedings of Machine Learning Research, vol. 151, pp. 680–720. PMLR (2022)
15. Safaryan, M., Islamov, R., Qian, X., Richtárik, P.: FedNL: making newton-type methods applicable to federated learning. In: Chaudhuri, K., Jegelka, S., Song, L., Szepesvári, C., Niu, G., Sabato, S. (eds.) International Conference on Machine Learning, ICML 2022, 17–23 July 2022, Baltimore, Maryland, USA. Proceedings of Machine Learning Research, vol. 162, pp. 18959–19010. PMLR (2022)

16. Singh, D., Tankaria, H., Yamada, M.: Nys-newton: Nystr\" om-approximated curvature for stochastic optimization. arXiv preprint arXiv:2110.08577 (2021)
17. Süli, E., Mayers, D.F.: An Introduction to Numerical Analysis. Cambridge University Press (2003)
18. Sun, Y., Shen, L., Tao, D.: Understanding how consistency works in federated learning via stage-wise relaxed initialization. CoRR abs/2306.05706 (2023)
19. Tan, A.Z., Yu, H., Cui, L., Yang, Q.: Towards personalized federated learning. CoRR abs/2103.00710 (2021)
20. Tan, H.H., Lim, K.H.: Review of second-order optimization techniques in artificial neural networks backpropagation. In: IOP conference series: materials science and engineering, vol. 495, p. 012003. IOP Publishing (2019)
21. Wang, J., Liu, Q., Liang, H., Joshi, G., Poor, H.V.: Tackling the objective inconsistency problem in heterogeneous federated optimization. In: Advances in Neural Information Processing Systems 33: Annual Conference on Neural Information Processing Systems 2020, NeurIPS 2020, December 6–12, 2020, virtual (2020)
22. Wang, S., Roosta-Khorasani, F., Xu, P., Mahoney, M.W.: GIANT: globally improved approximate newton method for distributed optimization. In: Advances in Neural Information Processing Systems 31: Annual Conference on Neural Information Processing Systems 2018, NeurIPS 2018, December 3–8, 2018, Montréal, Canada, pp. 2338–2348 (2018)
23. Wu, H., Wang, P.: Fast-convergent federated learning with adaptive weighting. CoRR abs/2012.00661 (2020)
24. Ye, M., Fang, X., Du, B., Yuen, P.C., Tao, D.: Heterogeneous federated learning: state-of-the-art and research challenges. CoRR abs/2307.10616 (2023)
25. Yurochkin, M., Agarwal, M., Ghosh, S., Greenewald, K.H., Hoang, T.N., Khazaeni, Y.: Bayesian nonparametric federated learning of neural networks. In: Chaudhuri, K., Salakhutdinov, R. (eds.) Proceedings of the 36th International Conference on Machine Learning, ICML 2019, 9–15 June 2019, Long Beach, California, USA. Proceedings of Machine Learning Research, vol. 97, pp. 7252–7261. PMLR (2019)
26. Zhao, Y., Li, M., Lai, L., Suda, N., Civin, D., Chandra, V.: Federated learning with Non-IID data. CoRR abs/1806.00582 (2018)

A Review of Approaches on Facets for Building IT-Based Career Guidance Systems

N. Chandra Shekar[1], P. Krishna Reddy[1(✉)], K. Anupama[1], and A. Amarender Reddy[2]

[1] International Institute of Information Technology, Hyderabad, India
chandra.shekar@research.iiit.ac.in

[2] School of Policy Support Research, ICAR-National Institute of Biotic Stress Management, Raipur, India

Abstract. Providing employment or jobs to job seekers is a significant challenge faced by several countries worldwide, including India. In addition to accelerating agricultural and industrial growth, governments are trying to provide education and skills to their citizens to tackle this challenge. In developing countries like India, due to industrial growth, thousands of job types or career options are available in the job market. However, most individuals, especially those from rural areas and underprivileged backgrounds, unware of the available career options and often choose sub-optimal career paths. We refer to this issue as the "Career Information Gap (CIG)." To address the issue of CIG, several countries have initiated national career services as mission-mode projects to assist individuals in selecting suitable careers or jobs. Notably, research efforts are being made to explore the progress in information technology (IT), such as the Internet, smartphones, and artificial intelligence, to build systems for reducing CIG. There is a scope to develop IT-based career guidance systems to address CIG. With this background, in this paper, we reviewed research on four facets of career guidance systems: theoretical modeling of a career, career guidance methods, ethical and fairness issues, and IT-based efforts. We hope this review will encourage researchers, especially those from the IT field, to collaborate with theoretical career modelers and counsellors to build scalable systems for tackling CIG to address the unemployment and underemployment problem in India and abroad.

Keywords: career guidance · career counselling · computer-assisted career guidance system · decision support systems · recommender system · unemployment · jobs · skills · national career service

1 Introduction

All individuals engaged in some form of work or activity as a means of livelihood. People are unique regarding their background, education, skills, and experience.

V. Goyal et al. (Eds.): BDA 2023, LNCS 14418, pp. 246–260, 2023.
https://doi.org/10.1007/978-3-031-49601-1_17

For a living, an individual makes an effort to do some job or activity. For a satisfactory living, an individual makes an effort to do a job or an activity that matches his or her interests [1]. As per the Oxford Dictionary [2], the term career is defined as "the series of jobs that a person has in a particular area of work, usually involving more responsibility as time passes". The goal of any government is to empower and enable all individuals to pursue careers (or jobs) that match their interests. Providing employment or jobs to the eligible population is a significant challenge faced by several countries worldwide, including India. Notably, India is facing a substantial challenge with unemployment and underemployment. Unemployment is when an individual does not have a job, but underemployment is when the individual is working at a lower capacity than his or her skills, capabilities, and qualifications. According to the Centre for Monitoring Indian Economy (CMIE) survey [3], India's unemployment rate is 8.11 per cent in April 2023. In developing countries like India, the employment problem manifests itself in the widespread prevalence of low-productivity work and underemployment rather than in high unemployment.

To tackle the challenge of unemployment and underemployment, besides accelerating agricultural and industrial growth, Indian government is also taking steps to provide education and skills to all the eligible populations. For a developing country like India, due to industrial and agricultural growth, hundreds of job types or career options are available in the job market. For better employment, the Indian government has started the national career service as a mission-mode project to assist individuals in selecting suitable careers or jobs. As a part of this effort, starting in 1973, the following programmes are being operationalized [4]: (i) Integrated Rural Development Programme (IRDP) to promote employment in rural areas; (ii) Training of Rural Youth for Self-Employment (TRYSEM) to provide skills training to unemployed rural youth; (iii) Rural Development and Self Employment Training Institute (RUDSETI) was set up to reduce youth unemployment through self-employment training; (iv) Jawahar Rozgar Yojana (JRY); (v) Pradhan Mantri Kaushal Vikas Yojana (PMKVY); and (vi) The Start Up India Scheme to provide employment opportunities and enhance skills across various sectors.

However, it has been reported that 93% of Indian students are aware of just seven types of careers [5], while more than 1000 job types are available globally [6]. On the other hand, India needs more career counsellors, as 93% of schools do not have professional career counsellors [7]. Also, in India, the counsellor-to-student ratio is significantly low, equal to 1:3,000, compared to the globally accepted counsellor-to-student ratio, which is equal to 1:250. In the Indian scenario, most individuals, especially those from rural areas and underprivileged backgrounds, are not aware of possible career options. Usually, an individual typically depends on parents, friends, and relatives for career guidance. However, they need to gain knowledge of career options due to the limited outreach, and information about the fast changes in the job market. Also, for an individual who lives in a village or remote area, getting updated and trustworthy information about new jobs is difficult, especially in remote areas where there might be less access to resources. In India, there is also an issue of an information gap about skills [8] due to poor quality education and a mismatch between available skills and employer demands. It was reported in 2013 that an additional 291 million individuals would need to be upskilled by 2022, as manufacturing requires 85 million, non-manufacturing requires 36 million, and the service sector requires 225 million. We call this issue the "Career Information Gap (CIG)". The cost of CIG may be as high as 10–15% of a country's Gross Domestic Product (GDP), which can be avoided with institutionalizing systems that can optimally match multiple skill sets of individuals with potential jobs.

It can be observed that the progress in the Internet, smartphones, and artificial intelligence is impacting all sections of society. Due to this progress in the Information Technology (IT) revolution, search engines, online platforms, mobile apps, and digital databases provide services to the population to improve lives. By exploring the progress in IT and research related to careers, there is scope to develop IT-based career guidance systems to address the issue of CIG. For building an IT-based system, it will be helpful for an IT researcher to understand the existing career-related research. With this background, in this paper, we discuss key career-related research by dividing it into four facets.

i **Modeling of a Career**: We reviewed a few key research papers about how a notion of career is modelled.
ii **Career Guidance Methods**: We discuss important existing methods and practices, including case studies and strategies, followed for career guidance.
iii **Ethical and Fairness Issues**: We discuss the approaches on the issues of ethical and fairness related to career.
iv **IT-based Methods**: We discuss the major IT-based efforts for career guidance and the corresponding challenges.

Notably, the career subject is multi-disciplinary, spanning disciplines such as Psychology, Sociology, Economics, Education, Business Management and Human Resources, Information Technology, Political Science, Anthropology and Cultural Studies, Gender Studies, and Neuroscience. The research contributions are published in the dedicated journals [9–13]. Broadly, these contributions cover multiple aspects such as theoretical frameworks, counseling theories, assessment techniques, technology-enabled career guidance, virtual reality interventions, mobile applications, gamification, user interface and user experience (UX) design, information visualization, and decision support systems. Case studies are also abundant, including controlled trials, longitudinal studies, meta-analyses, and qualitative research. Specialized focus areas in the research include school/college counseling, adult career transitions, special populations, international perspectives, and industry-specific research. Ethical and policy considerations are also being explored, with topics like access and equity, privacy and confidentiality, quality assurance, credentialing and certification, and government and institutional policies taking the forefront.

The scope of this paper is to give a review of research on essential facets for developing career guidance systems by extending the developments in IT.

We hope this review will encourage researchers, especially from the IT field, to conduct further research to build scalable systems for tackling CIG.

The rest of this paper is organized as follows: In the next section, we discuss approaches for modeling a career. In Sect. 3, we discuss approaches to career counseling methods. The approaches to ethical and fairness issues are discussed in Sect. 4. In Sect. 5, we discuss the IT-based efforts and the challenges. In Sect. 6, we provide a discussion about the utility of the approaches in building career guidance systems. The last section contains the conclusion.

2 Modeling of a Career

This section explains significant theories proposed to model the career notion.

As per John Holland's theory of career choice [1], the personalities of each person and the workplace environment influence the individual's career choices. It proposes that people choose careers that fit their personality type and allow them to use their skills and express their values. This theory identified six personality types: Realistic, Investigative, Artistic, Social, Enterprising, and Conventional. Based on this theory, the study in [14] examines how people can make long-lasting changes by combining personal and career counseling. The counseling has three parts: hearing about the person's life and career, putting those stories together, and planning the future together.

Albert Bandura's social cognitive theory [15] considers that people learn by observing others. The thoughts of others are crucial in shaping their respective personalities. This theory outlines four steps in learning: attention (noticing something), retention (remembering it), reproduction (being able to do it), and motivation (having a reason to do it). This theory is particularly relevant for those who want to understand and change their own behaviour. Based on this theory, an integrative model was proposed in [16], which ties together personal characteristics, career choices, and environmental circumstances to explain career motivation.

Frank Parsons, the vocational guidance movement's founder, introduced the talent-matching approach, later called the Trait and Factor Theory of Occupational Choice [17]. It emphasizes matching individuals to careers based on their aptitudes, interests, and abilities, and a good match leads to optimal performance and productivity. The three-part theory involves understanding one's traits, knowing the job market, and making rational decisions about the fit between the personal traits and the job requirements. A seven-stage process for career counsellors to guide clients in finding suitable careers was proposed. Based on this theory, a framework is proposed in [18], which traverses the landscape of career development methodologies, tools, and strategies, spotlighting the assessment of an individual's interests, necessities, and talents.

John Krumboltz's planned happenstance theory [19] proposes that indecision can be beneficial. It allows one to capitalize on unplanned events in one's career path. It emphasizes the importance of adapting to change in the rapidly shifting labour market by considering the impact of unpredictable factors on career choices. It also highlights the importance of ongoing learning, self-assessment, feedback, networking, work-life balance, and financial planning for successful career management by considering unplanned events.

Donald Super's career development theory [20] emphasizes the evolution of self-concept over time, considering that career planning is a lifelong process. As per this theory, an individual's occupational preferences, competencies, and life situations evolve through experience. The notion of a career is divided into five stages: growth (birth-14), exploration (15–24), establishment (25–44), maintenance (45–64), and decline (65 onwards). Each stage represents different developmental tasks and career and personal growth challenges. Based on this theory, the proposal in [21] advocates for a lifespan-oriented perspective on career exploration and placing importance on immediate experiences.

Table 1. Description of Career Theories

Career Theory	Description
Theory of Career Choice [1]	People tend to choose careers that fit their personality type and allow them to use their skills and express their values
Social Cognitive Theory [15]	People learn by observing others and that their thoughts are crucial in shaping their respective personalities
Trait and Factor Theory [17]	Individuals can be matched to careers based on their aptitudes, interests, and abilities, and a good match leads to optimal performance and productivity
Planned Happenstance theory [19]	Emphasises the importance of adapting to change in the rapidly shifting labour market and acknowledges the impact of unpredictable factors on career choices
Super's Career Development Theory [20]	Proposes that career planning is a lifelong process and the notion of a career is divided into five stages: growth (birth-14), exploration (15–24), establishment (25–44), maintenance (45–64), and decline (65 onwards)
Theory of Work Adjustment [22]	If a person's abilities, such as skills, knowledge, experience, and behaviours, align with the job or organisation's requirements, they are more likely to succeed and be seen as satisfactory
Theory of Circumscription and Compromise [23]	Career development involves narrowing down career choices based on personal and social factors

The theory of Work Adjustment [22] studies the alignment between an individual's attributes and the work environment. As per this theory, if a person's abilities, such as skills, knowledge, experience, and behaviours, align with the job or organization's requirements, the corresponding person is likely to succeed with satisfaction. Also, if a job's or organization's rewards match the individual's values and work-related desires, job satisfaction increases.

Gottfredson's theory of circumscription and compromise [23] suggests that career development involves narrowing down career choices based on personal and social factors. As individuals mature, they gradually exclude career options incompatible with their self-concept and social identity. However, they may need to compromise and adjust career preferences due to real-world limitations, emphasizing both personal agency and external influences in career development. Based on this theory, the analysis in [24] examines career challenges and presses for a cohesive framework in career counseling practises.

Summary: Table 1 provides a summary of different theoretical models, which are being proposed for modelling the notion of career. Each model captures a distinct aspect of an individual's development. Studying the existing models will help in developing effective IT-based career guidance systems.

3 Career Guidance Methods

The profession of career guidance or counseling has evolved as a significant profession. In this section, we discuss the major frameworks employed for career guidance.

The Client-Centered Approach [25] underscores the necessity of positioning the individual at the centre of career guidance. As per this approach, every individual has unique attributes, experiences, and preferences that must be acknowledged and understood in any guidance process. Cultural Sensitivity based approach [26] considers individuals' socio-cultural backgrounds. Recognizing that people come from varied backgrounds with different sets of values, norms, and experiences ensures that career guidance can consider the socio-cultural differences of individuals and provide career guidance to ensure inclusiveness in various career paths.

The Life Design approach [27] focuses on co-constructing future possibilities rather than solely reflecting on past experiences. On the other hand, the storytelling approach [28] emphasizes the importance of narratives in helping individuals make sense of their career journeys, enabling them to craft coherent career stories. Informed by a storytelling approach, a case study of a black South African student is presented in [29] and it advocates for culturally appropriate career counseling methods

It was stressed in [20] that career guidance should align with the client's expectations. For guidance to be practical, the career guidance must resonate with what clients' hope to achieve and where they envision themselves.

A Holistic Perspective [30] emphasizes the importance of viewing career guidance as more than just job placements. It considers personal aspirations, work-life balance, potential challenges, and the evolving nature of the job market. The importance of continuous learning is emphasized in [31]. As a part of continuous learning, adapting and updating one's skills becomes the cornerstone of career longevity and success due to the ever-changing dynamics of the modern workplace. A career management framework is proposed in [32] to assist career counsellors in guiding clients and helping individuals understand and improve the resources necessary for successful career development. As a practical tool, the Life-Career Rainbow is introduced in [33] for career education and counseling.

Table 2. Description of Career Counseling Methods

Career Counseling Method	Description
Client-Centered Approach [25]	Focuses on the uniqueness of each client, tailoring guidance based on individual attributes, experiences, and preferences
Cultural Sensitivity based approach [26]	Incorporates socio-cultural backgrounds to make guidance relatable and applicable to individuals from diverse cultures
Life Design approach [27]	Co-constructs future possibilities with the client, moving beyond reflecting solely on past experiences
Storytelling approach [28]	Utilizes narratives to help clients make sense of their career journeys, enabling the creation of coherent career stories
Aligning with Client's Expectations [20]	Ensures the guidance resonates with clients' goals, expectations, and envisioned future
Holistic approach [30]	Considers multiple facets of the client's life, including personal aspirations, work-life balance, and potential challenges
Continuous Pursuit of Knowledge [31]	Highlights the importance of continuous learning and skill updating in the modern, dynamic workplace
Contextual Action Theory [38]	Captures the role of a career counsellor in delving into both the conscious and unconscious thoughts and actions of clients
Career Management Framework [32]	Assists career counselors and individuals in understanding and improving resources for career development
Life-Career Rainbow as a Tool [33]	Introduces this tool for career education and counseling, offering a practical solution for career development

To underscore the significance of the methods mentioned above, we highlight several studies related to career guidance. An effort has been made in [34] to review the policies of organizations like the Organisation for Economic Cooperation and Development (OECD), European Commission, and World Bank to study the global trend towards lifelong learning and employability, and suggesting improvements in career guidance. The analyses in [35,36] advocate for aligning career counseling services with client expectations by emphasizing a consumer-driven approach. Inspired by life design counseling, the study in [37] highlights the benefits of personalized and diverse approaches, particularly for an abandoned adolescent in South Africa. Using Contextual Action Theory, the research presented in [38] underscores the crucial role of career counsellors in probing both the conscious and unconscious aspects of their clients' viewpoints. This comprehensive approach enables counsellors to better assist individuals in discerning and pursuing their desired career paths and life decisions. The study in [39] analyzes the influence of cultural and social-cognitive factors on career choices among Indian students. Informed by human capital theory, career progression factors are examined in [40], using a unique sample of American football players.

Summary: A brief summary of various career counseling methods is given in Table 2. Career guidance is about aligning the advice based on the client's requirements by seeing a career in the long-term perspective by considering the client's background. Overall, by studying the existing counseling frameworks and the corresponding studies, one can observe that the problem of providing career guidance or counseling is multidimensional. An IT-based or human-in-the-loop career guidance system aims to improve scalability and efficiency by matching the quality of service of the existing human-based career guidance or counseling systems. For this, it is necessary to understand and reflect on the multifaceted nature of the existing career guidance/counseling frameworks in IT-based career guidance systems.

4 Ethical and Fairness Issues

This section discusses studies about ethics and fairness regarding career guidance systems.

The fundamental principle of protecting a client's personal information is the basis of a trusted counseling relationship. This was made evident in [41], by carefully evaluating the results on betraying the client's trust. When confidentiality was violated, particularly in situations involving severe difficulties, the study's audio-taped counselor-client exchanges showed a noticeable reduction in perceived trustworthiness. These findings highlight that, even when breaking confidentiality is justified, severe ethical ramifications could undermine the fundamental trust in the counseling relationship. In order to keep the greatest standards of ethics and justice in counseling, the research serves as a forceful reminder of the importance of maintaining secrecy.

The study [42] presents a striking portrait of the ethical environment, highlighting crucial factors, including cultural Sensitivity, conformity to legal frameworks, and the imperative requirement for confidentiality. Additionally, it emphasizes the usefulness of standards established by reputable organizations like the National Career Development Association (NCDA). It also highlights the significance of values, norms, and ethical qualities in ethical counseling practices. The investigation in [43] examines the use of artificial intelligence (AI) in recruitment processes with a critical eye. It clarifies the possible ethical issues raised by AI, such as algorithmic biases and discrimination, and suggests practical ways to deal with them.

Summary: The studies demand a cooperative effort from all stakeholders (practitioners, researchers, and policymakers) to address ethical and fairness issues and strive towards establishing a just and equitable environment in the counseling and career advising industries. The design of effective IT-based career guidance systems must also incorporate the aspects of privacy and fairness.

5 IT-Based Methods

In this section, we discuss IT-based efforts for providing career guidance.

LinkedIn's personalized job recommendation system [44] analyses user input such as profiles, preferences, and interactions in addition to millions of structured job documents. The processing uses information retrieval, machine learning, statistical modeling, and recommender system algorithms. It mainly addresses the issue of providing users with timely, relevant employment recommendations that are in line with their profiles. Techniques for candidate selection are used to sort and order job advertisements effectively. In order to improve personalization, user interaction data is explored. The main task is finding relevant job posts that match users' skills and interests. The difficulties encountered by LinkedIn for job recommendation are discussed in [44]. One of the challenges is to suggest to consumers the appropriate number of positions so that companies receive the proper volume of applications. The difficulties include providing quick, tailored job recommendations, taking user behaviour into account, and adjusting to the reality that job ads are for specific opportunities, unlike book or movie recommendations.

A job recommendation system was proposed in [45], which considers both user interactions and job descriptions as the input. It employs a novel algorithm that constructs a recommendation graph, calculates similarity scores between jobs, and combines them to form weighted correlation scores. The output is a list of relevant job recommendations, with personalized rankings for active users based on recent interactions and tailored categories. In contrast, new and passive users receive recommendations based on their job preferences and locations.

In [46], the authors propose a framework to improve the skills given a set of tasks and difficulty level associated with each task. The framework models skill and skill difficulty and proposes a dynamic programming-based algorithm to recommend the skills to the individuals. It primarily focuses on comprehending and quantifying skill development.

A recommendation system called CareerRec, proposed in the [47], uses machine learning algorithms to help IT graduates choose their career options based on their skill sets. The method focused on predicting one of three career trajectories for graduates (developer, analyst, or engineer) using data from 2,255

IT sector professionals in Saudi Arabia. CareerRec considers technical capabilities (such as programming languages) and soft skills (like communication and logical thinking) when making recommendations.

Table 3. Description of IT-Based methods

IT-based system	Description
LinkedIn's job recommendation system [44]	Presents the details of LinkedIn's personalised job recommendation system by matching the user profiles with job profiles
A graph-based job recommendation [45]	A graph model is employed to process user's interaction data and job descriptions for personalized recommendations
Skill Improvement Suggestion [46]	The model suggests skills required for individuals to progress in their career and also estimates the difficulty of learning each skill
CareerRec approach [47]	Uses a combination of machine learning algorithms to help IT graduates select a career path based on their skills
Infinity System [48]	Explores the narrative and adaptability facets of career construction theory for enhancing career adaptability and planning,
Exploring Machine Learning for career path [49]	Conducts experimental study by comparing machine learning algorithm by analyzing students' academic details

The development of Computer-Assisted Career Guidance Systems (CACGS) is explored in [48] based on the so-called Infinity system. By leveraging the narrative tradition and adaptability facets of Career Construction Theory (CCT), Infinity is built as a holistic platform, enabling users to enhance their career planning skills. It is built based on trait-factor theory and person-environment fit models of CCT. The system recommends career-based personal assessment tests.

A machine learning-based career recommendation system is presented in [49]. The system recommends future professional trajectories and job placement eligibility based on patterns and connections extracted by analyzing the student data, including academic standing and personal traits. The machine learning algorithms such as K-means, decision trees, Naive Bayes, and Multilayer Perceptrons were employed to analyze the data.

Summary: A brief summary of various IT based systems is given in Table 3. It can be observed that not many efforts are being made in IT-based systems to address the career problem comprehensively. Research efforts are required to build a comprehensive career recommendation by extending insights from career-related theories, career counseling frameworks, and related studies.

6 Discussion

As reported in the introduction, India and several other countries are facing the issue of unemployment and underemployment. The studies report that career guidance systems could address unemployment and underemployment problems [50,51]. However, in India, the issue of a significant "Career Information Gap (CIG)" exists, as 93% of Indian students are aware of just seven types of careers [5]. At the same time, more than 1000 job types are available globally [6]. 93% of schools in India need professional career counsellors [7].

Further, in countries like India, the hierarchical organization of educational institutions creates a particular obstacle to fair career guidance. Students attending second-tier engineering institutions are frequently underprivileged because elite colleges have ample resources. In order to address this imbalance, a bottom-up strategy that takes into account the unique requirements and of individual students must replace the current generalized policy approach. Rather than imposing one-size-fits-all solutions, the emphasis should be on identifying and addressing students' particular needs in less privileged educational settings and customizing treatments to their particular surroundings. By doing this, we can make sure that career advice is a practical and readily available tool for all students, regardless of their educational background, enabling them to make well-informed and independent career decisions.

Conventional career counselling is based on an individualized, human-centred approach in which a qualified counsellor guides a client through different facets of their professional path while taking into account their intrinsic traits, emotional intelligence, cultural influences, and external dynamics. Every counseling session is customized to the client's needs and level of competence, with a focus on interpersonal communication and human insight. However, the scale of career counseling is limited by physical access and, beyond the reach of people in need, who are generally poor. With the advancement of IT, there is a scope for scaling up career counseling. There is a need, though, for the same level of personalization and depth to continue when we move into IT-enabled career guidance. While real-time data, flexibility, and efficiency are advantages of IT-enabled career counselling, the emotional depth, and thorough understanding of the traditional approach are frequently absent. For each potential seeker/client, the IT-based guidance needs to be evolved to answer some pertinent questions like

- What factors contribute to a mismatch between the job and core skill-sets of the individuals?
- What is the best match between multiple skill sets possessed by individuals and available jobs in the market?
- For a seeker/client, given a career stage with a skill-set, what are the next job options?
- For a seeker/client, given a career stage with a skill-set, what are the next business/start-up/occupation options?
- What skill sets are required for a job?

- What skill sets do individuals have?
- Why did a person fail in a specific job but succeed in another?

To answer these questions, IT-based systems must be built based on a proper theoretical and empirical understanding of career progression and counseling strategies. A mismatch between skillsets and occupation depresses wage growth in the current occupation; it also leaves a scarring effect-by stunting skill acquisition-that reduces wages in future occupations [52–57].

The need of the hour is a well-balanced strategy that would make the most of technical innovation and human empathy, providing people at all stages of their professional journeys with a thorough, flexible, and encouraging career counseling experience. We hope that the research covered in this paper on the issue of CIG and different facets related to career guidance systems will encourage collaboration between the researchers who work in the IT area and theoretical fields related to career progression, career counseling to address the challenges of building IT-based career guidance systems.

7 Conclusion

Unemployment and underemployment are twin challenges faced by all developing countries, including India. The countries have started national career services as mission-mode projects to assist individuals in selecting suitable careers or jobs. However, the issue of a significant career information gap persists in developing countries like India. Most individuals in India are aware of just seven types of careers, while more than 1,000 job types are available globally. India also needs more career counsellors. Notably, the progress in the Internet, smartphones, and artificial intelligence is impacting all sections of society. There is an opportunity to build scalable career guidance systems by exploring the progress in IT and Artificial Intelligence. With this background, we reviewed important theoretical and empirical studies on the various facets of building career guidance systems in this paper. We have discussed the theoretical basis for career modeling, career guidance methods, ethical and fairness issues, and IT-based methods. We concluded that the IT-based tools must be grounded on theoretical models and counseling frameworks to suggest optimal career choices. We hope this review will encourage researchers, especially from the IT field, to understand the complexity of career guidance systems and conduct further research to build scalable systems for tackling the career information gap.

References

1. Holland, J.L.: Making Vocational Choices: A Theory of Vocational Personalities and Work Environments. Psychological Assessment Resources, Inc. (1997)
2. Oxford Learners' Dictionaries, Oxford (2023). www.oxfordlearnersdictionaries.com. Accessed 22 Oct 2023
3. CMIE: Economic outlook www.economicoutlook.cmie.com. Accessed 4 May 2023

4. Roy, P., et al:. Use of ICT and advanced media for skill development in agriculture. In: ICT and Social Media for Skill Development in Agriculture, pp. 105–124. Today & Tomorrow's Printers and Publishers, New Delhi (2019)
5. Mindler.: Job option awareness among Indian students www.mindler.com/blog/career-coaching-choice-compulsion. Accessed 4 Feb 2019
6. List of over 1388 careers. www.careerplanner.com. Accessed Sep 2023
7. India Needs 15 Lakh Counsellors for 315 Million Students, Times of India (2019)
8. Why Career Guidance is Important for Students, India Today (2022)
9. Repetto, E.: International competencies for educational and vocational guidance practitioners: an IAEVG trans-national study. Int. J. Educ. Vocat. Guidance **8**, 135–195 (2008)
10. Hooley, T.: How the internet changed career: framing the relationship between career development and online technologies. J. Nat. Inst. Career Educ. Counselling (NICEC) **29**, 3–12 (2012)
11. Arup, V., Kumar, S., Sureka, R., Lim, W.M.: What do we know about career and development? insights from career development international at age 25. Career Dev. Int. **27**(1), 113–134 (2022)
12. Hirschi, A.: Whole-life career management: a counseling intervention framework. Career Dev. Q. **68**(1), 2–17 (2020)
13. Autin, K.L., Blustein, D.L., Ali, S.R., Garriott, P.O.: Career development impacts of COVID-19: practice and policy recommendations. J. Career Dev. **47**(5), 487–494 (2020)
14. Maree, J., Fabio, A.: Integrating personal and career counseling to promote sustainable development and change. Sustainability **10**, 4176 (2018)
15. Bandura, A., Walters, R.H.: Social Learning Theory, Vol. 1. Prentice Hall (1977)
16. London, M.: Toward a theory of career motivation. Acad. Manage. Rev. **8**(4), 620–630 (1983)
17. Parsons, F.: Choosing a Vocation. Houghton, Mifflin and Company (1909)
18. Brown, S.D., Lent, R.W.: Career Development and Counseling: Putting Theoryand Research to Work. Wiley (2004)
19. Krumboltz, J.D.: The happenstance learning theory. J. Career Assess. **17**(2), 135–154 (2009)
20. Super, D.E.: A life-span, life-space approach to career development. J. Vocat. Behav. **16**, 282–298 (1980)
21. Jiang, Z., Newman, A., Le, H., Presbitero, A., Zheng, C.: Career exploration: a review and future research agenda. J. Vocat. Behav. **110**, 338–356 (2018)
22. Brown, D., Brooks, L.: Career Choice and Development. Jossey-Bass (1996)
23. Gottfredson, L.S.: Circumscription and compromise: a developmental theory of occupational aspirations. J. Couns. Psychol. **28**, 545–579 (1981)
24. Cochran, L.: What is a career problem? Career Dev. Q. **42**(3), 204–215 (1994)
25. Rogers, C.R.: Client-centered Therapy: Its Current Practice. Implications and Theory. Constable, London (1951)
26. Leong, F., Blustein, D.: Toward a global vision of counseling psychology. Couns. Psychol. **28**, 5–9 (2000)
27. Savickas, M., et al.: Life designing: a paradigm for career construction in the 21st century. J. Vocat. Behav. **75**, 239–250 (2009)
28. Brott, P.: The storied approach: a postmodern perspective for career counseling. Career Dev. Q. **49**, 304–313 (2001)
29. McMahon, M., Watson, M., Chetty, C., Hoelson, C.: Examining process constructs of narrative career counseling: An exploratory case study. Br. J. Guidance Couns. **40**, 1–15 (2012)

30. Guichard, J.: Career guidance, education, and dialogues for a fair and sustainable human development. In: Inaugural Conference of the UNESCO Chair of Lifelong Guidance and Counseling, Poland (2013)
31. Collin, K., Van der Heijden, B., Lewis, P.: Continuing professional development. Int. J. Training Dev. **16**(3), 155–163 (2012)
32. Hirschi, A.: The career resources model: an integrative framework for career counsellors. Br. J. Guidance Couns. **40**, 1–15 (2012)
33. Super, D.E.: A life-span, life-space approach to career development. J. Vocat. Behav. **16**(3), 282–298 (1980)
34. Watts, A.G., Sultana, R.G.: Career guidance policies in 37 countries: contrasts and common themes. Int. J. Educ. Vocat. Guidance **4**, 105–122 (2004)
35. Lim, R.B.: Career counseling services: client expectations and provider perceptions, Ph. D. Thesis, Queensland University of Technology (2005)
36. Lim, R., Patton, W.: What do career counsellors think their clients expect from their services? are they right? Aust. J. Career Dev. **15**(2), 32–41 (2006)
37. Maree, J., Crous, S.: Life design career counseling with an abandoned adolescent: a case study. J. Psychol. Afr. **22**, 106–113 (2012)
38. Dyer, B., Pizzorno, M., Qu, K., Valach, L., Marshall, S., Young, R.: Unconscious processes in a career counseling case: an action-theoretical perspective. Br. J. Guidance Couns. **38**, 343–362 (2010)
39. Arulmani, G.: The internationalization of career counseling: bridging cultural processes and labour market demands in India. Asian J. Couns. **16**(2), 149–170 (2009)
40. Harris, C., Pattie, M., McMahan, G.: Advancement along a career path: the influence of human capital and performance. Hum. Resour. Manage. J. **25**(3), 102–115 (2014)
41. Merluzzi, T.V., Brischetto, C.S.: Breach of confidentiality and perceived trustworthiness of counselors. J. Couns. Psychol. **30**(2), 245–251 (1983)
42. Partner, M., Ajagbawa, H.: The role of ethics in career counseling in the 21st century. IOSR J. Humanit. Soc. Sci. **19**(1), 12–22 (2014)
43. Hunkenschroer, A., Lütge, C.: Ethics of AI-enabled recruiting and selection: a review and research agenda. J. Bus. Ethics **178**(7), 977–1007 (2022)
44. Kenthapadi, K., Le, B., Venkataraman, G.: Personalized job recommendation system at LinkedIn: practical challenges and lessons learned. In: Conference on Human Information Interaction & Retrieval (CHIIR 2017), pp. 346–347 (2017)
45. Shalaby, W., et al.: Help me find a job: a graph-based approach for job recommendation at scale. In: IEEE International Conference on Big Data, pp. 1544–1553 (2017)
46. Umemoto, K., Milo, T., Kitsuregawa, M.: Toward recommendation for upskilling: modeling skill improvement and item difficulty in action sequences. In: IEEE 36th International Conference on Data Engineering (ICDE), pp. 169–180 (2020)
47. Al-Dossari, H., Nughaymish, F., Al-Qahtani, Z., Alkahlifah, M., Alqahtani, A.: CareerRec: a machine learning approach to career path choice for information technology graduates. Eng. Technol. Appl. Sci. Res. J. **10**, 6589–6596 (2020)
48. Leung, S.A.: New frontiers in computer-assisted career guidance systems(CACGS): implications from career construction theory. Front. Psychol. **13**, 786232 (2022)
49. Sengupta, S., Banerjee, A., Chakrabarti, S.: Prediction of future career path using different machine learning models. In: Tavares, J.M.R.S., Chakrabarti, S., Bhattacharya, A., Ghatak, S. (eds.) Emerging Technologies in Data Mining and Information Security. LNNS, vol. 164, pp. 241–252. Springer, Singapore (2021). https://doi.org/10.1007/978-981-15-9774-9_23

50. Sampson, J.P.: Career Counseling and Services: A Cognitive Information Processing Approach. Graduate Career Counseling Series (2004)
51. Brown, S.D., Krane, N.E.R.: Four (or five) sessions and a cloud of dust: old assumptions and new observations about career counseling. In: Handbook of Counseling Psychology, pp. 740–766. Wiley (2000)
52. McGuinness, S., Pouliakas, K., Redmond, P.: Skills mismatch: concepts, measurement and policy approaches. J. Econ. Surv. **32**(4), 985–1015 (2018)
53. Mahboubi, P.: Bad fits: the causes, extent and costs of job skills mismatch in Canada. CD Howe Inst. Commentary **10**, 552 (2019)
54. Nedelkoska, L., Neffke, F., Wiederhold, S.: Skill mismatch and the costs of job displacement. In: Annual Meeting of the American Economic Association (2015)
55. Reddy, A.A.: Growth, structural change and wage rates in rural India. Econ. Polit. Wkly. **50**, 56–65 (2015)
56. Reddy, A.A.: Disparities in employment and income in rural Andhra Pradesh, India. Bangladesh Dev. Stud. **34**, 73–96 (2011)
57. Mehrotra, S., Gandhi, A., Sahoo, B.: Estimating India's skill gap on a realistic basis for 2022. Econ. Pol. Wkly **48**, 102–111 (2013)

Author Index

V. Goyal et al. (Eds.): BDA 2023, LNCS 14418, pp. 261–262, 2023.
https://doi.org/10.1007/978-3-031-49601-1

Printed in the United States
by Baker & Taylor Publisher Services